Springer Series on Environmental Management

Robert S. DeSanto, Series Editor

William H. Smith

Air Pollution and Forests

Interactions between Air Contaminants and Forest Ecosystems

With 60 Figures

Springer-Verlag

New York   Heidelberg   Berlin

William H. Smith
School of Forestry and Environmental Studies
Yale University
Greeley Memorial Laboratory
370 Prospect Street
New Haven, Connecticut 06511

*On the front cover: A view of the Northern Hardwood
Forest of New Hampshire as taken by the author from the
top of Red Hill (elevation 618 m), near Moultonborough.*

Library of Congress Cataloging in Publication Data

Smith, William H.   1939−
. Air pollution and forests.

(Springer series on environmental management) Includes
bibliographical references and index.
1. Air—Pollution—Environmental aspects. 2. Forest
ecology. 3. Plants, Effect of air pollution on. I. Title. II.
Series. QH545.A3S64   581.5′2642   81-257 AACR1

Printed in the United States of America

9 8 7 6 5 4 3 2 1

ISBN 0-387-90501-4 Springer-Verlag   New York
ISBN 3-540-90501-4 Springer-Verlag   Berlin   Heidelberg

To my family, friends, and students, for their encouragement and inspiration.

Series Preface

This series is dedicated to serving the growing community of scholars and practitioners concerned with the principles and applications of environmental management. Each volume will be a thorough treatment of a specific topic of importance for proper management practices. A fundamental objective of these books is to help the reader discern and implement man's stewardship of our environment and the world's renewable resources. For we must strive to understand the relationship between man and nature, act to bring harmony to it, and nurture an environment that is both stable and productive.

These objectives have often eluded us because the pursuit of other individual and societal goals has diverted us from a course of living in balance with the environment. At times, therefore, the environmental manager may have to exert restrictive control, which is usually best applied to man, not nature. Attempts to alter or harness nature have often failed or backfired, as exemplified by the results of imprudent use of herbicides, fertilizers, water, and other agents.

Each book in this series will shed light on the fundamental and applied aspects of environmental management. It is hoped that each will help solve a practical and serious environmental problem.

Robert S. DeSanto
East Lyme, Connecticut

Preface

This book was made possible by the research performed by a very large number of scientists. Their productivity and publication have allowed the author to construct this review. This volume provides a compendium of the most significant relationships between forests and air pollution under low, intermediate, and high dose conditions. Under conditions of low dose, the vegetation and soils of forest ecosystems function as important sources and sinks for air pollution. When exposed to intermediate dose, individual tree species or individual members of a given species may be subtly and adversely affected by nutrient stress, impaired metabolism, predisposition to entomological or pathological stress, or direct disease induction. Exposure to high dose may induce acute morbidity or mortality of specific trees.

At the ecosystem level, the impact of these various interactions would be very variable. In the low-dose relationship, pollutants are exchanged between the atmospheric compartment, available nutrient compartment, other soil compartments, and various elements of the biota. Depending on the nature of the pollutant, this transfer can be innocuous or stimulatory to the forest. Forest exposure to intermediate dose conditions may result in inimical influence. The ecosystem impact in this instance may include reduced productivity and biomass, alterations in species composition or community structure, increased insect outbreaks or microbial disease epidemics, and increased morbidity. Under conditions of high dose and concentrated mortality, ecosystem impacts may include gross simplification, impaired energy flow, and biogeochemical cycling, changes in hydrology and erosion, climate alteration, and major impacts on associated ecosystems as well as forest destruction.

The author hopes that this book will provide a strategy for a comprehensive introduction to the complex relationship between forest systems and air contaminants. An understanding of this relationship is essential for the protection and wise management of our forest resources.

William H. Smith
Yale University
New Haven, Connecticut

Contents

1
Introduction

Forests of the world are of enormous importance to mankind for wood, watershed, wildlife, recreation, wilderness, and aesthetic and amenity values. Over the next several decades improved understanding of forests, wilderness stewardship, improved natural forest management, and increased acreages of intensively managed artificial forests will significantly improve these values. During this same period, unfortunately, other human activities will exert a variety of stresses on forest ecosystems. One of the most important stresses of widespread significance is air pollution. This book is an exploration of the various interactions between air pollution and temperate forest ecosystems.

A. Air Pollution

Approximately 95% of the atmosphere of the earth occurs in an 8-12 km (5-7 mile) layer surrounding the earth, termed the troposphere. For the perspective of this book, an appropriate definition of air pollutants is *materials that occur in the troposphere in quantities in excess of normal amounts.* These materials may be solid, liquid, and gaseous in character and they may result from both natural and human (anthropogenic) processes. Natural sources of air pollution are diverse and include volcanic and other geothermal eruptions, forest fires, gases released from vegetation, wind blown soil and other debris, pollen, spores, and sea spray particles. Anthropogenic sources are also diverse, and unfortunately also typically very concentrated, and include a variety of combustion and industrial activities. The specific materials that contaminate the troposphere are as varied as the sources and have differential importance depending on their ability to influence natural processes or elements of the biota and human health or activities. Air pollutants of particular importance are presented in Table 1-1.

Human beings have been polluting the atmosphere for millennia and it is quite impossible to know the exact ranges of normalcy for tropospheric materials. In concept, however, unpolluted tropospheric air contained a variety of naturally generated particles plus gases in the approximate concentrations presented in Table 1-2. Gas concentrations in Table 1-2 and gas and particulate concentrations throughout this book will generally be expressed as micrograms per cubic meter of air (μg m^{-3}) (for gases, standard conditions of 25°C and 760 mm mercury are assumed). While it is impossible to specify trace gas concentrations of unpolluted air, global sources of air pollution vary greatly, and as a result, various portions of the troposphere may be designated "relatively clean" and "relatively polluted," respectively. A comparison of trace gas concentrations for air environments so designated is presented in Table 1-3. Suspended particles in United States tropospheric samples during 1966-1967, averaged 102 μg m^{-3} in urban areas and 45, 40, and 21 μg m^{-3} in nonurban proximate, intermediate, and remote areas, respectively (Corn, 1976).

Due to increasing awareness of the adverse consequences of air pollution, numerous countries of the temperate latitudes have promulgated air quality standards. In the United States, the Clean Air Act of 1970 directed the Environmental Protection Agency to establish air quality standards for particulates, sulfur dioxide, nitrogen dioxide, carbon monoxide, hydrocarbons, and ozone. A standard for lead was added in 1978. Primary and secondary standards were established. Primary standards are intended to protect human health, while secondary standards are intended to protect public welfare. The latter includes consideration of air pollution influence on vegetative health, materials weathering, and visibility. The National Ambient Air Quality Standards of the United States as amended through February 1979 are presented in Table 1-4. It is of constant interest to compare various dose-response relationships reviewed in this book with ambient air quality measurements and standards. Since air pollution investigations monitor and average air pollutant concentrations for variable time periods, it is helpful to realize that a generalized relationship between averaging time (length of monitoring period) and maximum average atmospheric concentration has been proposed (Table 1-5).

B. Forest Ecosystems

Forests introduce air pollutants to the troposphere and remove air pollutants from the troposphere. Forest development is both stimulated and inhibited by air pollution. In order to obtain a comprehensive impression of the various interactions between forests and air pollution it is useful to employ an ecosystem perspective.

An ecosystem is a unit of nature in which all the organisms of a given area interact with the physical environment to produce a flow of energy that leads to

Table 1-1. Materials of Importance That Pollute the Troposphere[a]

I. Particulate pollutants
   A. Primary
      1. Inorganic
         a. Variable (dust, asbestos, soil, salt)
         b. Chlorides
         c. Fluorides*
         d. Trace metals*
      2. Organic
         a. Spores*
         b. Pollen*
   B. Secondary
      1. Inorganic
         a. Sulfates*
         b. Nitrates*
      2. Organic
         a. Hydrocarbons*
         b. Aliphatic nitrates
         c. Carboxylic acids
         d. Dicarboxylic acids
II. Gaseous pollutants
   A. Primary
      1. Inorganic
         a. Oxides
            i. Carbon*
            ii. Sulfur*
            iii. Nitrogen*
         b. Halogens
            i. Chloride
            ii. Fluoride*
         c. Other
            i. Ammonia
            ii. Hydrogen sulfide
      2. Organic
         a. Hydrocarbons*
         b. Ketones
         c. Mercaptans
         d. Sulfides
         e. Halocarbons
   B. Secondary
      1. Inorganic
         a. Ozone*
      2. Organic
         a. Aldehydes
         b. Peroxyacetylnitrate and homologues
         c. N-nitroso compounds

[a]Materials released directly into the atmosphere are designated primary, while those synthesized in the atmosphere are termed secondary. Materials importantly associated with forest ecosystems are followed by an asterisk.

Table 1-2. Approximate Gaseous Composition of Unpolluted Tropospheric Air on a Wet Basis

Gas	$\mu g\ m^{-3}$
Nitrogen	9×10^8
Oxygen	3×10^8
Water	2×10^7
Argon	1×10^7
Carbon dioxide	5×10^5
Neon	1×10^4
Helium	8×10^2
Methane	$6\text{-}8 \times 10^2$
Krypton	3×10^3
Nitrous oxide	9×10^2
Xenon	4×10^2
Hydrogen	4×10^1

clearly defined trophic structure, biotic diversity, and exchange of materials between biotic and abiotic parts of the unit. Several components and processes characterize all ecosystems. Components include inorganic substances, organic compounds, producers, macroconsumers, microconsumers, and climate. Processes include energy flow, food chains, diversity patterns, biogeochemical cycles, development, evolution, and control (Odum, 1971). A diagram representing the major interactions between living and nonliving components of a terrestrial ecosystem is presented in Figure 1-1. A thorough explanation of ecosystem structure and function is beyond the scope of this book. Readers interested in a more complete discussion of the ecosystem concept are referred to the very concise and readable Chapter 2 in *Agricultural Ecology* by Cox and Atkins (1979).

The earth is covered by a mosaic of ecosystems. These ecosystems are connected and influence one another in a variety of ways. Temperate forest ecosystems occupy a position of prominence among all ecosystems. Temperate forests (1.8 billion ha) are second only to tropical forest ecosystems (2 billion ha) in size. Temperate forest ecosystems (200-400 tons ha^{-1}) are second only to rain

Table 1-3. Trace Gas Concentrations for Relatively "Clean" and "Polluted" Atmospheres

Gas	$\mu g\ m^{-3}$	
	Clean air	Polluted air
Carbon dioxide	57.6×10^4	72.0×10^4
Carbon monoxide	115	$46\text{-}80.5 \times 10^3$
Methane	920	1533
Nitrous oxide	450	?
Nitrogen dioxide	1.9	376
Ozone	39	980
Sulfur dioxide	0.5	524
Ammonia	7.0	14.0

Source: Urone (1976).

Table 1-4. National Ambient Air Quality Standards of the United States (as amended through February 1979)[a]

Pollutant	Standards ($\mu g\ m^{-3}$)	
	Primary	Secondary
Particulates (total suspended)		
annual	75	60
24-hr	260	150
Sulfur dioxide		
annual	80	
24-hr	365	
3-hr		1,300
Nitrogen dioxide		
annual	100	100
Carbon monoxide		
8-hr	10,000	10,000
1-hr	40,000	40,000
Hydrocarbons		
3-hr	160	160
Ozone		
1-hr	240	240
Lead		
3-month	1.5	

[a]Short-term standards (24 hr and less) are not to be exceeded more than once a year. Long-term standards are maximum permissible concentrations never to be exceeded.

forest ecosystems (400-500 tons ha^{-1}) in biomass. In terms of primary productivity, temperate forest ecosystems (5-20 tons $ha^{-1}\ yr^{-1}$) rank third behind only tidal zone (20-40 tons $ha^{-1}\ yr^{-1}$) and rain forest (10-30 tons $ha^{-1}\ yr^{-1}$) ecosystems. Unfortunately temperate forest ecosystems are also located in the zone of maximum air pollution because of their extensive distribution throughout the zone of primary urbanization and industrialization of the earth.

Table 1-5. Relationship between Averaging Time Employed and Relative Maximum Average Concentration in the Atmosphere

Averaging time	Relative maximum concentration (average) in the atmosphere
1 month	0.5
24 hr	1.0
8 hr	1.2
2 hr	1.8
1 hr	2.2
30 min	2.4
15 min	2.7
Single measurement[a]	3.3

Source: Stern (1964).
[a]Encompasses range up to 10 min.

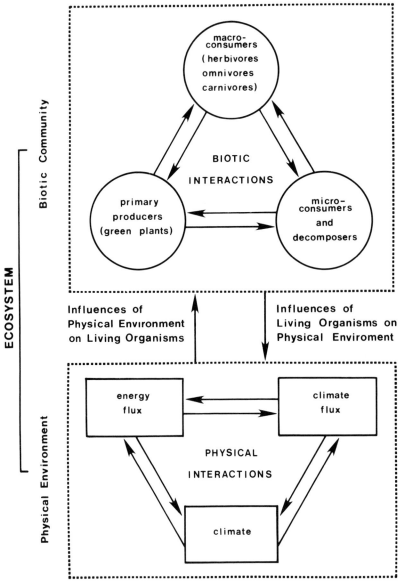

Figure 1-1. Major interactions between the biotic and abiotic components of a typical terrestrial ecosystem. (From Cox and Atkins, 1979.)

C. Interactions between Air Pollutants and Forest Ecosystems

We will not completely appreciate the interactions between air pollutants and forest ecosystems until we understand the influence of various air contaminants on the numerous components and processes of the forest. During the past several decades we have greatly advanced our awareness of this complex and important topic. This book attempts to summarize our understanding as we start the eighth decade of the twentieth century.

In 1972, the author was asked to review the relationship between air pollution and forests for the annual meeting of the American Association for the Advancement of Science. I approached this formidable assignment by dividing the interactions between air contaminants and forest ecosystems into three classes (Smith, 1974). Under conditions of low dose (Class I relationship) the vegetation and soils of forest ecosystems function as important sources and sinks for air pollutants. When exposed to intermediate dose (Class II relationship) individual tree species or individual members of a given species may be subtly and adversely affected by nutrient stress, impaired metabolism, predisposition to entomological or pathological stress, or direct disease induction. Exposure to high dose (Class III relationship) may induce acute morbidity or mortality of specific trees. At the ecosystem level the impact of these various interactions would be very variable. In the Class I relationship, pollutants would be exchanged between the atmospheric compartment, available nutrient compartment, other soil compartments, and various elements of the biota. Depending on the nature of the pollutant, the ecosystem impact of this transfer could be undetectable (innocuous effect) or stimulatory (fertilizing effect). If the effect of air pollution dose on some component of the biota is inimical then a Class II relationship is established. The ecosystem impact in this case could include reduced productivity or biomass, alterations in species composition or community structure, increased insect outbreaks or microbial disease epidemics, and increased morbidity. Under conditions of high dose and Class III relationship, ecosystem impacts may include gross simplification, impaired energy flow and biogeochemical cycling, changes in hydrology and erosion, climate alteration, and major impacts on associated ecosystems. While these classes of interaction are conceptual and artificial, and not necessarily discrete entities in time nor space, I have found them useful in teaching and my colleagues and I in organizing and establishing research priorities (Smith and Dochinger, 1975).

The information reviewed in this book concerns the temperate latitudes of the Northern Hemisphere, primarily the area between lat. $40°$ and $60°$N, because the primary release of anthropogenic air pollutants is within this zone and as a result the research on effects is concentrated in this area. The information reviewed is principally from North America. This reflects the relative availability and accessibility of the literature to the author. The literature review was completed in January 1980 and the book contains only a small amount of material published after that date. The author hopes, however, the information presented will provide assistance to the reader in understanding the extremely complex relationships between air pollutants and forest ecosystems.

References

Corn, M. 1976. Aerosols and the primary pollutants—nonviable particles. Their occurrence, properties, and effects. *In*: A. C. Stern (Ed.), Air Pollution. Vol. 1. Academic Press, New York, pp. 77-168.

Cox, G. W., and M. D. Atkins. 1979. Agricultural Ecology. Freeman, San Francisco, California, 721 pp.

Odum, J. A. 1971. Ecosystem structure and function. *In*: J. A. Wiens (Ed.), Proc. 31st Annual Biol. Colloquium, Oregon State Univ. Press, Corvallis, Oregon, pp. 11-24.

Smith, W. H. 1974. Air pollution—Effects on the structure and function of the temperate forest ecosystem. Environ. Pollut. 6:111-129.

Smith, W. H., and L. S. Dochinger. 1975. Air Pollution and Metropolitan Woody Vegetation. U.S.D.A. Forest Service. PIEFR-PA-1. Northeastern For. Exp. Sta., Upper Darby, Pennsylvania, 74 pp.

Stern, A. C. 1964. Summary of existing air pollution standards. J. Air Pollut. Control Assoc. 14:5-15.

Urone, P. 1976. The primary air pollutants—gaseous. Their occurrence, sources, and effects. *In*: A. C. Stern (Ed.), Air Pollution, Vol. 1. Academic Press, New York, pp. 23-75.

SECTION I

FORESTS FUNCTION AS SOURCES AND SINKS FOR AIR CONTAMINANTS—CLASS I INTERACTIONS

2
Role of Forests in Major Element Cycles: Carbon, Sulfur, and Nitrogen

Despite the fundamental and enormous importance of carbon, sulfur, and nitrogen to the biota, our appreciation of the cycling, residence, and flux of these elements among various ecosystems, oceans, and the atmosphere is incomplete. Despite our deficiencies efforts are being made to refine our understanding of major element cycles and to more accurately estimate global budgets. This is especially important for those interested in air pollution as compounds containing these elements are extremely significant air contaminants under certain circumstances. Forest ecosystems, because of their extensive distribution and their varied functions, play important roles in global element cycles. The imprecision of our estimates of global nutrient budgets is large and conclusions and predictions based on them must be qualified and cautious.

Carbon, sulfur, and nitrogen have analogous biogeochemical cycles. All three cycles contain a biological reduction step with a reduced residence period in the biota where these elements exist prior to reintroduction to the atmosphere (Deevey, 1973; Hitchcock and Wechsler, 1972). Since oxides and oxidized compounds of these three elements can be important atmospheric contaminants and since forest ecosystems are involved in the global release of these compounds, it is important to review the significance of forests as sources of carbon, sulfur and nitrogen oxides, and oxidized end products.

A. Carbon Pollutants

Carbon monoxide and carbon dioxide are both very important atmospheric contaminants. Human activities are responsible for the introduction of increasing quantities of these gases to the atmosphere. Carbon monoxide is particularly important because of its potent mammalian toxicity, while carbon dioxide is most significant because of its ability to regulate global temperature. Neither gas is thought to cause direct damage to vegetation at ambient concentrations presently monitored. Carbon monoxide has not been shown to produce acute effects on plants at concentrations below 100 ppm (11.5×10^4 μg m^3) for exposures from one to three weeks (U.S. Environmental Protection Agency, 1976). The threshold of carbon dioxide toxicity to plants is in such excess of ambient conditions as to be completely unimportant. The hypothesis that the increasing concentration of carbon dioxide in the atmosphere will result in elevated global temperatures, however, has enormous implications for the health of forest ecosystems and will be discussed in more detail later.

1. Atmospheric Increase of Carbon Dioxide

The carbon dioxide concentration of the global atmosphere has been estimated to have been approximately 290 ppm (5.2×10^5 μg m^{-3}) in the middle of the nineteenth century. Today the carbon dioxide concentration approximates 335 ppm (6.0×10^5 μg m^{-3}). One fourth of this increase is estimated to have occurred between 1967 and 1977. Evidence for this increase has come from careful carbon dioxide monitoring carried out over the last 20 years in Hawaii, Alaska, New York, Sweden, Australia, and the South Pole. The current rate of increase is about one ppm (1.8×10^3 ug m^3) per year, 1.5 ppm (2.7×10^3 μg m^3) in 1977-1978. In the year 2020, if the increasing rate continues, the carbon dioxide amount in the global atmosphere may be nearly two times the present value (Ember, 1978; Woodwell, 1978).

2. Carbon Cycle

Pool sizes, transfer patterns, and flux rates of the carbon cycle are not fully nor very accurately appreciated. Woodwell (1978) has presented estimates of the major carbon pools. The atmosphere contains 700×10^{15} g of carbon as carbon dioxide. The global biota contains 800×10^{15} g of reduced carbon. Organic matter of soil, primarily humus and peat, contains between 1000×10^{15} and 3000×10^{15} g of carbon. The largest carbon pool is the global ocean system with a carbon content of approximately $40,000 \times 10^{15}$ g.

A constant and critically important flux of carbon, as carbon dioxide, occurs between the biota and the surface waters of the ocean. A biogeochemical cycle for carbon is presented in Figure 2-1.

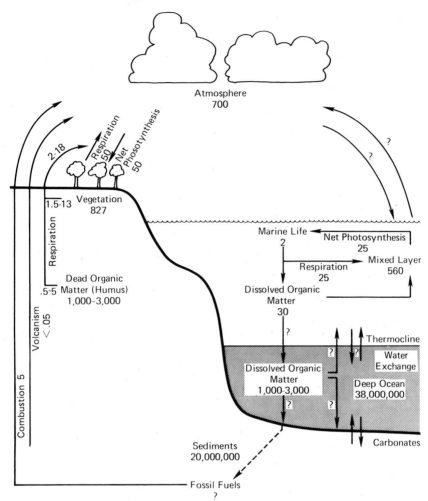

Figure 2-1. Major global carbon pools and annual exchange rates between them. Amounts are billions of tons. (From Woodwell, 1978.)

Forests occupy approximately one-third of the area of global terrestrial ecosystems (Reichle et al., 1973; Whittaker, 1975) and constitute from approximately 60% (Reichle et al., 1973) to 90% (Woodwell, 1978) of the total terrestrial carbon pool. Global net primary production and total mass of carbon for various ecosystems are listed in Table 2-1.

The natural net input of carbon dioxide to the atmosphere from vegetative systems is close to zero (equilibrium) when the systems exist in natural, undisturbed states. Gross input by forest fires, organic matter decomposition, and plant respiration is balanced by carbon dioxide uptake for photosynthesis in existing and newly established vegetation. Under conditions of widespread and rapid forest destruction, however, equilibrium conditions may be lost.

Table 2-1. Major Ecosystems of the Earth: Area, Net Primary Production, and Mass of Carbon

Plant community	Area (10^6 km^2)	Total net primary production of carbon $(10^{15} \text{ g yr}^{-1})$	Total plant mass of carbon (10^{15} g)
Tropical rain forest	17.0	16.8	344
Tropical seasonal forest	7.5	5.4	117
Temperate evergreen forest	5.0	2.9	79
Temperate deciduous forest	7.0	3.8	95
Boreal forest	12.0	4.3	108
Woodland and shrubland	8.5	2.7	22
Savanna	15.0	6.1	27
Temperate grassland	9.0	2.4	6.3
Tundra and alpine meadow	8.0	0.5	2.3
Desert scrub	18.0	0.7	5.9
Rock, ice, and sand	24.0	0.03	0.2
Cultivated land	14.0	4.1	6.3
Swamp and marsh	2.0	2.7	13.5
Lake and stream	2.0	0.4	0.02
Total continental	149	52.8	827
Open ocean	332	18.7	0.45
Upswelling zones	0.4	0.1	0.004
Continental shelf	26.6	4.3	0.12
Algal bed and reef	0.6	0.7	0.54
Estuaries	1.4	1.0	0.63
Total marine	361	24.8	1.74
Full total	510	77.6	828

Source: Woodwell (1978).

3. Relationship of Forests to Increasing Atmospheric CO$_2$ Levels

In 1954, G. Evelyn Hutchinson, presently professor emeritus of zoology at Yale University, suggested in a chapter in *The Earth as a Planet* edited by G. P. Kuiper that the increasing content of carbon dioxide of the atmosphere is due to the destruction of forests as well as the combustion of fossil fuels (Woodwell, 1978). George M. Woodwell of the Ecosystems Center, Marine Biological Laboratory, by providing and summarizing recent evidence, has actively advanced the hypothesis that forest destruction contributes to increasing atmospheric carbon dioxide levels. In addition to forest destruction, conversion of original-natural forests to you.   , a forests, forest fires and soil management practices may significantly contribute carbon dioxide to the atmosphere. The latter may be of the same order of magnitude as the input of fossil fuel carbon dioxide (Bolin et al., 1977).

One-third of the United States land area or 300×10^6 ha (741 million acres) consisted of forest ecosystems in 1979. This area is continually changing in response to clearing of forests to convert to agricultural activity, abandonment of agricultural land to forest development, and loss of forest systems to urban, industrial, or transportation use (Spurr and Vaux, 1976). It has been estimated that over the next 50 years the United States will witness the following net loss of forested land: 8×10^6 ha (20 million acres) to crop and pasture use and 3.2×10^6 ha (8 million acres) to development pressure (Spurr and Vaux, 1976). Adams et al. (1977) point out, however that the United States forest loss is quite insignificant when compared to the forest destruction in some of the developing countries such as Brazil where the minimum net loss of wood is estimated to be 3.5 tons per capita per year. Global forest destruction may be underestimated by overreliance on data from developed countries relative to the developing countries where forest loss to agriculture and firewood is particularly high (Adams et al., 1977).

Bolin et al. (1977) have concluded that changes in forest area in developed countries do not constitute a significant net flux of carbon to or from the atmosphere. For the developing countries, however, Bolin estimated that approximately 0.12×10^6 km^2 of natural forest is cleared and burned annually and that this loss has significantly added to the atmospheric carbon dioxide load during this century as indicated in Table 2-2.

Comparison of the ratios of carbon isotopes ^{12}C, ^{13}C, and ^{14}C has been useful to those studying the flux and sources of carbon dioxide in the atmosphere. The procedure is based on the fact that the ratios differ for the major sources of carbon. Fossil fuels do not contain ^{14}C and are elevated in ^{12}C relative to ^{13}C. The present biota contain ^{14}C and also has elevated ^{12}C compared to the atmosphere. Burning of fossil fuels liberates carbon deficient in ^{14}C and thus dilutes the ^{14}C of the atmosphere, the so-called Suess effect (Woodwell et al., 1978). Anthropogenic release of CO_2 from both fossil fuels and the biota reduces ^{13}C levels in the atmosphere, but ^{14}C is reduced only by the fossil fuel source. Using the ^{13}C content of woody tissues of trees, Stuiver (1978) has suggested that between 1850 and 1950 the biospheric release of carbon dioxide to the atmosphere was 120×10^9 tons of carbon while the release from fossil fuels combustion was only half this amount (60×10^9 tons). Stuiver cited forest cutting in the Lake States starting in 1870 and Pacific Northwest logging during the early 1900s as examples of deforestation contributing to the biospheric carbon dioxide release.

It has been suggested that increased atmospheric carbon dioxide may stimulate vegetative growth through enhancement of photosynthesis and that the biota may serve as a sink for carbon dioxide. Woodwell et al. (1978) considered this possibility and concluded that the available evidence does not support the

Table 2-2. Net Average Annual Input of Carbon (as Carbon Dioxide) into the Atmosphere and Accumulated Input since the Early Nineteenth Century due to Human Perturbation of Terrestrial Ecosystems

	Input (10^9 tons)	
Source	Present average annual	Accumulated
Reduction of forests		
Developed countries	0 ± 0.1	
Developing countries		45 ± 15
Forestation	-0.3 ± 0.1	
Deforestation	0.8 ± 0.4	
Use of fuel wood	0.3 ± 0.2	
Changes of organic matter in soil	0.3 ± 0.2	24 ± 15
Total	1.0 ± 0.6	70 ± 30

Source: Bolin (1977).

notion that the biota is or has been an appreciable sink for carbon. These authors supported the hypothesis that manipulation of terrestrial vegetation is a source rather than a sink for carbon dioxide and that the most probable range for global release from the biota is 4 to 8×10^{15} g of carbon annually. Contributions by various forest types are presented in Table 2-3.

Wong (1978) has estimated the atmospheric input of carbon dioxide from burning wood. He estimated input from forest fires in boreal and temperate regions resulting from both natural and human-related fires and fires in tropical regions caused by shifting cultivation. As indicated in Table 2-4, these sources resulted in a net input of carbon of approximately 1.5×10^{15} g carbon per year and this was solely from new tropical forest clearing.

An examination of global models for the natural carbon dioxide cycle has led Siegenthaler and Oeschger (1978) to conclude that we are deficient in our understanding of the role of the biosphere in the carbon cycle, that fertilization by

Table 2-3. Estimated Net Release of Carbon from Major Global Terrestrial Plant Communities

Plant community	Net release of carbon ($\times 10^{15}$ g)	% of total
Tropical forests	3.5	44
Temperate zone forests	1.4	18
Boreal forest	0.8	10
Other vegetation (including agriculture)	0.2	3
Total vegetation	5.9	75
Detritus and humus	2.0	25
Total land	7.9	100

Source: Woodwell et al. (1978).

Table 2-4. Area Burned and Net Input of Carbon Dioxide into the Atmosphere from Nonfossil Wood Burning

Ecosystem	Area burned (10^{10} m^2 yr^{-1})	Net carbon input (10^{15} g yr^{-1})
Boreal forest	1.2	0
Boreal nonwooded land	1.3	
Temperate forest	4.6	0
Temperate nonwooded land	4.6	
Tropical forest: new clearing		
FAO estimate	24.0	1.5
Rural population increase estimate	7.5	(0.7-2.2)
Tropical shifting cultivation: existing	300	0

Source: Wong (1978).

carbon dioxide and deforestation partly compensate one another, and in any case, that a sink for the large carbon dioxide release proposed by Woodwell and others cannot be found. Broecker et al. (1979) have employed several models to estimate the global carbon budget and have concluded that changes in forest biomass, relative to fossil fuel combustion, have not contributed significant carbon dioxide to the atmosphere.

4. Carbon Monoxide

Unlike carbon dioxide, the concentration of carbon monoxide does not appear to be increasing in "clean" atmospheres remote from excessive local input of carbon oxides. Carbon monoxide is input to the atmosphere in very approximately equal amounts from anthropogenic sources and from the biota on a global basis (Table 2-5). If the inputs estimated in Table 2-5 approximate the actual fluxes in nature, the role of forest ecosystems in the input of carbon monoxide

Table 2-5. Primary Sources of Atmospheric Carbon Monoxide

Source	CO (10^6 tons yr^{-1}) Total Min.	Max.	Northern Hemisphere Min.	Max.
Methane oxidation	400	4000	200	2000
Anthropogenic	600	1000	540	900
Biota				
Oceans	100	220	40	90
Terrestrial plants	20	200	14	140
Chlorophyll degradation	300	700	200	500

Source: Nozhevnikova and Yurganov (1978).

to the atmosphere may be significant. Since the global atmospheric concentration of carbon monoxide is not increasing, despite increasing combustion of fossil fuels, it must be assumed that an effective global carbon monoxide sink is operating. It will be suggested in Chapter 4 that the soils of forest ecosystems may play a particularly important role in this sink function.

B. Sulfur Pollutants

The atmospheric sulfur contaminants of primary interest to those concerned with vegetative and human health effects and environmental quality are currently sulfur dioxide and sulfates. The latter include sulfuric acid, metallic sulfates, and ammonium sulfate. Because of the role of sulfates in precipitation acidity and the increasing geographic area subject to precipitation of lowered pH, considerable interest is focused on the sulfate group. Sulfur containing air contaminants exert a major impact on agricultural and forest ecosystems.

1. Sulfur Cycle

In a recent review, Graedel (1979) has emphasized the uncertainties associated with quantifying and balancing the global sulfur cycle and with identifying the primary means of atmospheric sulfate formation.

Almost without exception, the traditional models of the sulfur cycle have indicated that approximately 50% of the sulfur in the atmosphere results from biological transformations of sulfur in soil and water ecosystems. The microbial activity within these natural ecosystems is presumed to volatilize sulfur in the form of hydrogen sulfide. The primary elements of the classic sulfur cycle are presented in Figure 2-2.

Much evidence exists to suggest that microbes produce hydrogen sulfide by two principal routes: sulfate reduction and organic matter decomposition. Sulfate reducers, *Desulfovibrio* and related bacteria, proliferate in swamp, mud, and poorly drained soils and employ sulfate as a terminal electron acceptor. An extremely large and diversified group of microorganisms, including aerobes, anaerobes, thermophiles, psychrophiles, bacteria, actinomycetes, and fungi decompose sulfur containing organic compounds and release hydrogen sulfide (Alexander, 1974).

The actual quantity of hydrogen sulfide produced by natural systems, while assumed by the traditional sulfur cycle to be substantial, is not directly measured and has been only crudely calculated by balancing the global sulfur cycle, to range from 58 to 110×10^6 tons of sulfur annually (Table 2-6).

A striking feature of the traditional sulfur cycle has been the considerable importance of natural ecosystems relative to human activities in placing sulfur compounds into the atmosphere. The natural release of sulfur to the atmosphere has been variously estimated to exceed anthropogenic input by two to 30 times (see Table 2-7). Marchesani et al. (1970) suggested that 0.07 ton of hydrogen

ATMOSPHERE

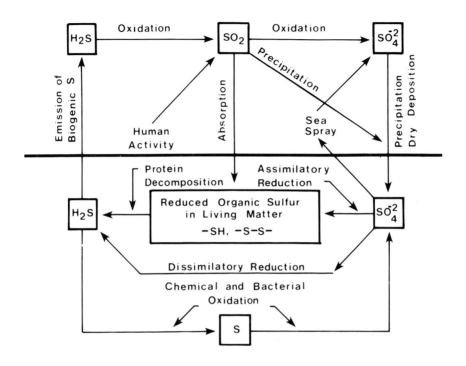

PEDOSPERE and HYDROSPHERE

Figure 2-2. Traditional elements and pathways of sulfur in the atmosphere and the biota. (From Hill, 1973)

Table 2-6. Estimates of Global Biogenic Production of Hydrogen Sulfide by Terrestrial Systems

Reference	Sulfur $(10^6 \text{ tons yr}^{-1})$
Junge (1963)	70
Eriksson (1963)	110
Robinson and Robbins (1968)	68
Friend (1973)	58
Kellogg et al. (1972)	90

Table 2-7. Global Transfer Rates of Sulfur to the Atmosphere

	Sulfur (10^6 tons yr^{-1})		
Sources	Terrestrial	Marine	Total
Human activity	50		50
Sea spray sulfate		43	43
Biogenic H_2S	90		90
Total			183

Source: Hill (1973).

sulfide was emitted per 1000 square miles daily from natural sources in the United States.

The atmospheric chemistry of the traditional sulfur cycle involves the oxidation of hydrogen sulfide to sulfur dioxide and the oxidation of sulfur dioxide to sulfates. Hydrogen sulfide may be oxidized by reactions involving atomic oxygen, molecular oxygen and ozone (Kellogg et al., 1972). Sulfur dioxide may be oxidized by reaction with OH, HO^2, and RO^2. The reaction with OH forms HSO^3 and is several orders of magnitude faster than the others (Wolff, 1979). Sulfates are also introduced directly to the atmosphere as sea salt particles from sea spray as indicated in Table 2-7.

2. New Sulfur Cycle Hypothesis

While there is little doubt that soil microbes can release hydrogen sulfide by reduction of sulfate and decomposition of organic compounds containing sulfur, there does not appear to be reliable, direct evidence that significant amounts of hydrogen sulfide are released from soils in natural ecosystems (Bremner and Steele, 1978). Microbially generated hydrogen sulfide may be rapidly converted in soil to metallic sulfides, for example, iron sulfide (Ayotade, 1977; Kittrick, 1976). When soils with high levels of sulfate and decomposable organic matter become poorly aerated (poorly drained) and are deficient in cations required to precipitate hydrogen sulfide, the latter gas may be released to the atmosphere.

An alternative and new hypothesis for sulfur input to the atmosphere from natural ecosystems, however, proposes that most of the sulfur volatilized from soils through microbial activity (Zinder and Brock, 1978) is in the form of organic compounds such as carbonyl sulfide, dimethyl sulfide, dimethyl disulfide, and methyl mercaptan (Bremner, 1977; Adams et al., 1979). The flux of sulfur gases from a variety of soils was given by Adams et al. (1979) to average 72 g sulfur m^{-2} yr^{-1} with a range of 0.002 to 152 g sulfur m^{-2} yr^{-1}. Most of the soils tested by Adams were tidal or poorly drained and, unfortunately, none was a forest soil. In addition to sulfur dioxide and hydrogen sulfide, recent advances in sulfur gas analysis demonstrate that carbonyl sulfide, carbon disulfide, dimethyl sulfide, and sulfur fluoride are components of tropospheric air (Bremner and Steele, 1978).

Banwart and Bremner (1976) and Bremner and Steele (1978) have thoroughly reviewed this new hypothesis for sulfur input to the atmosphere and conclude that most of the dimethyl sulfide originates via microbial degradation of methionine or dimethyl-β-propiothetin, most of the carbon disulfide through microbial degradation of cysteine or cystine, and at least some of the carbonyl sulfide by microbial decomposition of thiocyanates and isothiocyanates in plant materials. Dimethyl sulfide is presumed to be directly oxidized in the atmosphere to sulfate without the formation of a sulfur dioxide intermediate (Granat et al., 1976).

It is important to realize that this new hypothesis also has its critics (Maroulis and Bandy, 1977) and that much uncertainty still characterizes our appreciation of the sulfur cycle.

3. Input of Sulfur to the Atmosphere: Anthropogenic Sources Relative to Natural, Particularly Forest, Sources

Global sulfur budgets based on the new hypothesis concerning organic sulfur compounds in the atmosphere are relatively few. Estimates for natural source release of dimethyl sulfide and other volatile sulfur compounds has been presented by Hitchcock (1975) (see Table 2-8).

As indicated in Table 2-8, these data suggest a relatively minor input directly from vegetation and a relatively major release from the soil. The relative efficiency of forest soils compared to other soils for dimethyl sulfide release is not clear. It is clear, however, that the maximum estimated dimethyl sulfide release $(5.5 \times 10^{12}$ g S yr^{-1}) is modest compared with the anthropogenic release of approximately 65 $\times 10^{12}$ g S yr^{-1} (2% sulfate, 98% sulfur dioxide) (Granat et al., 1976). The release of dimethyl sulfide, however, may represent less than 10% of the total biogenic emission of sulfur (Bremner and Steele, 1978).

It is also possible that all models of the sulfur cycle have considerably overestimated the contribution of the biota. Granat et al. (1976) have suggested a total flux of reduced sulfur to the atmosphere of approximately 35-37 $\times 10^{12}$ g per year, less than one-half the estimate of several other authors.

In the Northern Hemisphere, where anthropogenically input sulfur has a residence time of very approximately two days, human sulfur inputs dominate natural fluxes. The current contribution to sulfur burden in submicron particles in the troposphere of the Northern Hemisphere is 0.17×10^{12} g natural relative to 0.23×10^{12} g human (Granat et al., 1976).

Table 2-8. Global Emissions of Dimethyl Sulfide from Natural Sources

Source	Sulfur (10^{12} g yr^{-1})
Marine algae	0.05
Fresh leaves	0.01
Senescent leaves	0.53
Soils	1.5-4.9
Total	2.1-5.5

Source: Granat et al. (1976).

C. Nitrogen Pollutants

The primary atmospheric contaminants containing nitrogen include ammonia, nitrogen oxides, and nitrates. Direct injury to humans or vegetation from gaseous ammonia is very infrequent, and has usually occurred in localized regions from accidental spills rather than continuous emissions to the atmosphere. Ammonia is important because of its reactions to form aerosols, especially ammonium sulfate. Nitrogen forms seven oxides that reflect each of its recognized oxidation states. Of these oxides, only nitrous oxide (N_2O), nitric oxide (NO), and nitrogen dioxide (NO_2) exist at measurable concentrations in the atmosphere and only nitrogen oxide and nitrogen dioxide are important air pollutants (Spedding, 1974). Nitrogen dioxide may directly effect vegetation in regions of high local release. Indirect impact on vegetation is much more important, however, because of the key role of oxides of nitrogen in forming generally phytotoxic oxidants, such as ozone and peroxyacetylnitrates. Since the primary sources of hydrogen ions that cause the phenomenon of acid precipitation are the strong mineral acids, sulfuric and nitric, nitrate is also an important air contaminant (Galloway et al., 1976).

1. Nitrogen Cycle

The global nitrogen cycle is extremely complex because of the large number and variable nature of the nitrogenous compounds that are involved, the myriad of processes operative, and the large regional and local deviations from generalized patterns. Söderlund and Svensson (1976) have thoroughly reviewed the various nitrogen cycles and have presented their own (Figure 2-3) that substantially differs in selected aspects from those of earlier efforts. Their model presents a net flow of ammonia and oxides of nitrogen from terrestrial to oceanic systems through the atmosphere. They conclude that natural fluxes of nitrogen oxides to the atmosphere may be smaller than generally suggested. This has the effect of increasing the relative importance of anthropogenic flows to the atmospheric compartment. These latter flows, estimated below, are the most accurately known fluxes of the nitrogen cycle. The greatest uncertainty of the cycle is associated with the quantification of denitrification processes that release molecular nitrogen and nitrous oxide into the atmosphere.

2. Sources of Ammonia

The sources of ammonia input to the atmosphere are varied and include minor release from various industries, coal combustion, fertilizer breakdown, volatilization from wild and domestic animal and human excreta, and the decomposition of organic matter in the soil. With the exception of the industrial release and coal combustion, the above sources are biological processes mediated by a variety of soil microorganisms. Alexander (1974) has listed alkaline pH, warm

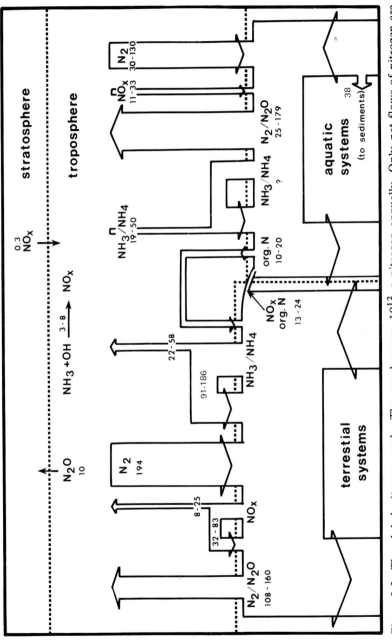

Figure 2-3. The global nitrogen cycle. The numbers are 10^{12} g nitrogen annually. Only net flows of nitrogen are presented. (From Söderlund and Svensson, 1976.)

temperatures, high rates of evaporation, and low cation exchange capacity of soils as all favoring microbial ammonia volatilization.

In the portions of the temperate zone subject to air masses contaminated by anthropogenic input, widespread acid aerosols would combine quickly with ammonia, and the latter would probably be found in particles rather than in the gaseous phase (Söderlund and Svensson, 1976).

3. Nitrogen Oxides

The high temperature combustion of fossil fuels for transportation, energy generation and the manufacture of petroleum products all contribute to the input of nitric oxide and nitrogen dioxide to the atmosphere. Söderlund and Svensson (1976) have presented an estimate of 19×10^{12} g of nitrogen as the annual anthropogenic input via nitrogen oxide release to the atmosphere. The use of higher combustion temperatures, probable in the future, will increase the atmospheric input of nitrogen oxides from fossil fuel burning. A less quantifiable, but probably larger amount of nitrogen oxides (exclusive of nitrous oxide) are released via several natural mechanisms including fixation by lightning (Junge, 1958; Ferguson and Libby, 1971), inflow from the stratosphere (Söderlund and Svensson, 1976), chemical conversion from ammonia in the troposphere (Crutzen, 1974; McConnell, 1973), and loss of gaseous nitric oxide from soils (Robinson and Robbins, 1975).

The global input of nitrogen to the atmosphere in the form of nitrogen oxides from natural sources may range from approximately equal to 4.5 (Söderlund and Svensson, 1976) to 15 (Rasmussen et al., 1975) times greater than the human input.

4. Role of Terrestrial Ecosystems, Particularly Forests, in the Nitrogen Cycle

The single, most important fixation of nitrogen on a global basis occurs in terrestrial ecosystems (Table 2-9). Of the total nitrogen fixed by these ecosystems, approximately equal amounts are fixed by agricultural, grassland, and forest systems (Table 2-10). Tables 2-9 and 2-10 establish the relative importance of forest ecosystems in the global fixation of nitrogen. In light of this importance, what is the significance of forests in the release of ammonia, nitrogen oxides and nitrates to the atmosphere?

5. Release of Ammonia

Living plants are generally not considered important sources of ammonia to the atmosphere. Farquhar et al. (1979), however, have presented evidence that corn plants release ammonia in normal air as they senesce. These investigators indicated that their observed rate of release was equivalent to 7 g of nitrogen ha^{-1}

Table 2-9. Global Fixation of Nitrogen

Process	Nitrogen (10^{12} g yr^{-1})
Biotic	
Terrestrial	139
Aquatic, pelagic	20-120
Abiotic	
Industrial	36
Combustion	19
Total	214-314

Source: Söderlund and Svensson (1976).

day^{-1} with a leaf area index of one. Comparable release of ammonia from tree leaves is undetermined.

Ammonia is generated during the mineralization of humus by a wide variety of bacteria and fungi. The acid pH, cool temperatures, low rates of evaporation, and high cation exchange capacity characteristic of many temperate forest soils would not favor the volatilization of ammonia. Kim (1973), however, monitored the ammonia released from a Korean pine forest, oak forest, and grassland from May through July. His results showed that an average of 3.4, 2.6, and 1.8 kg ammonia ha^{-1} $week^{-1}$ were evolved from the pine and oak forest soil and in the grassland, respectively.

The use of fertilizers containing organic nitrogen compounds can result in substantial volatilization from treated soils. This is particularly true for urea, a common fertilizer, as this compound is hydrolyzed rapidly by cosmopolitan and common urease containing heterotrophic microbes (Alexander, 1974). Urease activity has been reported for numerous forest soils (Wollum and Davey, 1975). The application of urea to forested ecosystems is receiving active research attention. In the United States, experimental applications have been made to Douglas fir (Miller and Reukema, 1974, 1977), western hemlock (DeBell, 1975; Webster et al., 1976) and West Virginia hardwoods (Aubertin et al., 1973) and in Canada to jack pine (Armson, 1972; Morrison et al., 1976). The rate of urea application utilized in these tests generally ranged from 113 to 555 kg ha^{-1} (100-500 pounds $acre^{-1}$). The quantity of ammonia lost by volatilization may be substantial and has been estimated to range from 18% to 75% of applied nitrogen (Volk, 1959, 1970). Many of the studies supporting high percentage loss, however, have been performed in laboratory environments. Hargrove and Kissel (1979) have com-

Table 2-10. Global Nitrogen Fixation by Terrestrial Ecosystems

Ecosystem	Nitrogen (10^{12} g yr^{-1})
Agricultural	44
Grassland	45
Forest	40
Other	10
Total	139

Source: Burns and Hardy (1975).

pared ammonia volatilization from surface urea applications in the laboratory and in the field, and found the latter small relative to the former.

While the use of fertilizer in forest management practice is miniscule compared to agricultural practice, it is important to realize that the interest in fertilizer use on forests is growing and that the forest systems receiving the greatest research interest are expansive. Large scale fertilization programs currently exist in Sweden and Finland (Armson et al., 1975).

6. Nitrogen Oxides and Nitrates

The abiotic and biotic processes of denitrification globally release large quantities of nitrogen oxides to the atmosphere. Söderlund and Svensoon (1976) have indicated that terrestrial ecosystems may release from 1 to 14×10^{12} g nitrogen oxide-nitrogen annually. Focht and Verstraete (1977) have estimated that terrestrial denitrification may generate an annual surface flux of approximately 3 kg nitrogen ha^{-1}.

Biological denitrification is a respiratory function that employs nitrate as a terminal electron acceptor. Subsequent nitrogen reductions may release nitrogen dioxide, nitrous oxide, nitric oxide, and dinitrogen. Denitrification is carried out by a diverse group of microbes including bacteria and fungi, autotrophs and heterotrophs, and aerobes and anaerobes in a variety of habitats including poorly drained and well-drained and acid and alkaline soils (Delwiche and Bryan, 1976; Wollum and Davey, 1975; Bremner and Blackmer, 1978). As in the case of ammonia release, fertilization may stimulate denitrification (Blackmer and Bremner, 1979; Hutchinson and Mosier, 1979). In Kim's (1973) study of Korean forest soils, the release of nitrogen dioxide averaged 0.21, 0.12, and 0.19 kg ha^{-1} $week^{-1}$ in a pine forest, oak forest, and grassland, respectively.

The modest evidence supporting the importance of denitrification has not come from forest soils. In view of the ubiquitous character of the denitrifying population and their range of habitats, however, it appears highly probable that forest soils do denitrify (Wollum and Davey, 1975). It is very clear that no conclusion can be provided concerning the importance of forest soils relative to other soils in the release of nitrogen oxides to the atmosphere. Once released to the atmosphere, nitrous oxide is assumed to be relatively inert, with the major portion destroyed by photodissociation in the upper troposphere and stratosphere. Nitric oxide is presumed to be either oxidized to nitrogen dioxide or photolyzed to molecular nitrogen. Nitrogen dioxide is primarily removed by precipitation, presumably in the form of nitric acid (HNO_3) (Rasmussen et al., 1975).

D. Summary

Carbon dioxide is increasing in the atmosphere at a rate of approximately 1 ppm (1.8×10^3 μg m^{-3}) per year. Currently the 700 billion tons of carbon in the atmosphere in the form of carbon dioxide are supplemented with an additional 2.3 billion tons annually resulting in a 3% increase every decade. Anthropogenic combustion of fossil fuels contributes five billion tons of carbon to the atmosphere annually (Ember, 1978). A comparable or greater amount of carbon may be released to the atmosphere via forest clearance and soil losses. Approximately 75% of carbon introduced to the atmosphere from all sources is removed by effective sink mechanisms. The most important sink is presumed to be the oceans with some potential contribution from new vegetation. The imperfections in our understanding of global carbon pools and cycles and carbon models preclude conclusive assessment of the precise role of forest ecosystems in the increased loading of carbon dioxide in the atmosphere. If forests are ultimately judged to have an important role in the carbon dioxide cycle, it is probable that manipulations of tropical forest ecosystems will prove to be much more important than changes in temperate or boreal systems.

Our appreciation of the sulfur cycle is undergoing major change as new developments in sulfur gas analysis reveal the increasing importance of organic sulfur volatiles from natural systems. Sulfur may be input to the atmosphere in amounts approximately 50-70 $\times 10^{12}$ g yr^{-1} from both human activities and natural systems. It is probable that soil metabolism generates most of the natural contribution and that much of it is in the form of organic sulfur gases, but it is not possible to specifically rank forest soils relative to other soils in this production.

The important nitrogenous gases of air quality interest, ammonia and oxides of nitrogen, are released in important amounts from terrestrial soils. Forest soils do not appear unique nor outstanding in the production of these gases. The size of forest ecosystems, however, may make their release comparable with agricultural systems. Adoption of broad scale forest fertilization programs would increase ammonia release and enhance denitrification. The role of forest ecosystems in nitrate generation is intimately linked to the nitrogen oxide cycle and as a result cannot be specifically appraised at the present time.

References

Adams, D. F., S. O. Farwell, M. R. Pack, and W. L. Barnesberger. 1979. Preliminary measurements of biogenic sulfur-containing gas emissions from soils. Air Pollut. Control Assoc. 29:380-383.

Adams, J. A. S., M. S. M. Mantovani, and L. L. Lundell. 1977. Wood versus fossil fuel as a source of excess carbon dioxide in the atmosphere: A preliminary report. Science 196:54-56.

Alexander, M. 1974. Microbial formation of environmental pollutants. Adv. Applied Microbiol. 18:1-73.

Armson, K. A. 1972. Fertilizer distribution and sampling techniques in the aerial fertilization of forests. Tech. Report No. 11, University of Toronto, Ontario, Canada, 27 pp.

Armson, K. A., H. H. Krause, and G. F. Weetman. 1975. Fertilization response in the northern coniferous forest. *In*: B. Bernier and C. H. Winget (Eds.), Forest Soils and Forest Land Management, Proc. Fourth North Amer. Forest Soils Conf., Laval Univ., Quebec, Les Presses de l'Université Laval, Quebec, Canada, pp. 449-466.

Aubertin, G. M., D. E. Smith, and J. H. Patric. 1973. Quantity and quality of streamflow after urea fertilization on a forested watershed: First-year results. *In*: Forest Fertilization Symp. Proc., U.S.D.A. Forest Service, Genl. Tech. Report NE-3, Upper Darby, Pennsylvania, pp. 88-100.

Ayotade, K. A. 1977. Kinetics and reactions of hydrogen sulfide in solution of flooded rice soils. Plant Soil 46:381-389.

Banwart, W. L., and J. M. Bremner. 1976. Evolution of volatile sulfur compounds from soils treated with sulfur containing organic materials. Soil Biol. Biochem. 8:439-443.

Blackmer, A. M., and J. M. Bremner. 1979. Stimulatory effect of nitrate on reduction of N_2O to N_2 by soil microorganisms. Soil Biol. Biochem. 11:313-315.

Bolin, B. 1977. Changes of land biota and their importance for the carbon cycle. Science 196:613-615.

Bolin, B., E. T. Degens, S. Kempe, and P. Ketner (Eds.). 1977. The Global Carbon Cycle. SCOPE Report 13. Wiley, New York, 491 pp.

Bremner, J. M. 1977. Role of organic matter in volatilization of sulfur and nitrogen from soils. *In*: Proceedings of Symposium on Soil Organic Matter Studies, Vol. 11, pp. 229-240, Braunschweig, Federal Republic of Germany Sept. 6-10, 1976. Internat. Atomic Energy Agency, Vienna.

Bremner, J. M., and A. M. Blackmer. 1978. Nitrous oxide: Emission from soils during nitrification of fertilizer nitrogen. Science 199:295-296.

Bremner, J. M., and C. G. Steele. 1978. Role of microorganisms in the atmospheric sulfur cycle. Adv. Microbial Ecol. 2:155-201.

Broecker, W. S., T. Takahashi, H. J. Simpson, and T. H. Peng. 1979. Fate of fossil fuel carbon dioxide and the global carbon budget. Science 206:409-418.

Burns, R. S., and R. F. W. Hardy. 1975. Nitrogen Fixation in Bacteria and Higher Plants. Springer-Verlag, New York, 189 pp.

Crutzen, P. J. 1974. Photochemical reactions initiated by and influencing ozone in unpolluted tropospheric air. Tellus 26:47-49.

DeBell, D. S. 1975. Fertilize western hemlock—yes or no? *In*: Global Forestry and the Western Role, Western Forestry and Conservation Association Proc. Portland, Oregon, pp. 140-143.

Deevey, E. S. 1973. Sulfur, nitrogen and carbon in the biosphere. *In*: G. M. Woodwell and E. V. Pecan (Eds.), Carbon and the Biosphere, Proceedings 24th Brookhaven Symposium in Biology, Upton, N.Y., May 16-18, 1972. Tech. Inform. Center, U.S. Atomic Energy Commission, pp. 182-190.

Delwiche, C. C., and B. A. Bryan. 1976. Denitrification. Annu. Rev. Microbiol. 30:241-262.

Ember, L. R. 1978. Global environmental problems: Today and tomorrow. Environ. Sci. Technol. 12:874-876.

Eriksson, E. 1963. The yearly circulation of sulfur in nature. J. Geophys. Res. 68:4001-4008.

Farquhar, G. D., R. Wetselaar, and P. M. Firth. 1979. Ammonia volatilization from senescing leaves of maize. Science 203:1257-1258.

Ferguson, E. E., and W. F. Libby. 1971. Mechanism for the fixation of nitrogen by lightning. Nature 229:37-38.

Focht, D. D., and W. Verstraete. 1977. Biochemical ecology of nitrification and denitrification. Adv. Microbial Ecol. 1:135-214.

Friend, J. P. 1973. The global sulfur cycle. In: S. I. Rasool (Ed.), Chemistry of the Lower Atmosphere. Plenum, New York, pp. 177-201.

Galloway, J. N., G. E. Likens, and E. S. Edgerton. 1976. Hydrogen ion speciation in the acid precipitation of the northeastern United States. In: L. S. Dochinger and T. A. Seliga (Eds.), Proc. First International Symp. on Acid Precipitation, U.S.D.A. Forest Service Genl. Tech. Report NE-23, Upper Darby, Pennsylvania, pp. 383-396.

Graedel, T. E. 1979. The oxidation of atmospheric sulfur compounds. In: Proc. MASS-APCA Technical Conf. on the Question of Sulfates, Philadelphia, Pennsylvania, April 13-14, 1978 (in press).

Granat, L., H. Rodhe, and R. O. Hallberg. 1976. The global sulphur cycle. In: B. H. Svensson and R. Söderlund (Eds.), Nitrogen, Phosphorus and Sulphur, SCOPE Report No. 7, Ecological Bulletins (Stockholm) 22:89-134.

Hargrove, W. L., and D. E. Kissel. 1979. Ammonia volatilization from surface application of urea in the field and laboratory. Soil Sci. Soc. Am. J. 43: 359-363.

Hill, F. B. 1973. Atmospheric sulfur and its link to the biota. In: G. M. Woodwell and E. V. Pecan (Eds.), Carbon and the Biosphere, Proc. 24th Brookhaven Symposium in Biology, Upton, N.Y., May 16-18, 1972. Tech. Inform. Center, U.S. Atomic Energy Commission, pp. 159-181.

Hitchcock, D. R. 1975. Biogenic contributions to atmospheric sulfate levels. Second Annual Conference on Water Reuse, Chicago.

Hitchcock, D., and A. E. Wechsler. 1972. Biogenic Cycling of Atmospheric Trace Gases. Final Report, NASA Contract HASW-2128, Arthur D. Little, Inc., Cambridge, Massachusetts.

Hutchinson, G. L., and A. R. Mosier. 1979. Nitrous oxide emissions from an irrigated cornfield. Science 205:1125-1127.

Junge, C. E. 1958. The distribution of ammonia and nitrate in rainwater over the United States. Trans. Am. Geophys. Union 39: 241-248.

Junge, C. E. 1963. Air Chemistry and Radioactivity. Academic Press, New York, pp. 59-74.

Kellogg, W. W., R. D. Cadle, E. R. Allen, A. L. Lazus, and E. A. Martell. 1972. The sulfur cycle. Science 175:587-596.

Kim, C. M. 1973. Influence of vegetation types on the intensity of ammonia and nitrogen dioxide liberation from soil. Soil Biol. Biochem. 5:163-166.

Kittrick, J. A. 1976. Control of Zn^{2+} in the soil solution by sphalerite. Soil Sci. Soc. Am. J. 40:314-317.

Marchesani, V. J., T. Towers, and H. C. Wohlers. 1970. Minor sources of air pollutant emissions. J. Air Pollut. Control. Assoc. 20:19-22.

Maroulis, P. J., and A. R. Bandy. 1977. Estimate of the contribution of biologically produced dimethyl sulfide to the global sulfur cycle. Science 196:647-648.

McConnell, J. C. 1973. Atmospheric ammonia. J. Geophys. Res. 75:7812-7821.

Miller, R. E., and D. L. Reukema. 1974. Seventy-five-year-old Douglas-fir on high quality site respond to nitrogen fertilizer. U.S.D.A. Forest Service, Res. Note PNW-281, Portland, Oregon, 8 pp.

Miller, R. E., and D. L. Reukema. 1977. Urea fertilizer increases growth of 20-year-old, thinned Douglas-fir on a poor quality site. U.S.D.A. Forest Service, Res. Note PNW-291, Portland, Oregon, 8 pp.

Morrison, I. K., F. Hegye, N. W. Foster, D. A. Winston, and T. L. Tucker. 1976. Fertilizing semimature jack pine (*Pinus banksiana* Lamb.) in northwestern Ontario: Fourth-year results. Report No. 0-X-240, Can. For. Service, Dept. Environment, Sault Ste. Marie, Ontario, Canada, 42 pp.

Nozhevnikova, A. N., and L. N. Yurganov. 1978. Microbiological aspects of regulating the carbon monoxide content in the earth's atmosphere. Adv. Microbial. Ecol. 2:203-244.

Rasmussen, K. H., M. Taheri, and R. L. Kabel. 1975. Global emissions and natural processes for removal of gaseous pollutants. Water, Air, Soil Pollut. 4:33-64.

Reichle, D. E., B. E. Dinger, W. T. Edwards, W. F. Harris, and P. Sollins. 1973. Carbon flow and storage in a forest ecosystem. *In*: G. M. Woodwell and E. V. Pecan (Eds.), Carbon and the Biosphere, Proc. 24th Brookhaven Symposium in Biology, Upton, N.Y., May 16-18, 1972. Tech. Inform. Center, U.S. Atomic Energy Commission, pp. 182-190.

Robinson, E., and R. C. Robbins. 1968. Sources, Abundance and Fate of Gaseous Atmospheric Pollutants. Final Report SRI Project PR-6755. Stanford Research Institute, Menlo Park, California.

Robinson, E., and R. C. Robbins. 1975. Gaseous atmospheric pollutants from urban and natural sources. *In*: S. F. Singer (Ed.), The Changing Global Environment. Reidel, Dordrecht, pp. 111-123.

Siegenthaler, U., and H. Oeschger. 1978. Predicting future atmospheric carbon dioxide levels. Science 199:388-395.

Söderlund, R., and B. H. Svensson. 1976. The global nitrogen cycle. *In*: B. H. Svensson and R. Söderlund (Eds.), Nitrogen, Phosphorus and Sulphur-Global Cycles. SCOPE Report No. 7. Ecological Bulletin No. 22. Royal Swedish Academy of Sciences, Stockholm, pp. 23-73.

Spedding, D. J. 1974. Air Pollution. Clarendon Press, Oxford, 76 pp.

Spurr, S. H., and H. J. Vaux. 1976. Timber: Biological and economic potential. Science 191:752-756.

Stuiver, M. 1978. Atmospheric carbon dioxide and carbon reservoir changes. Science 199:253-258.

U.S. Environmental Protection Agency. 1976. Diagnosing Vegetation Injury Caused by Air Pollution. Contract No. 68-02-1344. U.S.E.P.A., Air Pollution Training Institute, Research Triangle Park, North Carolina.

Volk, G. M. 1959. Volatile loss of ammonia following surface application of urea to turf or bare soils. Agron. J. 51:756-749.

Volk, G. M. 1970. Gaseous loss of ammonia from prilled urea applied to slash pine. Soil Sci. Soc. Am. Proc. 34:513-516.

Webster, S. R., D. S. DeBell, K. N. Wiley, and W. A. Atkinson. 1976. Fertilization of western hemlock. Proc. Western Hemlock Manage. Conf., Univ. Washington, Seattle, Washington, pp. 247-251.

Whittaker, R. H. 1975. Communities and Ecosystems. Macmillan, New York, 385 pp.

Wolff, G. T. 1979. The question of sulfates: A conference summary. J. Air Pollut. Control Assoc. 29:26-27.

Wollum, A. G., II, and C. B. Davey. 1975. Nitrogen accumulation, transformation transport. *In*: B. Bernier and C. H. Winget (Eds.), Forest Soils and Forest Land Management, Proc. Fourth North Amer. Forest Soils Conf., Laval Univ., Quebec, Les Presses de l'Université Laval, Quebec, Canada, pp. 67-106.

Wong, C. S. 1978. Atmospheric input of carbon dioxide from burning wood. Science 200:197-200.

Woodwell, G. M. 1978. The carbon dioxide question. Sci. Amer. 238:34-43.

Woodwell, G. M., R. H. Whittaker, W. A. Reiners, G. E. Likens, C. C. Delwiche, and D. B. Botkin. 1978. The biota and the world carbon budget. Science 199: 141-146.

Zinder, S. H., and T. D. Brock. 1978. Microbial transformations of sulfur in the Environment. Part II. *In*: J. O. Nriagu (Ed.), Ecological Impacts. Wiley, New York, pp. 445-466.

3

Forests as Sources of Hydrocarbons, Particulates, and Other Contaminants

In addition to whatever contribution forests may make to the atmospheric burden of carbon, sulfur, and nitrogen oxides, they are known to be important natural sources of hydrocarbons and particulates. Volatile hydrocarbons are released by a variety of woody plants during the course of normal metabolism. Pollen, the most significant particulate contaminant released by forests from the standpoint of human health, is also produced, of course, during normal reproductive metabolism. Hydrocarbon aerosols are viewed as an increasingly important particulate emission from forests. Forest burning, whether naturally occurring or artificially ignited, also produces hydrocarbons, particulates as well as carbon oxides. Even though forest fires may be a natural recurring event in most forest ecosystems, the pollutants generated by this process are not the result of normal metabolism but rather are generated by combustion of forest biomass. As a result, the latter are discussed in Section C.

A. Volatile Hydrocarbons

These organic gases vary greatly in chemical reactivity depending on structure. The paraffins, or aliphatic hydrocarbons, that contain the maximum number of hydrogen atoms (saturated) while not chemically inert are relatively unreactive and of limited importance as atmospheric components (Calvert, 1976). Alkenes (olefins) are open-chain hydrocarbons containing one or more carbon-carbon double bonds and alkynes contain one or more carbon-carbon triple bonds. Both groups do not contain the maximum number of hydrogen atoms and are termed unsaturated. Aromatic hydrocarbons contain ring systems (benzene structure) in which all carbon atoms are linked in a system of conjugated double bonds. Since all unsaturated organic gases are chemically reactive, many of them are important atmospheric contaminants.

Polluted urban atmospheres contain in excess of 100 different hydrocarbons, the most reactive of which are the olefins. The primary significance of these olefins is their important role in the synthesis of photochemical oxidants.

1. Sources of Hydrocarbons

A variety of anthropogenic and natural processes releases hydrocarbons to the atmosphere. The primary anthropogenic source in urban areas is motor vehicle exhaust (Heuss et al., 1974) with significant local contributions from incinerators (Davies et al., 1976) and industrial and other sources (Feldstein, 1974; Parsons and Mitzner, 1975). It has been estimated that the total global anthropogenic release of hydrocarbons may approximate 88×10^6 tons annually (Stern et al., 1973). Robinson and Robbins (1968) have estimated that the total annual emission of olefins and aromatics from the combustion of various fuels a decade ago was roughly 27×10^6 tons. This latter figure, 27×10^6, was the recent (1976) estimate of the Environmental Protection Agency for the anthropogenic hydrocarbon release for the United States alone exclusive of forest and agricultural burning (U.S. Environmental Protection Agency, 1977).

Globally natural biological processes may release volatile organic gases to the atmosphere in amounts many times the quantity released by the activities of human beings. Vegetation in general, and trees in particular, release various hydrocarbons to the atmosphere.

2. Forests as Sources of Hydrocarbons

In 1955, F. W. Went, then Director of the Missouri Botanical Garden, advanced the hypothesis that vegetation contributes to air pollution by the release of organic gases (Went, 1955). Went was particularly interested in the fate of plant products that were derivatives of isoprene. These compounds included terpenes,

$$H_2C = C - C - CH_2$$

with CH_3 above the first C and H below the second C.

isoprene unit

carotenoids, resin acids, rubber, and phytol. Went estimated that approximately 2×10^8 tons of organic volatile plant products were released to the global atmosphere annually (Went, 1960a,b).

Turk and D'Angio (1962) in their efforts to evaluate the sensory quality of "freshness" in natural area atmospheres studied the atmospheric composition along the coast in New Jersey, in a Connecticut state forest, and on Mount Washington in New Hampshire. They detected aromatic and unsaturated organic gases at both the coastal and forest sites and observed that the forest location was the richer of the two. Major et al. (1963) presented evidence that the leaves of several trees were capable of releasing the aldehyde α-hexanal from their leaves when injured at certain times during the growing season.

R. A. Rasmussen, a student of F. W. Went, stressed the importance of terpenes as important organic volatiles released from vegetation (Rasmussen, 1964). Terpenoid compounds are an extremely diverse group of organic materials built of isoprene units. Current wisdom suggests that isopentenyl pyrophosphate, and not isoprene itself, may be the actual building block of the terpenoids (Hess, 1975). Two, three, four, and more isopentenyl residues may be combined to form monoterpenoids (C_{10} compounds), sesquiterpenoids (C_{15} compounds), diterpenoids (C_{20} compounds), and triterpenoids (C_{30} compounds). Hydrocarbon terpenoids are specifically termed terpenes. Employing a gas chromatograph mounted in a mobile trailer laboratory, Rasmussen and Went (1965) studied volatile organic release in forest areas in North Carolina, Virginia, Missouri, and Colorado. They found three monoterpenes, α-pinene, β-pinene, and myrcene, along with isoprene particularly abundant. They also observed that concentrations varied with meteorological conditions and density of the vegetation. Terpene release from the vegetation was higher in the summer than winter. In the fall leaf litter became a major source of aromatic materials.

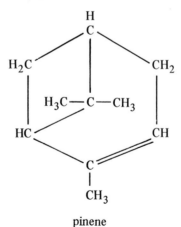

pinene

The terpenes characteristic of various tree species are quite variable and are under fairly rigid genetic regulation (Table 3-1). While considerable uncertainty has existed concerning the importance of the hemiterpene isoprene, Rasmussen (1970) presented evidence identifying isoprene as an important forest emission to the atmosphere. Numerous angiosperm species are efficient isoprene emitters. Unlike monoterpenes, isoprene is released from foliage only in the light.

Table 3-1. Monoterpenes Associated with Various Coniferous Species

Terpene	Douglas fir[a,b]	Western white pine[c]	Slash pine[d]
α-Pinene	X	X	X
Camphene	X	X	X
β-Pinene	X	X	X
Sabinene	X		
Δ-3-Carene	X	X	X
Myrcene	X	X	X
Limonene	X	X	X
β-Phellandrene	X		
γ-Terpinene	X		
Terpinolene	X		
α-Phellandrene			X
1,8-Cineole	X		
2-Hexenal	X		
Ethyl caproate	X		
cis-Ocimene	X		
Citronellal	X		
Linalool	X		
Fenchyl alcohol	X		
Bornyl acetate	X		
Sesquiterpene HC	X		
Terpinen-4-ol	X		
β-Caryophyllene	X		
Terpene alcohol	X		
Citronellyl acetate	X		
α-Terpineol	X		
Citronellol	X		
Geranylacetate	X		

[a] Radwan and Ellis (1975).
[b] Maarse and Kepner (1970).
[c] Hanover (1966).
[d] Squillace (1971).

In 1972, Rasmussen reviewed his own work and that of several others in an effort to inventory the sources of forest terpenes, estimate terpene emission rates, and calculate the significance of tree released hydrocarbons on a global basis. Six monoterpenes, α-pinene, camphene, β-pinene, limonene, myrcene, and β-phellandrene, were judged to be the major terpenic emissions from gymnosperm foliage. Numerous angiosperms, along with a few gymnosperms, were concluded to release isoprene. Table 3-2 indicates the relative importance of hemiterpene and monoterpene emitters for major United State forest types. Using a generalized release rate of 100 ppb hr^{-1}, Rasmussen estimated the relative importance of forest terpene emissions to the atmosphere on a global basis to approximate 175 × 10^6 tons of reactive materials. This is roughly six times the amount of hydrocarbons produced by human beings.

Table 3-2. Percentage of Monoterpene and Hemiterpene Emitters for Various Forest Regions of the United States

	% total U.S. forest area	% α-pinene emitters	% isoprene emitters
Eastern type groups			
Softwood types			
Loblolly-shortleaf pine	11	∿ 100	Some from oak and sweet-gum associates
Longleaf-slash pine	5	∿ 100	Some from oak and sweet-gum associates
Spruce-fir	4	∿ 75	25 from spruce, which also emits α pinene
White-red jack pine	2	∿ 90	10 from aspen trees
Subtotal	22	∿ 91	∿ 9
Hardwood types			
Oak-hickory	23	∿ 10	70, diluted by hickory, maple, and black walnut
Oak-gum-cypress	7	∿ 50	50 from plurality of oak, cottonwood and willow
Oak-pine	5	∿ 30	60, diluted by black gum and hickory associates
Maple-beech-birch	6	∿ 15	—, terpene foliages are hemlock and white pine
Aspen-birch	5	∿ 20	60, diluted by birch, α-pinene source is balsam fir and balsam poplar
Elm-ash-cottonwood	4	—	30 from cottonwood, sycamore, willow
Subtotal	50	∿ 21	∿ 45
Total	72	—	—
Western type groups			
Softwoods			
Douglas fir	7	∿ 100	—
Ponderosa pine	7	∿ 100	5 from aspen associates
Lodgepole pine	3	∿ 90	10 from Englemann spruce and aspen
Fir-spruce	3	∿ 100	40 from spruce trees
Hemlock-Sitka spruce	2	∿ 100	25 from Sitka spruce
White pine	1	∿ 100	5 from Englemann spruce
Larch	1	∿ 100	—
Redwood	0.5	∿ 100	—
Subtotal	24	∿ 98	∿ 12
Hardwoods	2	—	∿ 100 from aspen trees
Total	26		

Source: Rasmussen (1972).

In addition to terpenes the biota is the potential source of additional hydrocarbons to the atmosphere. Ethylene is released by numerous angiosperms during flowering and fruit maturation, and in response to injury and infection of certain tissues. Shain and Hillis (1972) have provided unique evidence that ethylene is also released by gymnosperms under certain conditions. Methane (nonreactive) is released from anaerobic ecosystems due to the activity of methanogenic bacteria (Mah et al., 1977; Dacey and Klug, 1979). These organisms are generally presumed unimportant in forest systems, unless the forests contain flooded soils for extended periods.

B. Particulates

1. Pollen

The medical significance of certain tree pollen makes pollen grains one of the most important air contaminants produced by woody vegetation. The time and method of opening of anthers or microstrobili, dehisence, and subsequent release of pollen are variable depending on tree family. Release in gymnosperms follows a parting of microstrobili sporophylls. Retraction of bract scales caused by dehydration frees pollen to be released by wind or shaking. All gymnosperm pollen, with the exception of cycads, is distributed following release by wind (anemophilous). Angiosperm pollen release follows one of numerous opening patterns of anther sac walls. Wall rupture is due to shrinkage occasioned by change in atmospheric humidity. In addition to wind, angiosperm pollen may be distributed following release by birds, bats, and insects (zoophilous/entomophilous) (Stanley and Linkskens, 1974).

a. Characteristics

The size of pollen grains among modern flowering plants varies from approximately 5 X 2.4 mm in *Myosotis* to 200 mm or greater in certain Cucurbitaceae and Nyctaginaceae (Erdtman, 1969). Wind pollinated species generally have a grain size in the range of 17 to 58 mm, while insect or animal pollinated species are typically larger or smaller (Stanley and Linskens, 1974). Size, volume, and weight variations of various tree pollens are presented in Table 3-3. The elaborate and variable shape and sculpturing of pollen allowing identification to species and the resistance of the walls of pollen grains allowing persistence overtime has made them extremely valuable in ecological, geological, ethnological, and medical studies. Excellent keys and descriptions of tree pollen are provided by Moore and Webb (1978), Wodehouse (1945) and Nilsson et al. (1977).

b. Production

The production and release of tree pollen is very variable and dependent on species, strategy of pollen transfer, time of year, numbers of anthers, microsporophylls and flowers produced, climate, tree age, and health. Entomophilus species (for example, *Acer, Malus*) have appreciably lower pollen production per anther

Table 3-3. Pollen Dimensions, Volume, and Weight for Selected Tree Species

| Species | Dimensions (μm) | | | Volume (10^{-9} cm^3) | Weight (10^{-9} g) |
	Length	Width	Height		
Silver fir	97.8	102.9	62.7	499.4	251.6
Cephalonica fir	97.1	98.6	86.2	422.6	212.2
Norway spruce	85.8	80.5	66.3	278.2	110.8
Scotch pine	41.5	45.9	36.0	35.5	37.0
European larch	76.0	72.0	50.0	180.2	176.3
Douglas-fir	84.8	81.1	54.8	219.2	188.8
Sugar maple	32.5	23.6	24.6	16.5	6.6
Horsechestnut	31.0	16.4	18.2	4.8	0.9
European alder	26.4	22.8	13.7	4.4	1.4
European birch	10.1	10.1	16.8	2.9	0.8
European beech	55.1	40.5	41.1	50.3	26.0
English oak	40.8	26.1	21.5	13.3	5.7
European cut-leaf linden	40.5	40.1	20.6	15.0	6.5
Rock elm	33.4	32.7	17.7	12.8	6.8

Source: Stanley and Linskens (1974).

than anemophilous species (for example, *Fagus*). Some representative estimates of pollen production for various species are presented in Tables 3-4 and 3-5. The variableness of pollen production may follow a predictable pattern in certain species. *Fraxinus* and *Ulmus* have a high yield approximately every third year, while American beech may give high yields every other year (Stanley and Liskens, 1974).

Hyde (1951) emphasized the importance of the influence of climate on pollen formation by monitoring seven tree genera over a six-year period in Great Britain (Table 3-6). These data suggest substantial influence of climatic variation on pollen production.

Table 3-4. Pollen Production by Reproductive Parts of Various Trees

Species	Number of pollen grains anther^{-1}	Number of pollen grains flower^{-1}	Number of pollen grains catkin^{-1}
Norway maple	1,000	8,000	
Apple	1,400-6,250		
European ash	12,500		
Common juniper		400,000	
Scotch pine		160,000	
Norway spruce		600,000	
European white birch			6,000,000
European alder			4,500,000
English oak			1,250,000
European beech			175,000

Source: Erdtman (1969).

Table 3-5. Volume of Pollen Yield by Reproductive Parts of Various Trees

			ml		
Genus	Catkin⁻¹	Inflorescence⁻¹	Strobili⁻¹	Flower⁻¹	Head⁻¹
Alnus	.04				
Betula	.01-.2				
Fagus		.01-.016			
Larix			.01-.014		
Liquidambar				.2-.3	
Pinus			.7-.35		
Platanus					.08
Populus	.5-1				
Pseudotsuga			.007-.037		
Ulmus				.001-.004	

Source: U.S.D.A. Forest Service (1974).

Of greatest interest to those interested in air contamination by pollen is the amount released per tree per unit time. Some estimations are presented in Table 3-7.

c. Distribution

Movement of pollen is critically important in air quality considerations. While animal disseminated pollen is moved in predictable patterns and for limited distances, anemophilous pollen may be moved considerable distances in complex patterns. The latter has been intensively studied and even a summary treatment is beyond the scope of this book. Pollen can spread over wide distances. Distance traveled is a function of sedimentation rate which is dependent on specific gravity, size, form, and degree of clumping (Stanley and Liskens, 1974). For various trees, therefore, dispersion distance may vary considerably (Table 3-8). Tauber (1965, 1967) has proposed a model for pollen dispersion from a forest that has three primary components. Trunk space component is the pollen that falls vertically from the tree and shrub canopies. Since air movement is slower below the canopy than above, most pollen in the trunk space component is deposited in

Table 3-6. Annual Variation in Pollen Productivity

	Average catch	Catch by year as % of 1943-1948 average					
Genus	1943-1948	1943	1944	1945	1946	1947	1948
Pinus	386	92	121	72	127	40	146
Alnus	385	45	226	120	55	34	118
Fraxinus	675	330	25	24	159	54	8
Fagus	273	23	140	11	143	8	275
Betula	620	57	160	42	106	22	214
Ulmus	4579	84	87	180	50	48	150
Quercus	2776	58	90	42	280	10	117

Source: Hyde (1951).

Table 3-7. Annual Pollen Production per Tree for Selected Species

Species	No. pollen grains ($\times 10^6$)[a]	kg[b]
European beech	409	0.15
Sessile oak	654	—
European hornbeam	3149	—
Norway spruce	5481	0.4
European birch	5562	0.03
European filbert	5603	0.05
Littleleaf linden	5603	—
Scotch pine	6462	0.12
Alder	7239	0.05

[a] From Erdtman (1969).
[b] From Brooks (1971).

relatively short distances. Canopy component is the pollen that escapes from the tree crowns and enters the faster moving air above the canopy. A portion of this component may be transported to high altitudes and be carried for considerable distances. Rain component refers to those pollen grains that function as nuclei for water droplet formation. This component may also result in long distance pollen distribution.

Anderson (1967) analyzed the pollen spectra of forest floor mors to study the distribution of tree pollen in a mixed deciduous forest in southern Denmark. For the species studied (*Betula, Quercus, Alnus, Fagus, Tilia,* and *Fraxinus*—canopy height, 20 m) moss analyses indicated surprisingly short dispersal distances for the majority of the pollen within the forest, that is, less than 20-30 m. Perhaps this largely represented the trunk space component.

Trees near the edge of forest stands or isolated trees (urban/suburban condition), however, may disseminate pollen great distances because they may release into locally turbulent atmospheric conditions. Pollen from trees upslope may tend to concentrate in valleys below (Silen and Copes, 1972).

Extensive surveys of contemporary pollen distribution have been conducted in numerous parts of North America and Europe (for example, Moore and Webb, 1978). Canadian studies suggest that long-distance transport of pollen of deciduous trees into boreal and tundra regions does not appear to be very significant. Input of *Picea* and *Pinus* pollen into deciduous forest regions, however, is measurable (Lichti-Federovich and Ritchie, 1968). Moore and Webb (1978) have con-

Table 3-8. Average Pollen Dispersion Distance at a Windspeed of 10 m sec^{-1}

Species	km
Norway spruce	22.2
Scotch pine	267.8
English oak	199.0
Alder	546.7
European filbert	267.8

Source: Stanley and Linskens (1974).

cluded that under appropriate meteorological conditions pollen grains may be carried thousands of kilometers from their point of origin.

d. Human Health Aspects

Human allergic response can result from exposure to a wide variety of allergenic materials including fungal spores, animal hair, feathers, dust, arthropod parts, insect and reptile venom, drugs, certain foods, and pollen. Allergic response to the latter, termed hayfever or pollinosis, is generally recognized as the most prevalent and important of all allergies (Stanley and Linskens, 1974). In sensitive individuals, hayfever symptoms, which include sneezing, watery eyes, nasal obstruction, itchy eyes and nose, and coughing, typically occur within minutes of exposure to allergenic pollen. Clinically hayfever is described as allergic rhinitis or conjunctivitis. Hyde et al. (1978) has suggested that between 10% and 20% of the population of the United States manifest recurrent or persistent allergic rhinitis. The cost to United States industry for lost wages due to airborne pollen allergies is in excess of $500 million annually.

Allergic rhinitis in a given individual may result from exposure to the pollen of any one of several hundred plant species. *Ambrosia* (ragweed) and *Phleum pratense* (timothy grass) pollen, both very widely distributed throughout the north temperate zone, elicit widespread allergic responses and cause approximately half the pollinosis cases in numerous urban areas. In this zone there is considerable temporal variation in disease associated with pollen production and distribution. Ragweed pollinosis is most prevalent from mid-August through late September while grass pollinosis is especially common from late spring through early July.

e. Tree Related Hayfever

For hayfever sufferers throughout the north temperate zone, pollinosis symptoms are frequently initiated in early spring (actually February through May, depending on location) when a variety of tree species releases pollen. A large number of tree species produces allergenic pollen (Table 3-9).

Based on adult skin test evidence, Lewis and Imber (1975) have ranked various tree groups relative to the ability of their pollen to incite rhinitis and asthma in North America. They suggest box elder, willow, and hickory pollen elicited the

Table 3-9. Principal Tree Genera with Species That Produce Allergenic Pollen

Acer	*Juniperus*
Alnus	*Morus*
Betula	*Olea*
Broussonetia	*Platanus*
Carpinus	*Populus*
Carya	*Quercus*
Casuarina	*Salix*
Cupressus	*Ulmus*
Fraxinus	

greatest allergenic reactivity. Oak, sycamore, poplar, maple, and birch proved moderately allergenic while elm, cottonwood, and white ash were lowest in producing adult skin reactions.

For any given location in the temperate zone, the most important tree species producing allergenic pollen is highly variable. Numerous regional studies are available (for example, Anderson et al., 1978; Newmark, 1978; Lewis and Imber, 1975). Rubin and Weiss (1974) list major tree pollens by state and Wodehouse (1945) by region.

Fortunately gymnosperm pollen is generally not allergenic. Pollen from *Abies* does not appear to cause symptoms while pollen from *Pinus* only rarely is involved in pollinosis (Newmark, 1978). There is evidence that pollen from certain *Juniperus* and *Cupressus* species is allergenic (Yoo et al., 1975).

2. Hydrocarbon Aerosol

The ability of various hydrocarbons to form particulate organic compounds in the atmosphere is a complex function of their ambient concentration, gas-phase reactivity, and ability to form products with appropriate physical characteristics for gas-to-aerosol conversion. Cyclic olefins (for example, cyclopentene) are probably the most significant class of urban organic aerosol precursors (National Academy of Sciences, 1977). Went (1960a) was the first to suggest that terpenoids from vegetation could serve as precursors for organic particulate formation. He concluded that laboratory and field evidence demonstrated that organic volatiles, specifically α-pinene, could form fine particles when subjected to high light intensities in the presence of nitrogen dioxide as a "light-absorbing catalyst" (Went, 1964). The size of these particles approximated 10^{-7} to 10^{-5} cm.

Organic aerosols formed from hydrocarbon precursors released from vegetation were judged to be responsible for the blue haze commonly observed in the Great Smoky Mountains of North Carolina and Tennessee, the sagebrush area in the western United States, the eucalypt forests of Australia, and the forested tropics (Went, 1960a,b). In 1964, Went estimated that organic gases released annually by vegetation may occasion the formation of 500 million tons of submicroscopic particulate matter globally. A mobile laboratory was used to obtain additional evidence for the hypothesis that terpenes were activated to agglomerate into fine particles by a photochemical process (Went et al., 1967).

The fate of gaseous olefins, for example, α-pinene, β-pinene, limonene, myrcene and isoprene, released to the atmosphere, however, remains largely undetermined. Rasmussen and Holdren (1972) have presented evidence that individual monoterpene hydrocarbons are present in the low parts-per-billion range in rural air. Aerosol formation from terpenes, however, has not been thoroughly investigated (National Academy of Sciences, 1977). Only one terpene, α-pinene, has received research attention, and this is very limited.

The formation of condensation nuclei from irradiated mixtures of 0.1 ppm (188 μg m^{-3}) nitrogen dioxide and 0.5 ppm (2780 μg m^{-3}) α-pinene was reported by Ripperton and Lillian (1971). Schwartz (1974) has intensively studied the aerosol products from reaction of nitrogen oxides and α-pinene. O'Brien et al. (1975) examined the aerosol-forming behavior of a representative group of hydrocarbons under standardized conditions and observed that α-pinene caused a large amount of light scattering (Table 3-10). The National Academy of Sciences, Subcommittee on Ozone and Other Photochemical Oxidants, has concluded that low volatility compounds are probably formed from other olefinic terpenes (for example, β-pinene, limonene), as in the case of α-pinene (Schwartz observations), and that these materials together probably constitute a major fraction of the blue haze aerosols formed naturally over forested areas (National Academy of Sciences, 1977). Weiss et al. (1977) do not agree, however, and have suggested that the rural haze observed in the Midwest and Southeast results from particulate sulfates rather than organic compounds. These investigators employed humidity-controlled nephelometry during 1975 to obtain information on the dominant optical scattering species in Michigan, Missouri, and Arkansas. They concluded at all sites that submicrometer-sized sulfate particles dominated during all wind and synoptic conditions. They judged that sulfate aerosol was regionally extensive and inferred the importance of determining the fraction of the sulfate due to natural versus anthropogenic sources.

Knights et al. (1975) have presented evidence for the occurrence of α-pinene aerosol products in urban air. It is not clear, however, whether the terpenes were released from urban tree foliage, industrial solvents (for example, turpentine), or were transported into the city from rural forested areas.

Table 3-10. Aerosol Formation from Selected Hydrocarbons

Hydrocarbon	Maximum light scattering ($\beta_{scat} \times 10^4$ m^{-1})
Glutaraldehyde	0
Ethylbenzene	1
Mesitylene	1
2,6-Octadiene	1
1-Octene	1
trans-4-Octene	1
5-Methyl-1-hexene	1
2,6-Dimethylheptane	1
1-Heptene	1
α-Xylene	8
1,5-Hexadiene	40
Cyclohexene	90
2-Methyl-1,5-hexadiene	110
1,6-Heptadiene	160
1,7-Octadiene	180
α-Pinene	180

Source: O'Brien et al. (1975).

C. Forest Fires

The burning of forests or forest debris represents a special case where forest systems supply air contaminants to the atmosphere. Fires can be conveniently divided into natural (wild) fires and prescribed (management) fires. Prescribed fires are artificially set and controlled and are intended to fulfill forest management objectives. Included in the latter are one or more of the following: (1) disposal of logging residue (slash), (2) forest fuel reduction to minimize the influence of wildfires, (3) control of unwanted vegetation, (4) reduction of microbial or insect pests, (5) facilitation of crop tree regeneration, and (6) improvement of wildlife habitat.

Hall (1972) presents an overview discussion of the various contaminants produced by forest burning. The visible smoke is very largely water condensed on particulate matter. Darley et al. (1976) and Komarek et al. (1973) have provided scanning electron microscope photomicrographs of these particles which consist largely of carbon. Aerosols, condensation products of terpenoids, phenols, and aldehydes are partially adsorbed on the carbon. A variety of hydrocarbons and carbon oxides are among the most important gaseous components produced.

1. Prescribed Fires

a. Tree Residue Burning (Slash Fires)

In the northwestern sections of the United States and in western Canada, burning is widely employed to dispose of logging residue. Harvest of old-growth stands of western conifers generates excessive quantities of nonmarketable tree debris. The only economically sound disposal strategy for decades has been burning (Figure 3-1.)

Approximately 10 years ago, roughly 81,000 ha (200,000 acres) of logging slash were burned annually west of the Cascade Range in the States of Washington and Oregon (Fritschen et al., 1970; Cramer and Westwood, 1970). This acreage has been decreasing as tree utilization has improved, as harvesting has involved progressively less old-growth timber, and as clean air laws have been implemented. Total forest and range land burned by prescription annually in the West currently approximates one million acres (Cooper, 1976). Individual slash fires average approximately 4-16 ha (10-40 acres) aflame with fuel loadings of 125-495 tons ha^{-1} (50-200 tons $acre^{-1}$). Smoke may be lifted in convective columns 500 to 1700 m above these fires (Cramer and Graham, 1971; Hedin and Turner, 1977).

Two types of slash burning have been practiced: broadcast and pile. The former leaves the logging debris in place while the latter concentrates it into piles. Broadcast fires burn at a relatively low temperature while pile fires burn at a relatively high temperature (Figure 3-2). Fritschen et al. (1970) analyzed the pollutants produced by broadcast and pile fires set experimentally in Douglas fir slash in Washington in order to explore the hypothesis that pile fires,

Figure 3-1. Broadcast slash burning in a clearcut block in a western forest ecosystem. (Photograph courtesy of U.S.D.A. Forest Service.)

because they are higher temperature fires, have more complete combustion and produce fewer emissions than broadcast burns. Particulates, at ground level, increased to roughly 10 times background immediately downwind from the broadcast burn. Smoke plume particulates in the vicinity of the fire reduced visibility to 0.5 km, but at a distance of 19 km from the fire, visibility was at the level found over Seattle. High carbon monoxide and carbon dioxide concentrations found at the fire site decreased rapidly to ambient levels in horizontal and vertical directions. Hydrocarbon analyses revealed low concentrations of 25 components. The most important were low molecular weight hydrocarbons and alcohols including ethylene, ethane, propene, propane, methanol, and ethanol. Several unsaturated components were detected but the amounts were small. Relative to the broadcast fire, the pattern of combustion and emission of the pile fire was more uniform and the carbon monoxide:carbon dioxide ratio lower, but otherwise the differences were not substantial.

Sandberg et al. (1975) sampled the emissions from ponderosa pine logging slash. Artificial fuel beds were prepared from material collected from the San Bernardino National Forest in California. The beds were the equivalent of a 125 ton ha^{-1} (50 ton acre^{-1}) fuel loading and similar in size and distribution to

Figure 3-2. Windrow burning of pine and hardwood slash in central Georgia. (Photograph courtesy of U.S.D.A. Forest Service.)

actual logging slash. Emission factors are presented in Table 3-11. Of the hydrocarbons, 15-40% were composed of methane and ethylene. Ethane and acetylene were the next most important. Photochemically important compounds represented only a minor portion of the hydrocarbon fraction (Sandberg et al., 1975).

The percentage of Douglas fir and western larch slash which was converted by controlled burns to various air contaminants was determined in the field at the Lolo National Forest, Montana, by Malte (1975). His results are presented in Table 3-12.

In an investigation of the contribution of various forms of agricultural burning to California air quality, Darley et al. (1966) included native woody brush in their artificial burnings. Considerable differences in emission were observed depending on whether the brush was "green" or dry (Table 3-13). The yield of olefins in the hydrocarbon fraction of the burned brush was equal to the ethene and saturates plus acetylenes combined.

Table 3-11. Emission Factors for Ponderosa Pine Slash

Air contaminant	kg ton^{-1} of fuel
Carbon monoxide	66
Hydrocarbons	3.6
Particulates	4

Source: Sandberg et al. (1975).

Table 3-12. Emission Factors for Douglas Fir and Western Larch Slash

Air contaminant	kg ton^{-1} of fuel
Carbon dioxide	1130
Carbon monoxide	129
Nitrogen dioxide	7.7
Particulates	6.3

Source: Malte (1975).

Adams et al. (1976) reported on field studies of particulate distribution from broadcast slash burns in the Flathead National Forest, Montana. Daily 24-hour hi-vol-suspended particulate concentrations measured were significantly higher at the three downwind sampling sites on prescribed fire days relative to nonfire days. Particulate emission factors were observed by these investigators to vary according to fuel size class (Table 3-14).

b. Management Burning (Nonslash Fires)

In warm and dry areas of the temperate forest, for example, the southern portion of the United States, litter may accumulate and constitute a significant fire hazard in production forests. As a result prescribed fire is used to reduce the potential damage by wildfire. Managed fire is also employed to protect the commercially important, subclimax pine species from hardwood incursion. Other benefits from artificial fires in the South frequently include site preparation, disease control, improvement of wildlife habitat, and occasionally slash disposal. Over 81×10^4 ha (two million acres) (Mobley, 1976) and six million tons of fuel (Dietrich, 1971) are prescribed burned annually in the South (Figure 3-3).

Georgia leads all states in United States in acreage burned for agricultural and forestry purposes with approximately 41×10^4 ha (1 million acres) artificially burned annually. Ward and Elliott (1976) have estimated that these Georgia fires produce 29,000 tons of particulate matter yearly.

Typically prescribed fires in the South consume about seven tons of fuel ha^{-1} (three tons acre^{-1}), with a range of roughly 2.5 to 27 tons (Cooper, 1976; Pharo, 1976). Published emission figures for contaminants produced by these burns are less prevalent than those available for managed fires in the West. Darley et al. (1976) have test burned dry loblolly pine needles sent from Georgia and determined that head fires produced between 24 to 31 kg of particulates ton^{-1} of fuel combusted.

Table 3-13. Emission Factors for California Woody Brush

Air contaminant	kg ton^{-1} of fuel	
	Green	Dry
Carbon dioxide	693	1168
Carbon monoxide	61	29
Hydrocarbons	12	2

Source: Darley et al. (1966).

Table 3-14. Particulate Emission Factors for Douglas Fir, Larch, and Spruce of Various Sizes

Tree material	kg ton^{-1} of fuel
Larch needles	5.8
Larch twigs 0-1 cm	9.3
Larch twigs 1-2.5 cm	3.3
Larch needles and twigs	5.1
Douglas fir needles	4.8
Douglas fir twigs 0.1 cm	3.8
Douglas fir twigs 1-2.5 cm	3.3
Douglas fir needles and twigs	2.8
Duff	7.5
Spruce	3.7

Source: Adams et al. (1976).

2. Wild Fires

Fires of natural origin have historically been extremely common throughout the temperate forests (Figure 3-4). The majority of these was initiated by lightning. Our understanding of the importance of natural fires to the structure and function of forest ecosystems continues to develop. In the United States, prescribed

Figure 3-3. A low intensity management or prescribed burn in a southeastern loblolly pine stand. (Photograph courtesy of U.S.D.A. Forest Service.)

Figure 3-4. Wolf Creek wildfire which burned 41 ha (100 acres) of the Ochoco National Forest. (Photograph courtesy of U.S.D.A. Forest Service.)

burning and other management strategies have substantially reduced the acreage burned by natural fires.

Wildfires differ substantially from managed fires. In the South a wildfire may consume three times as much fuel as a prescribed burn (22 tons vs 7 tons ha^{-1}). Combustion experiments at the U.S.D.A. Forest Service, Southern Forest Fire Laboratory, Macon, Georgia, have indicated that emissions from simulated wildfires are greater than those produced by simulated prescribed fires. Smoke production may be ten times greater in wildfire situations. Annual particulate production for prescribed fires in the South has been estimated to be 8 kg ton^{-1} of fuel while the wildfire estimate is 26 kg ton^{-1}. Emission of nitrogen oxides may also be greater in wildfires relative to controlled fires (Cooper, 1976).

3. Forest Fires and Ozone

Air contaminants of forest fire origin that have received the greatest research attention include particulates and the oxides of carbon. Oxidants and nitrogen oxides have received only minor study. In 1972 during the course of air monitoring flights over prescribed burns of Western Australia, Evans et al. (1974) detected elevated concentrations of ozone at the top of some smoke plumes. It is interesting to note that typical controlled burns in Western Australia may involve areas of 4000 ha and carry a fuel load of 4 tons ha^{-1}. After more thorough analysis, Evans et al. (1977) concluded that high-intensity burns for forest clearing purposes in Australia can produce ozone concentrations in excess of 100 ppb (196 μg m^{-3}) within one hour. Specific data from representative fires are presented in Table 3-15. The authors stressed that the ozone content of the plume at the point where the plume reaches the ground would be appreciably less than the upper plume maxima.

D. Summary

From an air quality perspective, volatile hydrocarbons represent the most important gaseous emission released directly to the atmosphere by forest trees during the course of normal metabolism. An average of 70% of the trees in various United States forests release reactive hydrocarbons. On a global basis, reactive hydrocarbons released by trees may exceed the total amount released by anthropogenic sources by a factor of 5-6. The most important compounds released include α-pinene, camphlene, β-pinene, limonene, myrcene, β-phellandrene, isoprene, and perhaps ethylene.

The most important particulate contaminants released by forests, in terms of direct human health significance, are the various pollens. Conifer pollen is distributed by wind for considerable distances but fortunately is generally nonallergenic. Deciduous pollen, which frequently is allergenic, may also be distributed considerable distances by wind and other vectors. While annual pollen production varies greatly due to several factors an average production of 0.10 kg tree^{-1} may be used for gross estimates of total particulate production. The medical evidence relating to human disease caused by pollen in terms of history, scope, abundance, and confidence of the data exceeds information available for the relationship between any other air contaminant and human health.

Hydrocarbon aerosols are another important particulate contaminant resulting from tree metabolism. Terpenes, particularly, α-pinene, released from tree foliage may react in the atmosphere to form submicrometer particles. The information on the production, transport, and persistence of these particles in natural environments is limited and their role in haze formation of forest regions is not completely clear, but probably is significant.

Table 3-15. Highest Ozone Concentrations in the Plume from Three West Australian Prescribed Forest Fires

Fire no.	Peak ozone ppb	Distance downwind of fire (km)
1	65 (127 μg m^{-3})	29
7	75 (147 μg m^{-3})	15
13	60 (118 μg m^{-3})	27

Source: Evans et al. (1977).

Burning of forests or forest debris by natural or managed fires is a special case where forests may contribute contaminants to the atmosphere. These fires can contribute locally and regionally significant amounts of particulates (Griffin and Goldberg, 1979), carbon oxides, and hydrocarbons to the atmosphere. Limited evidence suggests nitrogen oxides and ozone should be added to this list. Contaminant additions to the atmosphere from wildfires in most regions of temperate forests are less than they were several decades ago due to improved fire management practices. Burning in tropical forests, however, is widespread and may have important global air quality implications. The local and regional importance of contaminants released from controlled forest burning has been reduced again due to improvement in fire management especially in greater appreciation of meteorological conditions optimal for maximum effluent dispersion.

The U.S. Environmental Protection Agency has estimated that forest wildfires and managed burning in the United States contributed the following percentages of the total estimated national emissions from 1970 to 1976: 4% of the carbon monoxide, 3% of the particulates, 2% of the hydrocarbons, and 0.6% of the oxides of nitrogen (Table 3-16).

On a global basis carbon monoxide estimates from forest fires range from 11×10^6 tons yr^{-1} (Robinson and Robbins, 1968) to 6×10^7 tons yr^{-1} (Seiler, 1974). Wong estimated the gross input of global carbon dioxide to be 5.7×10^9 tons of carbon yr^{-1} (Wong, 1978). Fahnestock (1979) argues this latter figure may be as much as four times too high.

Table 3-16. United States Emission Estimates for Forest Wildfires and Managed Burning, 1970 through 1976

Year	10^6 tons yr^{-1}			
	Particulates	Nitrogen oxides	Hydrocarbons	Carbon monoxide
1970	0.5	0.1	0.7	3.5
1971	0.7	0.2	1.0	5.1
1972	0.5	0.1	0.7	3.6
1973	0.4	0.1	0.6	2.9
1974	0.5	0.1	0.6	3.9
1975	0.3	0.1	0.4	2.3
1976	0.6	0.2	0.8	4.8

Source: U.S. Environmental Protection Agency (1977).

References

Adams, D. F., R. K. Koppe, and E. Robinson. 1976. Air and surface measurements of constituents of prescribed forest slash smoke. *In*: Air Quality and Smoke from Urban and Forest Fires. National Academy of Sciences, Washington, D.C., pp. 105-147.

Anderson, E. F., C. S. Dorsett, and E. O. Fleming. 1978. The airborne pollens of Walla Walla, Washington. Ann. Allergy 41:232-235.

Anderson, S. T. 1967. Tree pollen rain in a mixed deciduous forest in South Jutland (Denmark). Rev. Palaeobot. Palynol. 3:267-275.

Brooks, J. 1971. Some chemical and geochemical studies on sporopollenin. In: J. Brooks, P. R. Grant, M. Muir, P. van Gijzel, and G. Shaw (Eds.), Sporopollenin. Academic Press, New York, pp. 351-407.

Calvert, J. G. 1976. Hydrocarbon involvement in photochemical smog formation in Los Angeles atmosphere. Environ. Sci. Technol. 10:256-262.

Cooper, R. W. 1976. The trade-offs between smoke from wild and prescribed forest fires. In: Air Quality and Smoke from Urban and Forest Fires. National Academy of Sciences, Washington, D.C., pp. 19-26.

Cramer, O. P., and H. E. Graham. 1971. Cooperative management of smoke from slash fires. J. For. 69:327-331.

Cramer, O. P., and J. N. Westwood. 1970. Potential impact of air quality restrictions on logging residue burning. U.S.D.A. Forest Service, Pac. S.W. Forest and Range Exp. Sta., Res. Paper PSW-64, 12 pp.

Dacey, J. W. H., and M. J. Klug. 1979. Methane efflux from lake sediments through water lilies. Science 203:1253-1254.

Darley, E. F., F. R. Burleson, E. H. Mateer, J. T. Middleton, and V. P. Osterli. 1966. Contribution of burning of agricultural wastes to photochemical air pollution. J. Air Pollut. Control Assoc. 11:685-690.

Darley, E. F., S. Lerman, G. E. Miller, Jr., and J. F. Thompson. 1976. Laboratory testing for gaseous and particulate pollutants from forest and agricultural fuels. In: Air Quality and Smoke from Urban and Forest Fires. National Academy of Sciences, Washington, D.C., pp. 78-89.

Davies, I. W., R. M. Harrison, R. Perry, D. Ratnayaka, and R. A. Wellings. 1976. Municipal incinerator as source of polynuclear aromatic hydrocarbons in environment. Environ. Sci. Technol. 10:451-453.

Dieterich, J. H. 1971. Air quality aspects of prescribed burning. In: Proc. Prescribed Burning Symposium. U.S.D.A. Forest Service, Southeastern For. Exp. Sta., Asheville, North Carolina, pp. 139-151.

Erdtman, G. 1969. Handbook of Palynology. Hofner, New York, 486 pp.

Evans, L. F., N. K. King, D. R. Packham, and E. T. Stephens. 1974. Ozone measurements in smoke from forest fires. Environ. Sci. Technol. 8:75-76.

Evans, L. F., I. A. Weeks, A. J. Eccleston, and D. R. Packham. 1977. Photochemical ozone in smoke from prescribed burning of forests. Environ. Sci. Technol. 11:896-900.

Fahnestock, G. R. 1979. Carbon input to the atmosphere from forest fires. Science 204:209-210.

Feldstein, M. 1974. A critical review of regulations for the control of hydrocarbon emissions from stationary sources. J. Air Pollut. Control Assoc. 24: 469-478.

Fritschen, L., H. Bovee, K. Buettner, R. Charlson, L. Monteith, S. Pickford, J. Murphy, and E. Darley. 1970. Slash fire atmospheric pollution. U.S.D.A. Forest Service, Pac. Northwest For. and Range Exp. Sta., Res. Paper No. PNW-97, 42 pp.

Griffin, J. J., and E. D. Goldberg. 1979. Morphologies and origin of elemental carbon in the environment. Science 206:563-565.

Hall, J. A. 1972. Forest fuels, prescribed fire, and air quality. U.S.D.A. Forest Service, Pacific Northwest Forest and Range Exp. Sta., Portland, Oregon, 44 pp.

Hanover, J. W. 1966. Genetics of terpenes. 1. Gene control of monoterpene levels in *Pinus monticola*. Dougl. Heredity 21:73-84.

Hedin, A., and T. Turner. 1977. What is burned in a prescribed fire? Department of Natural Resources Note No. 16, Olympia, Washington, 7 pp.

Hess, D. 1975. Plant Physiology. Springer-Verlag, New York, 333 pp.

Heuss, J. M., G. J. Nebel, and B. A. D'alleva. 1974. Effects of gasoline aromatic and lead content on exhaust hydrocarbon reactivity. Environ. Sci. Technol. 8:641-647.

Hyde, H. A. 1951. Pollen output and seed production in forest trees. Quart. J. For. 45:172-175.

Hyde, J. S., N. V. Aroda, C. M. Kumar, and B. S. Moore. 1978. Chronic rhinitis in the pre-school child. Ann. Allergy 41:216-219.

Knights, R. L., D. R. Cronn, and A. L. Crittenden. 1975. Diurnal patterns of several components of urban particulate air pollution. Paper No. 3, Pittsburgh Conference on Analytical Chemistry and Applied Spectoscopy, Cleveland, Ohio, March 3, 1975.

Komarek, E. V., B. B. Komarek, and T. C. Carlysle. 1973. The Ecology of Smoke Particulates and Charcoal Residues from Forest and Grassland Fires: A Preliminary Atlas. Miscell. Publica. No. 3, Tall Timbers Research Sta., Tallahassee, Florida, 75 pp.

Lewis, W. H., and W. E. Imber. 1975. Allergy epidemiology in the St. Louis, Missouri, Area. III. Trees. Ann. Allergy 35:113-119.

Lichti-Federovich, S., and J. C. Ritchie. 1968. Recent pollen assemblages from the western interior of Canada. Rev. Paleobot. Palynol. 7:297-344.

Maarse, H., and R. E. Kepner. 1970. Changes in composition of volatile terpenes in Douglas fir needles during maturation. J. Agr. Food Chem. 18:1095-1101.

Mah, R. A., D. M. Ward, L. Baresi, and T. L. Glass. 1977. Biogenesis of methane. Annu. Rev. Microbiol. 31:309-341.

Major, R. T., P. Marchini, and A. J. Boulton. 1963. Observations on the production of α-hexanal by leaves of certain plants. J. Biol. Chem. 238:1813-1816.

Malte, P. C. 1975. Pollutant production from forest slash burns. Bulletin No. 339, College of Engineering, Washington State Univ. Pullman, Washington, 32 pp.

Mobley, H. E. 1976. Summary of state regulations as they affect open burning. *In*: Air Quality and Smoke from Urban and Forest Firest. National Academy of Sciences, Washington, D.C., pp. 206-212.

Moore, P. D., and J. A. Webb. 1978. An Illustrated Guide to Pollen Analysis. Wiley, New York, 133 pp.

National Academy of Sciences. 1977. Ozone and Other Photochemical Oxidants. National Academy of Sciences, Washington, D.C., 719 pp.

Newmark, F. M. 1978. The hay fever plants of Colorado. Ann. Allergy 40:18-24.

Nilsson, S., J. Praglowski, L. Nilsson, and N. O. Kultur. 1977. Atlas of Airborne Pollen Grains and Spores in Northern Europe. Ljungforetagen, Stockholm, Sweden, 159 pp.

O'Brien, R. J., J. R. Holmes, and A. H. Bockian. 1975. Formation of photochemical aerosol from hydrocarbons. Environ. Sci. Technol. 9:568-576.

Parsons, J. S., and S. Mitzner. 1975. Gas chromatographic method for concentration and analysis of traces of industrial organic pollutants in environmental air and stacks. Environ. Sci. Technol. 9:1053-1058.

Pharo, J. A. 1976. Aid for maintaining air quality during prescribed burns in the South. U.S.D.A. Forest Service, Res. Paper No. SE-152, Southeastern For. Exp. Sta., Asheville, North Carolina, 11 pp.

Radwan, M. A., and W. D. Ellis. 1975. Clonal variation in monoterpene hydrocarbons of vapors of Douglas-fir foliage. For. Sci. 21:63-67.

Rasmussen, R. A. 1964. Terpenes: Their analysis and fate in the atmosphere. Ph.D. Thesis, Washington, Univ., St. Louis, Missouri.

Rasmussen, R. A. 1970. Isoprene: Identified as a forest-type emission to the atmosphere. Environ. Sci. Technol. 4:667-671.

Rasmussen, R. A. 1972. What do the hydrocarbons from trees contribute to air pollution? J. Air Pollut. Control Assoc. 22:537-543.

Rasmussen, R. A., and M. W. Holdren. 1972. Analyses of C_5 to C_{10} hydrocarbons in rural atmosphere. Paper No. 72-19, presented at 65th Annual Meeting of the Air Pollution Control Association, Miami Beach, Florida, June 18-22, 1972.

Rasmussen, R. A., and F. W. Went. 1965. Volatile organic material of plant origin in the atmosphere. Proc. Nat. Acad. Sci. U.S. 53:215-220.

Ripperton, L. A., and D. Lillian. 1971. The effect of water vapor on ozone synthesis in the photo-oxidation of alpha-pinene. J. Air Pollut. Control Assoc. 21:629-635.

Robinson, E., and R. C. Robbins. 1968. Sources, Abundance and Fate of Gaseous Atmospheric Pollutants. Final Report SRI Project PR-6755. Stanford Research Institute, Menlo Park, California.

Rubin, J. M., and N. S. Weiss. 1974. Practical Points in Allergy. Medical Examination, New York, 208 pp.

Sandberg, D. V., S. G. Pickford, and E. F. Darley. 1975. Emissions from slash burning and the influence of flame retardant chemicals. J. Air Pollut. Control Assoc. 25:278-281.

Schwartz, W. 1974. Chemical Characterization of Model Aerosols. EPA-650-3-74-011. Battelle Memorial Institute, Columbus, Ohio, 129 pp.

Seiler, W. 1974. The cycle of atmospheric CO. Tellus 26:116-135.

Shain, L., and W. E. Hillis. 1972. Ethylene production in *Pinus radiata* in response to Sirex-Amylostereum attack. Phytopathology 62:1407-1409.

Silen, R. R., and D. L. Copes. 1972. Douglas-fir seed orchard problems—A progress report. J. For. 70:145-147.

Squillace, A. E. 1971. Inheritance of monoterpene composition in cortical oleoresin of slash pine. For. Sci. 17:381-387.

Stanley, R. G., and H. F. Linskens. 1974. Pollen Biology Biochemistry Management. Springer-Verlag, New York, 307 pp.

Stern, A. C., H. C. Wohlers, R. W. Boubel, and W. P. Lowry. 1973. Fundamentals of Air Pollution. Academic Press, New York, 492 pp.

Tauber, H. 1965. Differential pollen dispersion and the interpretation of pollen diagrams. Danm. Geol. Anders. II R. 89:1-69.

Tauber, H. 1967. Investigations of the mode of pollen transfer in forested areas. Rev. Palaeobot. Palynol. 3:277-287.

Turk, A., and C. J. D'Angio. 1962. Composition of natural fresh air. J. Air Pollut. Control Assoc. 12:29-33.

U.S.D.A. Forest Service. 1974. Seeds of Woody Plants in the United States. Agr. Handbook No. 450, U.S.D.A., Forest Service, Washington, D.C., 833 pp.

U.S. Environmental Protection Agency. 1977. National Air Quality and Emissions Trends Report, 1976. U.S.E.P.A., EPA-450/1-77-002, Research Triangle Park, North Carolina.

Ward, D. E., and E. R. Elliott. 1976. Georgia rural air quality: Effect of agricultural and forestry burning. J. Air Pollut. Control Assoc. 26:216-220.

Weiss, R. E., A. P. Waggoner, R. J. Charlson, and N. C. Ahlquist. 1977. Sulfate aerosol: Its geographical extent in the midwestern and southern United States. Science 195:979-980.

Went, F. W. 1955. Air pollution. Sci. Amer. 192:62-72.

Went, F. W. 1960a. Blue hazes in the atmosphere. Nature 187:641-643.

Went, F. W. 1960b. Organic matter in the atmosphere and its possible relation to petroleum formation. Proc. Nat. Acad. Sci. U.S. 46:212-221.

Went, F. W. 1964. The nature of Aitken condensation nuclei in the atmosphere. Proc. Nat. Acad. Sci. U.S. 51:1259-1267.

Went, F. W., D. B. Slemmons, and H. N. Mozingo. 1967. The organic nature of atmospheric condensation nuclei. Proc. Nat. Acad. Sci. U.S. 58:69-74.

Wodehouse, R. P. 1945. Hayfever Plants. Chronica Botanica Co., Waltham, Massachusetts, 245 pp.

Wong, C. S. 1978. Atmospheric input of carbon dioxide from burning wood. Science 200:197-199.

Yoo, T., E. Spitz, and J. L. McGerity. 1975. Conifer pollen allergy: Studies of immunogencity and cross antigenicity of conifer pollens in rabbit and man. Ann. Allergy 34:87-93.

4

Forests as Sinks for Air Contaminants: Soil Compartment

Air contaminants may be removed from the atmosphere by a variety of mechanisms. The primary processes are precipitation scavenging, chemical reaction, dry deposition (sedimentation), and absorption (impaction) (Rasmussen et al., 1974). Loss via precipitation may occur in two ways: "rainout" which involves both absorption and particle capture by falling raindrops. Primary and secondary contaminants are subject to a large number of chemical reactions in the atmosphere that may ultimately transform them into an aerosol or oxidized or reduced product. Attachment by aerosols and subsequent deposition on the surface of the earth is termed dry deposition. Absorption by water bodies, soils, or vegetation at the surface of the earth is an additional extremely important removal process.

Components of the ecosystems of the earth that remove pollutants from the atmospheric compartment and store, metabolize, or transfer them may be termed "sinks" (Warren, 1973). Forest ecosystems in general, and temperate forest ecosystems in particular, are locally, regionally, and globally important sinks for a wide variety of atmospheric contaminants. Soil and vegetation surfaces represent the major sink for pollutants introduced into terrestrial ecosystems (Little, 1977).

The transfer of contaminants from the atmospheric compartment to the surfaces of soil or vegetation is expressed as a flux (pollutant uptake) rate and is given as a weight of pollutant removed by a given surface area per unit time. Actual determinations of flux rates (sink strengths) are extremely complex and involve an appreciation of atmospheric conditions (wind, turbulence, temperature, humidity), pollutant nature and concentration, sink surface conditions (geometry, presence or absence of moisture), and other parameters. Awareness of the following simple relationship between flux rate and pollutant concentration will assist the reader in evaluating the literature dealing with sink function:

$$F \text{ (pollutant uptake, flux)} = \nu \text{ (proportionality constant or deposition velocity)} \times C \text{ (pollutant concentration)}$$

typical units: F, mg cm^{-2} sec^{-1}
ν, cm sec^{-1}
C, mg cm^{-3}

The deposition velocity parameter is generally a function of the surface of the sink (soil or vegetation) and whether it is wet or dry. It may be thought of as the rate at which an absorbing surface "cleans" a pollutant from the air. If the deposition velocity of a pollutant is 1.0 cm sec^{-1}, it suggests that the surface is completely removing the pollutant from a layer of air 1.0 cm thick each second, with the "clean" layer immediately replaced by a "new" contaminated layer. Keep in mind that both the proportionality constant and pollutant concentration must be known in order to estimate pollutant uptake. This summary relationship between flux rate and pollutant concentration and associated jargon has been developed by those primarily concerned with the transfer of particles from the atmosphere to natural or other surfaces. The concept, however, is widely employed in air pollution literature to express the transfer of particulate and gaseous pollutants from the atmosphere to various terrestrial and aquatic sinks.

The principal repository in terrestrial ecosystems for air contaminants of anthropogenic origin is soil. By virtue of their distribution and physical and chemical characteristics, forest soils may be particularly efficient short- and long-term sinks. For selected trace metals and gases the evidence supporting the importance of soils is considerable.

A. Forest Soils as Particulate Sinks

Particles are transferred from the atmosphere to forest soils directly by dry deposition and precipitation scavenging and indirectly via leaf and twig fall. A very large number of human activities generate small particles (0.1-5 μm) with high concentrations of trace metals. Depending on weather conditions these particles may remain airborne for days or weeks and be transported 100 to 1000 km from their source. The evidence that forest soils may be the ultimate or temporary repository for the trace elements associated with these particles is substantial. Soils have a very high affinity for heavy metals, particularly the clay and organic colloidal components (John et al., 1972; Lagerwerff, 1967; Stevenson, 1972; Petruzzelli et al., 1978; Somers, 1978; Korte et al., 1976; McBride, 1978; Zunino and Martin, 1977a,b).

1. Lead

Lead is naturally present, in small amounts, in soil, rocks, surface waters, and the atmosphere. Due to its unique properties it has been an element widely useful to humans. This utility has resulted in greatly elevated lead concentrations in certain ecosystems. Locations where lead is being mined, smelted, and refined, where industries are consuming lead, and in urban-suburban complexes the environmental lead level is greatly elevated. It is widely agreed that a primary source in these latter sites is the combustion of gasoline containing lead additives. Other important sources include coal combustion, refuse and sludge incineration, burning or attrition of lead-painted surfaces, and industrial processes. Since the vast majority of atmospheric lead particles are less than 0.5 μm (Smith, 1976), they have been widely distributed to all parts of the earth.

The input of lead, its cycling within forest ecosystems, its transfer in food chains, its rate of loss to downstream aquatic systems, and its residence in soil has received considerable research attention.

Lead deposited from the atmosphere will vary greatly depending on area and regional source strengths. In urban areas, with abundant motor vehicle use, the lead flux rate may exceed 3000 g ha^{-1} yr^{-1} (Chow and Earl, 1970) while areas distant from cities and industrial or power generating sources may have flux rates less than 20 g ha^{-1} yr^{-1} (Chow and Johnstone, 1965).

The lead concentration of the upper soil horizons of unmineralized and uncontaminated areas ("baseline" level) is generally given as approximately 10-20 μg g^{-1} of dry soil. Analyses of forest soils from throughout temperate zone forests, however, frequently show elevated lead amounts associated with the soil compartment. Lead may be added to the soil as components of organic compounds in plant debris. The divalent cationic nature of lead added via precipitation, dry fallout, throughfall, or stemflow may cause the lead to be bound to organic exchange surfaces (Zimdahl and Skogerboe, 1977) abundant in the forest floor. Subsequent reaction of lead with sulfate, phosphate, or carbonate anions may reduce its solubility and impede its downward migration in forest soil profiles. Lead deposited from the atmosphere may have a residence time approximating 5000 yr in the surface organic soil horizons (Benninger, 1975), and long-term concentration increases can be predicted as long as inputs to forest soils exceed outputs.

The author and colleague T. G. Siccama have been involved in an intensive biogeochemical study of several trace metals including lead in the northern hardwood forest (Siccama and Smith, 1978; Smith and Siccama, 1980). The study is being conducted on the Hubbard Brook Experimental Forest in central New Hampshire (elevation 230-1010 m). This forest is typified by an unbroken canopy of second-growth northern hardwoods (sugar maple, American beech, and yellow birch) with patches of red spruce and balsam fir, particularly at higher elevations and along the valley bottoms. The annual lead flux to the forest is 266 g ha^{-1} yr^{-1}. The lead output from the system in streamwater approximates 6 g ha^{-1} yr^{-1} (Smith and Siccama, 1980). The extraordinary disparity in these input-output figures is accounted for by the accumulation of lead in the soil compartment.

The parent material of the Hubbard Brook soil is compact unsorted acidic glacial till derived primarily from local gneissic bedrock. Profiles are strongly developed with typical podsol horizonation. A forest floor of the mor type with well-developed L, F, and H horizons of unincorporated organic matter overlies an A2 of 2-6 cm. The average lead concentration in the forest floor was determined to be 89 μg g^{-1} with total lead of the forest floor averaging 9 kg ha^{-1} (Figure 4-1).

Reiners et al. (1975) also have investigated lead retention by forest soils in the White Mountains of New Hampshire in an area they estimated had a lead

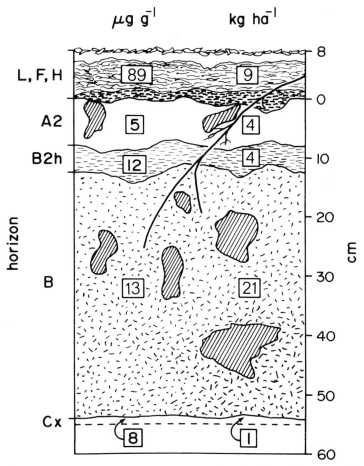

Figure 4-1. Lead distribution in a typical northern hardwood forest soil profile in central New Hampshire. Concentrations and amounts are based on acid extraction of ashed (475°C) samples and do not include lead in mineral crystal complexes not extracted by this procedure. (From Smith and Siccama, 1980.)

flux rate of approximately 200 g ha^{-1} yr^{-1}. Samples were collected from Mt. Moosilauke that represented various vegetative zones from the northern hardwood forest (~700 m) to the alpine tundra (~1400 m). Cores were removed from the organic layer (L = 01, plus F + H = 02) of the forest floor and analyzed for lead. Table 4-1 presents the soil lead concentrations obtained for the various vegetative types and compares them with various concentrations published in the literature.

These New Hampshire investigations were performed in an area reasonably remote from strong lead sources. The results, however, reveal high lead concentrations in the soil relative to other sites similarly remote from lead sources. The authors feel their data is consistent with the hypothesis that the New England mountains are frequently exposed to air masses from industrial and urban areas along with high winds, precipitation, and cloudiness and that the combination of these factors plus horizontal interception at higher elevations contributes to higher rates of aerosol deposition than are likely in other regions.

Van Hook et al. (1977) have examined lead cycling in eastern Tennessee's oak forests in the vicinity (14 km) of coal-fired electric generating facilities. The esti-

Table 4-1. Lead Concentrations of the Organic Horizon of Mt. Moosilauke, New Hampshire, Forest Soils Compared with Various Published Lead Concentrations

Area	Pb (μg g^{-1})
Minimum—all sites, layers	11
Maximum—all sites, layers	336
Weighted average, all layers	
Hardwoods	35
Spruce-fir	79
Fir forest	145
Fir krummholz	120
Alpine tundra	38
Mt. Moosilauke bedrock	46
World average (presumably inorganic)	10
Connecticut, U.S.A.	10-15
New Zealand	15
Missouri, U.S.A.	17
West England ⁻	42
Wales	42
Wales	45
Salt marsh, Connecticut	38
Heath, Sweden (above ground litter)	66, 80
Heath, Sweden (below ground litter)	40, 45
Spruce needle litter, Sweden	
Least decomposed	27, 34, 78
Intermediate	49, 61, 102
Most decomposed	66, 105
O$_2$, Pennsylvania	240-355

Source: Reiners et al. (1975).

mated lead input to the forest was given as 286 g ha^{-1} yr^{-1}. For four forest types examined, the litter layer was fractioned into two horizons: original leaf form still discernible (O_1) and original form lost (O_2). The lead concentrations and amounts in the litter are presented in Table 4-2. This study calculated the standing pools of lead in the vegetative components of the study forests. From these data it was observed that the litter (O_1 and O_2 horizons), which constituted 13% of the total forest organic matter, contained 71% of the lead contained within the ecosystem. Movement of lead from the O_2 litter to the underlying soil horizons was concluded to be high and estimated at 182 g ha^{-1} yr^{-1}.

As part of the International Biological Program, Heinrichs and Mayer (1977) gathered soil lead data from beech and spruce forests in Germany that are representative of ecosystems widely distributed in central Europe. The authors suggested that the sampled forests were in a "relatively unpolluted" environment yet the measured lead flux in precipitation below the beech canopy was 365 g ha^{-1} yr^{-1} and below the spruce canopy was 756 g ha^{-1} yr^{-1} (adjacent open field lead input in precipitation given as 405 g ha^{-1} yr^{-1}). Lead concentrations and amounts in the soils of these forests are presented in Table 4-3. All studies summarized to this point were conducted in rural sites in locations considered by their authors to be relatively remote from excessive lead contamination. Despite this fact it is obvious that considerable quantities of lead are accumulating in the soil and that this compartment is serving as an important sink for this trace metal. For locations subject to excessive lead input, for example, urban, industrial, or roadside locations, the accumulation of lead in the soil sink is even more impressive.

Parker et al. (1978) have examined lead distributions in an oak forest in a heavily urbanized and industrialized section of East Chicago, Indiana. Lead flux via precipitation and dry fall out was determined to be 815 g ha^{-1} annually. The lead concentrations for various soil horizons are given in Table 4-4.

Soil samples analyzed from sites 1-2 km from a lead smelter in Kellogg, Idaho, averaged 4640 μg g^{-1} at the surface (0-2 cm) (Ragaini et al., 1977). Soil samples collected from 65 major street intersections in urban locations in southern Ontario revealed a high of 21,000 μg g^{-1} lead in the upper 5 cm of soil (Linzon et al., 1976). These soils were not collected from forest sites but woody vegetation did occur in the general area.

Table 4-2. Lead Concentration and Amount Associated with the Litter of Eastern Tennessee Mixed Deciduous Forests

Litter	Yellow poplar		Chestnut oak		Oak-hickory		Pine	
	Conc. (μg g^{-1})	Amt. (g ha^{-1})	Conc. (μg g^{-1})	Amt. (g ha^{-1})	Conc. (μg g^{-1})	Amt. (g ha^{-1})	Conc. (μg g^{-1})	Amt. (g ha^{-1})
O_1	31	200	27	190	25	220	31	340
O_2	42	320	51	940	35	630	37	580

Source: Van Hook et al. (1973).

Table 4-3. Lead Concentration and Amount for Various Soil Horizons of Central German Beech and Spruce Forests

Horizon (cm)	Conc. (μg g^{-1})	Amt. (kg ha^{-1})
0-10	61	70.2
10-20	21	25.0
30-40	12	15.6
40-50	12	17.8

Source: Heinrichs and Mayer (1977).

Roadside environments are grossly contaminated with lead as motor vehicles combusting gasoline containing lead alkyls release approximately 80 mg of lead per km driven. The size of the roadside ecosystem approximates 3.04×10^7 ha (118,000 square miles) in the United States. Since much of this ecosystem contains woody vegetation, it is important to consider roadside soil lead burden. If 20μg g^{-1} lead is accepted as a baseline lead concentration for uncontaminated soils, it can be seen that soil samples taken within a few meters of the road surface of a heavily traveled highway may range to more than 30 times baseline. At 10 m distance from the roadway, however, the lead level is typically only 5-15 times baseline. At approximately 20 m distance, several studies suggest that a constant level of soil lead is achieved and the influence of the roadway is lost (Smith, 1976). Invariably, investigations of roadside soil lead conclude that the lead concentrations are positively correlated with traffic volume and negatively correlated with perpendicular distance from the roadway. Significant variations from strict correlation occur, however, due to numerous additional variables. Since lead accumulates over time in the soil, road age is important. Soil lead adjacent to an old road with lower traffic volume may exceed soil lead adjacent to a young road with higher traffic volume. If prevailing winds blow normal to the highway, significantly higher soil lead may be found on the lee side of the road. Other important variables causing deviation from generalized correlations include, of course, soil type, vehicle types, topography, and vegetative cover.

Detection of trends in rates of accumulation or loss of trace metals in soils over long periods of time is generally impossible due to the paucity of quantitatively obtained and systematically retained soil samples collected from the same site at different times. Reported changes in lead in mineral soils has generally been limited to measurements of changes in concentration. There have been no reports documenting changes in amount as well as concentration in areas not

Table 4-4. Lead Concentration for Various Soil Horizons from an Urban Oak Forest in East Chicago, Indiana

Soil horizon	Lead (μg g^{-1})
Surface litter (O horizon)	400
A1 (0-2.5 cm)	463
A1 (2.5-14 cm)	140
B (14-25 cm)	8

Source: Parker et al. (1978).

associated with major local sources of metals such as smelters or highways. A search by our laboratory for preserved soil samples from New England forests revealed a unique opportunity to analyze quantitatively sampled and carefully preserved forest floor material collected in 1962 by D. L. Mader, University of Massachusetts, as part of a study of forest floor characteristics. Lead contents were determined for L, F, and H layers of the forest floor of 10 white pine stands in central Massachusetts collected from the same place, in 1962 and 1978. Total lead content was found to have increased significantly. Average lead concentration increased in all layers but the increases were not statistically significant primarily due to the dilution effect of the concurrent increase in the mass of the forest floor. The observed net increase in lead of 30 mg m^{-2} yr^{-1} was approximately 80% of the estimated total annual input of this element via precipitation in this region during the 16 year period (Siccama et al., 1980). The total amount of lead in the forest floor of the pine stands (12 kg ha^{-1}) was greater than the amount in the forest floor of the northern hardwood forest in New Hampshire (9 kg ha^{-1}).

When a trace metal such as lead is added to the soil it may be (1) absorbed on soil particle exchange sites, (2) precipitated as an insoluble compound, (3) leached to lower depths in the soil profile, (4) lost to the atmosphere, (5) metabolized by soil fauna or microbes, or (6) absorbed by plant roots (compare Figure 4-2). In the case of lead and forest soils, it is apparent that mechanisms 3 through 6 are generally unimportant, that mechanisms 1 and 2 prevail, and as a result the soil compartment, particularly the organic forest floor, is an important sink for atmospheric lead. The flux of lead from the atmosphere to the forest floor may approximate 200-400 g ha^{-1} yr^{-1} in the forest soils of the eastern United States.

The threshold concentration of soil lead for phytotoxicity approximates 600 μg g^{-1} (Linzon et al., 1976). Since most forest ecosystems are well below this threshold, except those in selected urban, industrial, and roadside sites, and since evidence for tree root uptake of lead is meager, this biologically nonessential element is judged not to be responsible for direct impairment of tree health.

2. Other Trace Elements

Forest soils may constitute a sink for a variety of trace elements, in addition to lead. Elements judged to have particular potential for biological significance if accumulated in terrestrial ecosystems include cadmium, nickel, thallium, copper, fluorine, vanadium, zinc, cobalt, molybdenum, tungsten, mercury, and selenium (Van Hook and Shults, 1977). Under certain conditions manganese, chlorine, chromium, and iron might be appropriate additions to this list. Numerous of these elements differ substantially from lead in their interaction with plants. Some, for example, iron, chlorine, manganese, zinc, copper, and molybdenum, are required by vegetation in small amounts for normal growth and development. Others, for example, cadmium, nickel, thallium, and tungsten, are very mobile in plants. Cadmium, nickel, fluorine, thallium, vanadium, mercury, and copper have high potential for phytotoxicity.

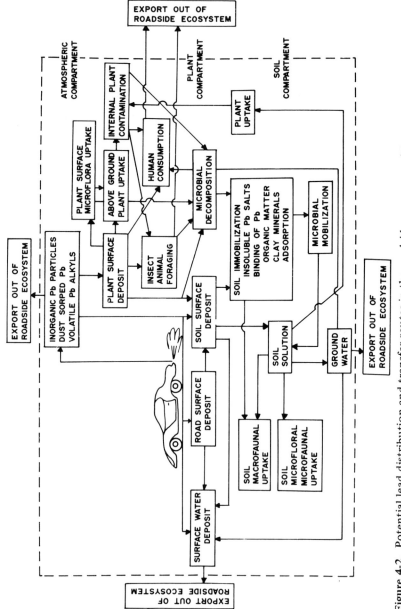

Figure 4-2. Potential lead distribution and transfer routes in the roadside ecosystem. (From Smith, 1976)

Information concerning the ability of forest soils to act as a sink for these elements is very limited. A comparison of Tables 4-5 and 4-6, which contain trace element concentrations for selected temperate forest soils, with Table 4-7, which contains baseline trace element concentrations for a variety of soils in relatively unpolluted areas and from regions lacking major mineral deposits, suggests that certain forest soils may be accumulating elevated levels of zinc, cadmium, copper, nickel, manganese, or iron. This is obviously the case for forests within several kilometers of an urban area, for example, the East Chicago situation, or metal smelter, for example the Palmerton environment. It is most significant, however, that it also appears that forest soils are accumulating elevated concentrations of these trace elements even when located many kilometers from a primary source. The residence time of heavy metals in forest soils is of fundamental importance in judging sink efficiency. Tyler (1978) has examined the leachability of manganese, zinc, cadmium, nickel, vanadium, copper, chromium, and lead from two organic spruce forest soils. One soil was from a site grossly polluted by copper and zinc from a metal smelter and the other from an unpolluted forest location. Both soils were treated in lysimeters with artificial rainwater, acidified to pH 4.2, 3.2, and 2.8. The results of this experiment are presented in Table 4-8. At pH 4.2, relatively commonly measured in the northeastern United States and in northern Europe, the residence times are impressive, particularly in the polluted soil. The high buffer capacity of the latter soil is cited by Tyler as conferring protection against leaching. While extrapolation of lysimeter data to natural soils can only be made with considerable risk, Tyler's experiment is very informative.

It is concluded that, in addition to lead, forest soils, in particular the organic forest floor, function with variable efficiency as a sink for zinc, cadmium, copper, nickel, manganese, vanadium, and chromium associated with particulates input from the atmosphere. Judgments concerning the relative efficiency and importance of the retention of these various metals must await greater understanding of their dynamics in a larger range of soil types.

3. Other Particulates

For several forest ecosystems in the United States the annual deposition of sulfate, expressed as sulfate-sulfur (SO_4^{2-}-S) is between 10 and 20 kg ha^{-1} (Likens et al., 1977; Shriner and Henderson, 1978; Swank and Douglass, 1977). While extremely limited information is available on the biogeochemical cycle of sulfur in forest ecosystems, Shriner and Henderson (1978) have examined sulfur cycling in various eastern Tennessee forest types in a region they estimated had an annual average deposition of 18.1 kg ha^{-1} sulfate. Eighteen percent of the deposition occurred as dry particulate fallout, while the bulk (82%) was dissolved in rainfall. Of this input, 64% or 11.5 kg ha^{-1} was lost from the forest in streamflow. The retention of 6.6 kg ha^{-1} (net annual accumulation) in this Tennessee forest greatly exceeds the sulfate retention estimated for New Hampshire forests (1.1 kg ha^{-1}; Likens et al., 1977) but approximates the estimate for North Carolina forests (7.6 kg ha^{-1}; Swank and Douglass, 1977).

Table 4-5. Trace Element Concentrations and Amounts for Various Forest Soils

Element	Location and sample site	Conc. ($\mu g\ g^{-1}$)	Amt. ($g\ ha^{-1}$)	Reference
Zinc	White Mountains, N.H., soil organic horizons			Reiners et al. (1975)
	Minimum—all sites, layers	19		
	Maximum—all sites, layers	169		
	Weighted average, all layers			
	Hardwoods	100		
	Spruce-fir	53		
	Fir forest	90		
	Fir krummholz	73		
	Alpine tundra	29		
	Eastern Tennessee forests			Van Hook et al. (1977)
	Soil litter layer O_1			
	Yellow poplar	58	380	
	Chestnut oak	42	290	
	Oak-hickory	48	420	
	Pine	56	610	
	Soil litter layer O_2			
	Yellow poplar	130	980	
	Chestnut oak	110	2000	
	Oak-hickory	125	2200	
	Pine	59	930	
	Central German beech and spruce forests, various soil depths			Heinrichs and Mayer (1977)
	0-10 cm	38	44×10^3	
	10-20 cm	50	60×10^3	
	30-40 cm	55	72×10^3	
	40-50 cm	49	73×10^3	
	East Chicago oak forest (urban)			Parker et al. (1978)
	Surface litter (O horizon)	915		
	A1 horizon (0-2.5 cm)	2456		
	A1 horizon (2.5-14 cm)	711		
	B horizon (14-25 cm)	103		
	Central Sweden spruce forest			Tyler (1972)
	Needle litter (O_1)	530		
	Humus (O_2)	2300		
	Palmerton, Pennsylvania, within 1 km of zinc smelter			Buchauer (1973)
	Humus (O_2)	135×10^3	13.5×10^6	
Cadmium	Eastern Tennessee forests			Van Hook et al. (1977)
	Soil litter layer O_1			
	Yellow poplar	1.0	8	
	Chestnut oak	0.8	15	
	Oak-hickory	0.2	4	
	Pine	0.6	9	

Table 4-5 (continued)

Element	Location and sample site	Conc. (μg g^{-1})	Amt. (g ha^{-1})	Reference
Cadmium	Central German beech and spruce forests, various soil depths			Heinrichs and Mayer (1977)
	0-10 cm	.056	64	
	10-20 cm	.060	71	
	30-40 cm	.074	96	
	40-50 cm	.096	142	
	East Chicago oak forest (urban)			Parker et al. (1978)
	Surface litter (O horizon)	4.7		
	A1 horizon (0-2.5 cm)	10.4		
	A1 horizon (2.5-14 cm)	3.8		
	B horizon (14-25 cm)	0.3		
	Central Sweden spruce forest			Tyler (1972)
	Needle litter (O$_1$)	24		
	Humus (O$_2$)	44		
	Palmerton, Pennsylvania, within 1 km of zinc smelter			Buchauer (1973)
	Humus (O$_2$)	1750	18×10^4	
Copper	Central German beech and spruce forests, various soil depths			Heinrichs and Mayer (1977)
	0-10 cm	24	28×10^3	
	10-20 cm	18	21×10^3	
	30-40 cm	23	30×10^3	
	40-50 cm	17	25×10^3	
	East Chicago oak forest (urban)			
	Surface litter (O horizon)	76		
	A1 horizon (0-2.5 cm)	119		
	A1 horizon (2.5-14 cm)	30		
	B horizon (14-25 cm)	3		
	Central Sweden spruce forest			Tyler (1972)
	Needle litter (O$_1$)	260		
	Humus (O$_2$)	660		
	Palmerton, Pennsylvania, within 1 km of zinc smelter			Buchauer (1973)
	Humus (O$_2$)	2000		
Nickel	Central German beech and spruce forests, various soil depths			Heinrichs and Mayer (1977)
	Needle litter (O$_1$)	26		
	Humus	36		
	Central Sweden spruce forest			Tyler (1972)
	Needle litter (O$_1$)	26		
	Humus (O$_2$)	36		

Table 4-5 (continued)

Element	Location and sample site	Conc. ($\mu g\ g^{-1}$)	Amt. ($g\ ha^{-1}$)	Reference
Manganese	Eastern Tennessee forests, various soil depths			Van Hook et al. (1973)
	Litter	2000		
	0-40 cm	700		
	40-80 cm	700		
	80-120 cm	75		
	120-160 cm	25		
	$>$ 160 cm	2000		
	Central German beech and spruce forests, various soil depths			Heinrichs and Mayer (1977)
	0-10 cm	330	380×10^3	
	10-20 cm	700	833×10^3	
	30-40 cm	730	949×10^3	
	40-50 cm	740	110×10^4	
Chromium	Eastern Tennessee forests, various soil depths			Van Hook et al. (1973)
	Litter	2		
	0-40 cm	5		
	40-80 cm	17		
	80-120 cm	2		
	120-160 cm	5		
	$>$ 160 cm	50		
	Central German beech and spruce forests, various soil depths			Heinrichs and Mayer (1977)
	0-10 cm	52	60×10^3	
	10-20 cm	59	70×10^3	
	30-40 cm	56	73×10^3	
	40-50 cm	51	76×10^3	
Chlorine	Eastern Tennessee forests, various soil depths			Van Hook et al. (1973)
	Litter	50		
	0-40 cm	55		
	40-80 cm	50		
	80-120 cm	150		
	120-160 cm	150		
	$>$ 160 cm	50		

Shriner and Henderson (1978) estimated that 65% of the net annual sulfur accumulation is located in the soil. Unlike trace metal accumulation, these authors conclude that 92% of the sulfur is located in mineral soil horizons, with only 3% contained in the organic forest floor.

Table 4-6. Trace Element Concentrations and Amounts for Central German Beech and Spruce Forests at Various Soil Depths

Soil depth	Iron		Cobalt		Vanadium		Mercury		Thallium	
	Conc. ($\times 10^5$ $\mu g\,g^{-1}$)	Amt. (tons ha^{-1})	Conc. ($\mu g\,g^{-1}$)	Amt. (kg ha^{-1})	Conc. ($\mu g\,g^{-1}$)	Amt. (kg ha^{-1})	Conc. ($\mu g\,g^{-1}$)	Amt. (g ha^{-1})	Conc. ($\mu g\,g^{-1}$)	Amt. (g ha^{-1})
0-10	1.6	18.4	7	8.1	55	63	0.12	138	0.5	610
10-20	1.9	22.6	11	13.1	62	74	0.08	100	0.3	381
30-40	1.9	24.7	12	15.6	62	81	0.05	65	0.2	247
40-50	2.0	29.6	13	19.2	63	93	0.02	27	0.2	340

Source: Heinrichs and Mayer (1977).

Table 4-7. Baseline or General Trace Element Concentrations for Soils in Unmineralized and Uncontaminated Areas

Element	Conc. in soil ($\mu g\ g^{-1}$)	
	Mean	Range
Cadmium	0.3-0.6	0.01-0.7
Chlorine	100	
Chromium	60	5-300
Cobalt	8	1-40
Copper	20	2-100
Fluorine	200	30-300
Iron	38,000	$7000\text{-}55 \times 10^4$
Manganese	550	100-4000
Mercury	0.05	0.001-0.5
Molybdenum	2	0.2-5
Nickel	14	<5-700
Selenium	0.1-2.0	<0.04-1200
Tungsten	1	
Vanadium	80	15-100
Zinc	50	25-65

Sources: Cannon (1974); Bowen (1966); Connor and Shacklette (1975).

B. Forest Soils as Sinks for Atmospheric Gases

Our understanding of the capacity of various soils to function as sinks for gaseous air contaminants is very limited. This general topic has received only meager research attention over the last ten years. Obviously soils have considerable capacity to absorb a variety of gases from the atmosphere and to incorporate and transform them in or on the soil through a large number of microbial, other biological, physical, and chemical processes. Specific information is lacking, however, on the relative importance of source versus sink function, the capacities and rates of various soils for absorption, residence and reaction rate times, the influence of soil physical (mineral and organic matter content, structure, porosity) and chemical (pH, moisture content, exchange capacity) properties and climate on removal rates, and the significance of soil management practices.

Bohn (1972) reviewed the literature on soil sink function and presented some generalizations. He concluded that soils will absorb organic gases more rapidly and in larger quantities with increasing molecular weight and with greater numbers of nitrogen, phosphorus, oxygen, sulfur, and other functional group substitutions in the compound. Absorption of low molecular weight and less substituted organic gases was judged to be dependent on the development of an appropriate microbial population. Soil removal of inorganic gases was concluded to involve primarily chemical and physical processes. The author observed that the literature regarding soil removal of reducing gases (oxides of carbon, sulfur and nitrogen, hydrocarbons, aldehydes) was modest while the information regarding oxidizing gases (ozone, peroxy compounds, chlorine) was nonexistent.

Table 4-8. Residence Times of Heavy Metals in Contaminated and "Clean" Organic Spruce Soils (Sweden) Artificially Leached with Acidified Precipitation Equivalent to an Annual Percolation of 150 liters m$^{-2 a}$

Metal	pH	Control soil	Polluted soil
Mn	4.2	3	30-40
	3.2	1.5	4
	2.8	0.5	1.5
Zn	4.2	7	9
	3.2	2	3
	2.8	0.8	1.2
Cd	4.2	6	20
	3.2	3	4-5
	2.8	1.3	1.7
Ni	4.2	5	15
	3.2	2	4-5
	2.8	2	2
V	4.2	17	2
	3.2	25-30	6-7
	2.8	9	9
Cu	4.2	13	80-120
	3.2	11	18-20
	2.8	9	6
Cr	4.2	20	100-150
	3.2	18-20	50-70
	2.8	15	50-70
Pb	4.2	70-90	>200
	3.2	40-50	>100
	2.8	20	17

Source: Tyler (1978).
[a] Data are number of years necessary for a 10% decrease of metal contamination.

Unfortunately the literature addressing the ability of forest soils to remove atmospheric gaseous contaminants is especially small. The following discussion will draw heavily on evidence obtained from situations involving agricultural or other nonforest soils.

1. Carbon Monoxide

Carbon monoxide is formed in all combustion processes as a result of the incomplete oxidation of carbon and as a result anthropogenic production is locally and regionally enormous. In excess of 6×10^{14} g of carbon monoxide is annualy discharged into the atmosphere (Seiler, 1974). The primary contributors of combustion carbon monoxide are the United States, Europe, and Japan. As a result most of the anthropogenic emissions are concentrated in the temperate latitudes of the northern hemisphere. As a consequence, a key feature of the global distribution of carbon monoxide is a higher concentration in the northern than in the southern hemisphere.

Carbon monoxide contents are at a maximum in the northern hemisphere during winter and spring and are at a minimum during the summer. Over the past two decades winter concentrations have tended to increase while summer levels have remained constant. Despite the geographic and seasonal variations, the available information supports the conclusion that global carbon monoxide concentrations have remained relatively constant in "clean" atmospheres. Available data further support the conclusion that during the warm season atmospheric levels of carbon monoxide are regulated by intense natural sink function (Nozhevnikova and Yurganov, 1978).

A large number of potential natural sinks have been proposed: (1) absorption by oceans, (2) oxidation to carbon dioxide by OH^- in the troposphere, (3) migration to the stratosphere followed by photochemical reaction, (4) reaction with animal hemoprotein, (5) fixation by higher plants, and (6) absorption by soil. In light of available evidence, only the latter two hypotheses can be said to be of general importance.

Vegetative oxidation of carbon monoxide to carbon dioxide (Ducet and Rosenberg, 1962) and fixation as serine (Chappelle and Krall, 1961) have been described. Employing ^{14}CO Bidwell and Fraser (1972) observed uptake by leaves of bean plants under both light and dark conditions. Numerous other non-tree species were tested for carbon monoxide uptake at low gas concentration in the light. Using their bean plant data, the authors estimated a summer removal capacity of 12-120 kg km^{-2} day^{-1}, globally $3-30 \times 10^8$ tons yr^{-1} (six month growing season). With the risks of extrapolation from in vitro work with greenhouse plants to the global scene aside, judgments concerning the significance of vegetation as a carbon monoxide sink remain difficult to make. Inman and Ingersoll (1971) failed to observe any capability of several plants, including seedlings of Monterey and knobcone pine and mimosa, to remove carbon monoxide from the atmosphere. Also the role of plants as producers of carbon monoxide (Nozhevnikova and Urganov, 1978) must be more accurately understood before the role of vegetation as a sink for carbon monoxide can be judged appropriately. While vegetation may play some role in the maintenance of carbon monoxide in the natural atmosphere, soils are concluded to be the most important removal agent.

The first evidence supporting the importance of soil as a sink for carbon monoxide was presented in 1926 and this study along with many that have followed indicate that soil microorganisms are responsible for the removal.

Inman et al. (1971) conducted preliminary experiments with a greenhouse potting mixture and found that the test soil could deplete carbon monoxide in an experimental atmosphere [containing 120 ppm (13.8×10^4 $\mu g\, m^{-3}$) CO] to near zero within 3 hr. Treatment of the soil with steam sterilization, antibiotics, salt, and anaerobic conditions all prevented carbon monoxide uptake and indicated the importance of biological processes. A variety of soils from California, Hawaii, and Florida were brought into the laboratory and tested for their ability to remove carbon monoxide (Table 4-9). Inman and co-workers generalized from their results that cultivated soils were less active than natural soils and that higher

Table 4-9. Carbon Monoxide Uptake from Test Atmospheres [80-130 ppm $(9.2\text{-}1.5 \times 10^4 \ \mu g \ m^{-3})$ CO] by Various Soils at $25°C$

Vegetation	Location	mg hr^{-1} m^{-2} of soil
Coast redwoods	CA	16.99
Oak	CA	15.92
Coast redwoods	CA	14.39
Ponderosa pine	CA	13.89
Grass-legume pasture	CA	11.94
Grapefruit	CA	11.48
Grass meadow	CA	10.52
Forest	HI	9.90
Chaparral	CA	6.46
Oak stubble	CA	6.23
Chaparral	CA	4.31
White fir	CA	3.48
Cotton (fallow)	CA	2.82
Almond weeds (fallow)	FL	2.65

Source: Inman et al. (1971).

organic matter and lower pH soils were the most active. These observations clearly support the potential importance of forest soils!

In order to improve the confidence of their observations, Inman and others (Ingersoll et al., 1974) outfitted a mobile laboratory and field tested soils in most major vegetative regions in the United States. Field testing was accomplished by covering a square meter of undisturbed soil and covering vegetation with a gas tight chamber. The carbon monoxide uptake rate showed considerable variation in the field ranging from 7.5 to 109 mg $hr^{-1} \ m^{-2}$. As in the previous work, cultivated soils were invariably lower than natural soils. Natural soil uptake rates are presented in Table 4-10. Ingersoll et al. (1974) concluded that the potential rates of carbon monoxide uptake by the soils of the United States and the world were 505 million and 14.3 billion tons per year, respectively (Tables 4-11 and 4-12). Forest soils are indicated to be of particular significance. If Ingersoll's estimates approximate the natural condition, it must be concluded that temperate and tropical forests play extraordinarily important roles as sinks for global carbon monoxide.

Seiler (1974) has determined carbon monoxide uptake rates for several European soils (location and vegetation unspecified) and calculated an average flux rate of 1.5×10^{-11} g $cm^{-2} \ sec^{-1}$ at $15°C$. This rate is very approximately an order of magnitude less than Ingersoll's average. Seiler explained the discrepancy by indicating that Inman's group had employed initial carbon monoxide concentrations of 100 ppm ($11.5 \times 10^4 \ \mu g \ m^{-3}$) while his laboratory had employed 0.20 ppm ($230 \ \mu g \ m^{-3}$) judged to be the normal ambient concentration in unpolluted areas of Europe. Seiler's estimate for global soil removal was 4.5×10^8 tons per year.

Table 4-10. Carbon Monoxide Uptake Rates Determined under Field Conditions (Corrected for Annual Average Temperature at Test Site) at 100 ppm CO for Various Soils under Natural Vegetation in North America

Vegetation	Location	mg hr^{-1} m^{-2} of soil
Flood plain forest	LA	26.2
Deciduous	NY	26.3
Mixed deciduous	OH	21.3
Oak-hickory forest	OH	21.3
Coastal forest	OR	14.1
Grassland	TX	11.9
Oak-hickory	MO	10.2
Boreal forest	Alberta	6.8
Boreal forest	Alberta	4.2
Grassland	KA	3.6
Grassland	Saskatchewan	2.7
Bluestem prairie	KA	2.3

Source: Ingersoll et al. (1974).

The question of inadequate account of carbon monoxide evolution by soils has been raised by Smith et al. (1973). This deficiency could contribute to important overestimations of sink function. Perhaps the actual capacity for carbon monoxide removal lies between the estimates of Inman's and Seiler's groups or perhaps it approximates one or the other. Whatever the case, it can be concluded that forest soils play a role of very significant importance as a repository

Table 4-11. Potential Carbon Monoxide Uptake Rates of the Soils of the Conterminous United States

Soil-vegetation type	Area (mi^2)	CO uptake (tons yr^{-1} mi^{-2})	Total CO uptake (tons $\times 10^6$ yr^{-1})
Pasture	616,674	179.3	110.57
Deciduous forest	339,485	254.8	86.50
Appalachian forest	265,519	313.7	84.17
Coastal forest	329,390	200.7	83.93
Cropland	468,000	86.0	40.25
Southern flood plain forest	58,782	595.0	34.95
Montane forest	87,150	242.0	21.10
Sagebrush steppe	264,867	76.7	20.32
Sagebrush	145,047		
Desert scrub	220,723	52.2	19.09
Southern mixed forest	48,644	86.5	4.21
Paved roads	28,100	0	0
Covered area	26,500	0	0
Lakes, rivers	78,267	0	0
Total	2,977,128		505.12

Source: Ingersoll et al. (1974).

Table 4-12. Potential Carbon Monoxide Uptake Rates of the Soils of the World

Soil-vegetation type	Area (10^6 mi^2)	Average CO uptake ($\times$ 10 tons yr^{-1} mi^{-2})	Total CO uptake ($\times$ 10^6 tons yr^{-1})
Tropical rain forest	6.55	805.6	5,277.5
Tropical grassland	3.45	886.9	3,062.5
Tropical deciduous forest	1.78	1,105.9	1,969.6
Montane forest	4.66	175.7	817.8
Taiga forest	4.71	127.5	600.9
Desert	10.86	52.2	567.0
Mixed and broadleaf forest	1.32	254.8	410.0
Agricultural	4.60	86.0	395.4
Pasture	1.84	179.3	329.7
Temperate grassland	3.10	86.5	268.4
Steppe	3.45	76.7	264.5
Southern pine forest	0.29	505.8	145.2
Tundra	1.15	123.0	141.3
Covered by ice, water, roads, structures	9.20	0	0
Total	56.96		14,250.0

Source: Ingersoll et al. (1974).

for atmospheric carbon monoxide. Since the soil removal rate increases with increasing levels of carbon monoxide, forest ecosystems in and around urban and industrial areas may be especially important sinks.

The specific components of the soil microflora involved in the removal of carbon monoxide are not completely understood and a thorough discussion of this topic is beyond the scope of this book. Inman and Ingersoll (1971) reported the identification of 16 fungal isolates belonging to the *Penicillium, Aspergillus, Mucor, Haplosporangium* and *Mortierella* genera that were capable of removing atmospheric carbon monoxide. As Nozhevnikova and Yurganov (1978) point out, however, all microbes with hemoproteins and cytochrome oxidase or other carbon monoxide reacting enzymes will fix this gas to some extent. The microbes of importance in this sink function are the organisms for which oxidation of carbon monoxide serves as a source of energy. Bacteria that oxidize carbon monoxide belong to a physiological group termed carboxydobacteria and belong to genera that appear to be related to the genus *Pseudomonas*. They are gram-negative, aerobic species that are considered to be widely distributed in nature. Their abundance in urban substrates may reflect a more favorable environment created by elevated carbon monoxide levels (Nozhevnikova and Yurganov, 1978). Their particular abundance in forest soils is unfortunately not clear.

2. Sulfur Dioxide and Hydrogen Sulfide

The sulfur gases currently recognized as important components of tropospheric air include sulfur dioxide (SO_2), hydrogen sulfide (H_2S), carbonyl sulfide (COS), carbon disulfide (CS_2), dimethyl sulfide ($CH_3 SCH_3$), and sulfur hexafluoride (SF_6) (Bremner and Stelle, 1978). Until very recently it was generally assumed that the primary and most important forms of sulfur in the atmosphere were hydrogen sulfide, sulfur dioxide, and sulfates. The information available concerning soil as a sink for atmospheric sulfur has dealt almost exclusively with the latter two.

As previously discussed in this chapter, sulfate will be added to soil largely by precipitation and will become part of the soluble sulfur content held by soil colloids. Soils also have a large capacity to quickly absorb sulfur dioxide from the atmosphere. Factors that tend to increase the soil uptake of sulfur dioxide include fine texture, high soil organic matter content, high pH, presence of free $CaCO_3$, high soil moisture content, and the presence of soil microorganisms (Nyborg, 1978).

Since Alway et al. (1937) presented the first evidence that soils can absorb sulfur dioxide, a large number of studies have followed. Smith et al. (1973) studied the capability of six soils, from Oregon, Iowa, and Saskatchewan with variable chemical and physical properties and found that removal of sulfur dioxide and hydrogen sulfide was much more rapid than the removal of carbon monoxide. It is not clear if any of the soils were from forest ecosystems, but the pH (4.8) and organic carbon percent (9.38) suggest that the Astoria, Oregon, soil may have been (Table 4-13). Clearly soil moisture favors uptake of sulfur dioxide, presumably due to the high solubility of this gas in water. The Astoria soil appears quite average in its ability to remove sulfur dioxide and hydrogen sulfide. The authors cautioned that the uptake rates of Table 4-13 should not be judged to be maxima for natural soils as the conversion to sulfate in vivo would presumably create more absorption sites in natural soils. Bohn (1972) has

Table 4-13. Sorption Capacity of Several Soils for Sulfur Dioxide and Hydrogen Sulfide Determined under Laboratory Conditions (100 ppm Gas Concentration at $23^{\circ}C$)

	mg g^{-1} of soil			
	Air-dry		Moist	
Soil	SO_2	H_2S	SO_2	H_2S
Astoria	8.9	62.9	37.1	51.6
Weller	15.3	61.0	31.9	58.1
Okoboji	10.2	52.9	31.4	44.5
Thurman	1.1	15.4	9.3	11.0
Regina	13.2	65.2	50.4	62.5
Harpster	10.2	46.6	66.8	40.6
Mean	9.8	50.7	37.8	44.7

Source: Smith et al. (1973).

suggested that the absorption rate of hydrogen sulfide slows with high soil moisture contents perhaps due to slow diffusion rates in water-filled pores. The fact that hydrogen sulfide absorption capacity increases with higher soil pH may reduce the importance of forest ecosystems in removing this gas from the atmosphere. Despite the relatively large number of papers addressing the soil removal of these two gases (Bremner and Steele, 1978; Moss, 1975), estimates of global and regional removal amounts by soil are relatively few (Rasmussen et al., 1974). Available data suggest that deposition velocities (v) for sulfur dioxide are generally in the range of 0.2 to 0.7 cm sec^{-1} (Rasmussen et al., 1974) which is less than those for vegetation (Chapter 5). Eriksson (1963) has estimated that the global removal of sulfur dioxide by soil equals 25×10^9 kg sulfur dioxide-sulfur annually. Abeles et al. (1971) examined the capacity of soil collected from Waltham, Massachusetts, to remove sulfur dioxide under laboratory conditions [100 ppm (26.2×10^4 μg m^{-3}) gas] and extrapolated their results to suggest that United States soils may be capable of removing 4×10^{13} kg of sulfur dioxide (2×10^{13} kg sulfur dioxide-sulfur) per year.

The mechanism of soil removal of sulfur dioxide is not completely understood but is generally concluded to involve both microbial and chemical mechanisms. Since autoclaving only partially attenuated the ability of their Massachusetts soil to removal sulfur dioxide, Abeles et al. (1971) concluded that the major removal mechanism was chemical rather than microbial. Sterilization of test soils has led others to conclude that microbial metabolism of sulfur dioxide may be relatively unimportant (Smith et al., 1973; Ghiorse and Alexander, 1976). Using a soil of unclear origin and $^{35}SO_2$, however, Craker and Manning (1974) studied uptake by soil bacteria and fungi and found that fungi were capable of removing sulfur dioxide. The relative importance of biological and nonbiological removal strategies will be determined as more sensitive and specific analytical techniques for sulfur forms become available. Present evidence encourages the conclusion that sorption of sulfur dioxide by soils involves the formation of sulfite and sulfate and that sorption of hydrogen sulfide involves formation of metallic sulfides and elemental sulfur (Bremner and Steele, 1978).

Moss (1975) presented one of a few investigations that have systematically examined sulfur uptake in natural ecosystems adjacent to excessive sulfur release. He investigated plant and soil sulfate burdens in the vicinity of industrial centers in Sheffield, England, and Welland, Ontario. In the latter situation, the sulfate accumulation was observed to be greater in grassland relative to forest sites at equivalent distance downwind from the source. Why this was so was unclear and highlights our limited understanding of actual sink capabilities in natural environments.

a. Other Sulfur Gases

The gas chromatographic studies of Bremner and Banwart (1976) showed that air-dry and moist soils had the capacity to sorb dimethyl sulfide, dimethyl disulfide, carbonyl sulfide, and carbon disulfide, but did not sorb sulfur hexafluoride. The first four gases were removed more efficiently by moist soils. Soil sterilization indicated that soil microorganisms were partially responsible for the removal of these gases. As the rates of removal were substantially less than those for sulfur dioxide or hydrogen sulfide, however, the authors concluded that while soils may constitute a sink for low levels of dimethyl sulfide, dimethyl disulfide, carbonyl sulfide, and carbon disulfide they would not effectively reduce elevated levels of these gases in areas of high anthropogenic emission.

3. Nitrogen Oxides and Other Nitrogen Gases

In the atmosphere nitric oxide is either oxidized to nitrogen dioxide or photolyzed to nitrogen gas. Nitrogen dioxide reacts photochemically or is removed by precipitation, primarily in the form of nitric acid. Nitric oxide and nitrogen dioxide may also be removed by soils. Nitric oxide is oxidized to nitrogen dioxide in soil, but the former gas does not persist in acid soils as long as it does in basic soils (Bohn, 1972).

Working with Waltham, Massachusetts, soil, Abeles et al. (1971) found that the removal rate for nitrogen dioxide was slower than the removal rate for sulfur dioxide. Twenty-four hours were required to reduce nitrogen dioxide concentrations from 100 to 3 ppm (18.8×10^4 to 56.4×10^2 μg m^{-3}) in test atmospheres. Extrapolation of these laboratory experiments allowed these authors to suggest that the soils of the United States may be capable of removing 6×10^{11} kg of nitrogen dioxide per year.

Both Ghiorse and Alexander (1976) and Smith and Mayfield (1978) have documented rapid absorption of nitrogen dioxide by both sterile and nonsterile soil. In the latter study, soil from an uncultivated grassland in Ontario absorbed 99% of the nitrogen dioxide introduced into a test vessel at 25°C in 15 min.

Mechanisms of nitrogen dioxide uptake may involve reaction with soil cations to form $NaNO_2$ or KNO_2, reaction with soil water to form HNO_2 and HNO_3, binding with organic matter, or persistence as a gas in interparticle soil spaces (Smith and Mayfield, 1978).

Since ammonia would probably be present in the atmosphere in the form of $(NH_4)_2SO_4$ rather than NH_3 in all but the most unusual environments, direct soil absorption of gaseous NH_3 is probably not important. Where ambient conditions might expose soils to high ammonia levels, however, evidence suggests that acid soils are particularly efficient removal agents (Rasmussen et al., 1974).

The specific capabilities that forest soils may have to remove nitrogenous gases must await further experimentation.

4. Hydrocarbons

Hydrocarbons are generally not soluble in water, and as a result, soil uptake where it is important is concluded to be primarily microbial. The light hydrocarbon from motor vehicles most actively removed by soil is ethylene (Zimmerman and Rasmussen, 1975). Abeles et al. (1971) observed that Maryland soil samples removed ethylene more slowly than other soils removed sulfur and nitrogen dioxides. Soil removal of ethylene, mediated by various microorganisms, was calculated by Abeles' group to approximate 7×10^9 kg of ethylene annually in the United States. Smith et al. (1973) determined that the soil flux rate for acetylene was from 0.24 to 3.12×10^{-1} mole g^{-1} day^{-1}. Their test soil with the lowest pH and highest organic matter (forest soil) was the most active of all soils tested for acetylene removal.

5. Oxidants

There is limited evidence that soils function as a sink for atmospheric ozone (Rasmussen et al., 1974). Aldaz (1969) concluded that the soil and vegetation of the surface of the Earth represent a major sink for this gas and estimated the capacity of this sink to be within the range of 1.3 to 2.1×10^{12} kg ozone yr^{-1}.

Most reports of ozone removal have examined plant uptake or plant and soil uptake combined. Turner et al. (1973) have tested the sink capacity of a freshly cultivated sandy loam devoid of vegetation. Their results, which were recorded under field conditions, showed the flux rate of ozone removal varied from 3 to 12×10^{11} mole cm^{-2} sec^{-1}, making bare soil in the author's judgment an important sink for ozone.

6. Other Gases

Fang (1978) examined the uptake of mercury vapor by five Montana soils by exposing them to a test atmosphere containing 75.9 μg metallic ^{203}Hg vapor m^{-3} for 24 hr. The soil with the highest organic matter content had the highest mercury uptake. While mercury vapor is currently only an extremely localized problem, more than 90% of the mercury contained in coal is vaporized during combustion. Even with relatively low mercury concentrations in coal, widespread or large volume coal combustion may increase the significance of soil retention of this gas.

C. Summary

Forest soils are important sinks for a variety of air contaminants. Retention of particulate lead by organic materials in the forest floor is a most dramatic example. The flux of lead to temperate forest ecosystems downwind of industrial, urban, or roadside sources may approximate 200-400 g ha^{-1} yr^{-1}, with much of this lead accumulating in the forest floor. Certain forest soils may also serve as a sink for additional trace metals including zinc, cadmium, copper, nickel, manganese, vanadium, and chromium. The efficiency of sink function for the latter metals is generally substantially less than for lead, but may be important, particularly in forest systems close to primary sources.

Forest soils remove pollutant gases from the atmosphere via several microbial, chemical, and physical processes. Forest soils function as an especially efficient sink for carbon monoxide and may play a dominant role in regulating the concentration of this gas in the atmosphere. Other gases that may be significantly removed by forest soils include sulfur dioxide, ammonia, some hydrocarbons, and mercury vapor.

References

Abeles, F. B., L. E. Craker, L. E. Forrence, and G. R. Leather. 1971. Fate of air pollutants: Removal of ethylene, sulfur dioxide and nitrogen dioxide by soil. Science 173:914-916.

Aldaz, L. 1969. Flux measurements of atmospheric ozone over land and water. J. Geophys. Res. 74:6943-6946.

Alway, F. J., A. W. Marsh, and W. J. Methley. 1973. Sufficiency of atmospheric sulfur for maximum crop yields. Soil Sci. Soc. Am. Proc. 2:229-238.

Benninger, L. K., D. M. Lewis, and K. K. Turekian. 1975. The use of natural Pb-210 as a heavy metal tracer in the river-estuarine system. In: T. M. Church (ed.), Marine Chemistry and Coastal Environment, American Chemical Society Symposium Series No. 18, pp. 201-210, American Chemical Society, Washington, D.C.

Bidwell, R. G. S., and D. E. Fraser. 1972. Carbon monoxide uptake and metabolism by leaves. Can. J. Bot. 50:1435-1439.

Bohn, H. L. 1972. Soil absorption of air pollutants. J. Environ. Qual. 1:372-377.

Bowen, H. J. M. 1966. Trace Elements in Biochemistry. Academic Press, New York, 241 pp.

Bremner, J. M., and W. L. Banwart. 1976. Sorption of sulfur gases by soils. Soil Biol. Biochem. 8:79-83.

Bremner, J. M., and C. G. Steele. 1978. Role of microorganisms in the atmospheric sulfur cycle. Adv. Microb. Ecol. 2:155-201.

Buchauer, M. J. 1973. Contamination of soil and vegetation near a zinc smelter by zinc, cadmium, copper, and lead. Environ. Sci. Technol. 7:131-135.

Cannon, H. L. 1974. Natural toxicants of geologic origin and their availability to man. In: P. L. White and D. Robbins (Ed.), Environmental Quality and Food Supply. Futura, New York, pp. 143-163.

Chappelle, E. W., and A. R. Krall. 1961. Carbon monoxide fixation by cell-free extracts of green plants. Biochem. Biophys. Acta 49:578-580.

Chow, T. J., and J. L. Earl. 1970. Lead aerosols in the atmosphere: Increasing concentration. Science 169:577-580.

Chow, T. J., and M. S. Johnstone. 1965. Lead isotopes in gasoline and aerosols of Los Angeles Basin, California. Science 147:502-503.

Connor, J. J., and H. T. Shacklette. 1975. Background Geochemistry of Some Rocks, Soils, Plants, and Vegetables in the Conterminous United States. U.S.D.I., Geological Survey Professional Paper 574-F, Washington, D.C., 168 pp.

Craker, L. E., and W. J. Manning. 1974. SO_2 uptake by soil fungi. Environ. Pollut. 6:309-311.

Ducet, G., and A. I. Rosenberg. 1962. Leaf respiration. Ann. Rev. Plant Physiol. 13:171-200.

Eriksson, E. 1963. The yearly circulation of sulfur in nature. J. Geophys. Res. 68:4001-4008.

Fang, S. C. 1978. Sorption and transformation of mercury vapor by dry soil. Environ. Sci. Technol. 12:285-288.

Ghiorse, W. C., and M. Alexander. 1976. Effect of microorganisms on the sorption and fate of sulfur dioxide and nitrogen dioxide in soil. J. Environ. Qual. 5:227-230.

Heinrichs, H., and R. Mayer. 1977. Distribution and cycling of major and trace elements in two central European forest ecosystems. J. Environ. Qual. 6:402-407.

Ingersoll, R. B., R. E. Inman, and W. R. Fisher. 1974. Soils potential as a sink for atmospheric carbon monoxide. Tellus 26:151-158.

Inman, R. E., and R. B. Ingersoll. 1971. Uptake of carbon monoxide by soil fungi. J. Air Pollut. Control Assoc. 21:646-657.

Inman, R. E., R. B. Ingersoll, and E. A. Levy. 1971. Soil: A natural sink for carbon monoxide. Science 172:1229-1231.

John, M. K., H. H. Chuah, and C. J. Vandaerhoven. 1972. Cadmium and its uptake by oats. Environ. Sci. Technol. 6:555-557.

Korte, N. E., J. Skopp, W. H. Fuller, E. E. Niebla, and B. A. Alessii. 1976. Trace element movement in soils: Influence of soil physical and chemical properties. Soil Sci. 122:350-359.

Lagerwerff, J. V. 1967. Heavy metal contamination of soils. *In*: N. C. Brady (Ed.), Agriculture and the Quality of Our Environment. Amer. Assoc. Adv. Sci., Public No. 85, Washington, D.C., pp. 343-364.

Likens, G. E., F. H. Bormann, R. S. Pierce, J. S. Eaton, and N. M. Johnson. 1977. Biogeochemistry of a Forested Ecosystem. Springer-Verlag, New York, 146 pp.

Linzon, S. N., B. L. Chai, P. J. Temple, R. G. Pearson, and M. L. Smith. 1976. Lead contamination of urban soils and vegetation by emissions from secondary lead industries. J. Air Pollut. Control Assoc. 26:650-654.

Little, P. 1977. Deposition of 2.75, 5.0 and 8.5 μm particles to plant and soil surfaces. Environ. Pollut. 12:293-305.

McBride, M. B. 1978. Transition metal bonding in humic acid: An ESR study. Soil Sci. 126:200-209.

Moss, M. R. 1975. Spatial patterns of sulfur accumulation by vegetation and soils around industrial centres. J. Biogeography 2:205-222.

Nozhevnikova, A. N., and L. N. Yurganov. 1978. Microbiological aspects of regulating the carbon monoxide content in the earth's atmosphere. Adv. Microbial Ecol. 2:203-244.

Nyborg, M. 1978. Sulfur pollution and soils. *In*: J. O. Nriagu (Ed.), Sulfur in the Environment. Part II. Ecological Impacts. Wiley, New York, pp. 359-390.

Parker, G. R., W. W. McFel, and J. M. Kelly. 1978. Metal distribution in forested ecosystems in urban and rural northwestern Indiana. J. Environ. Qual. 7:337-342.

Petruzelli, G., G. Guidi, and L. Lubrano. 1978. Organic matter as an influencing factor on copper and cadmium adsorption by soils. Water, Air, Soil Pollut. 9:263-269.

Ragaini, R. C., H. R. Ralston, and N. Roberts. 1977. Environmental trace metal contamination in Kellogg, Idaho, near a lead smelting complex. Environ. Sci. Technol. 11:773-781.

Rasmussen, K. H., M. Taheri, and R. L. Kabel. 1974. Sources and Natural Removal Processes for Some Atmospheric Pollutants. U.S. Environmental Protection Agency, Publica. No. EPA-650/4-74-032, U.S.E.P.A., Washington, D.C., 121 pp.

Reiners, W. A., R. H. Marks, and P. M. Vitousek. 1975. Heavy metals in subalpine and alpine soils of New Hampshire. Oikos 26:264-275.

Seiler, W. 1974. The cycle of atmospheric CO. Tellus 26:116-135.

Shriner, D. S., and G. S. Henderson. 1978. Sulfur distribution and cycling in a deciduous forest watershed. J. Environ. Qual. 7:392-397.

Siccama, T. G., and W. H. Smith. 1978. Lead accumulation in a northern hardwood forest. Environ. Sci. Technol. 12:593-594.

Siccama, T. G., W. H. Smith, and D. L. Mader. 1980. Changes in lead, zinc, copper, dry weight and organic matter content of the forest floor of white pine stands in central Massachusetts over 16 years. Environ. Sci. Technol. 14:54-56.

Smith, E. A., and C. I. Mayfield. 1978. Effects of nitrogen dioxide on selected soil processes. Water, Air, Soil Pollut. 9:33-43.

Smith, K. A., J. M. Bremner, and M. A. Tabatabai. 1973. Sorption of gaseous atmospheric pollutants by soils. Soil Sci. 116:313-319.

Smith, W. H. 1976. Lead contamination of the roadside ecosystem. J. Air Pollut. Control Assoc. 26:753-766.

Smith, W. H., and T. G. Siccama. 1980. The Hubbard Brook Ecosystem study: Biogeochemistry of lead in the northern hardwood forest. J. Environ. Qual. (in press).

Somers, G. F. 1978. The role of plant residues in the retention of cadmium in ecosystems. Environ. Pollut. 17:287-295.

Stevenson, F. J. 1972. Role and function of humus in soil with emphasis on adsorption of herbicides and chelation of micronutrients. Bioscience 22:643-650.

Swank, W. T., and J. E. Douglass. 1977. Nutrient budgets for undisturbed and manipulated hardwood forest ecosystems in the mountains of North Carolina. In: Watershed Research in Eastern North America. Smithsonian Inst., Edgewater, Maryland, pp. 343-364.

Turner, N. C., S. Rich, and P. E. Waggoner. 1973. Removal of ozone by soil. J. Environ. Qual. 2:259-264.

Tyler, G. 1972. Heavy metals pollute nature, may reduce productivity. Ambio 1:53-59.

Tyler, G. 1978. Leaching rates of heavy metal ions in forest soil. Water, Air, Soil, Pollut. 9:137-148.

Van Hook, R. I., and W. D. Shults. 1977. Effects of Trace Contaminants from Coal Combustion. Proc. Workshop, Aug. 2-6, 1976, Knoxville, Tenn., U.S. E.R.D.A. Publica. No. 77-64, U.S. Energy Research and Development Administration, Washington, D.C., 79 pp.

Van Hook, R. I., W. F. Harris, and G. S. Henderson. 1977. Cadmium, lead and zinc distributions and cycling in a mixed deciduous forest. Ambio 6:281-286.

Van Hook, R. I., W. F. Harris, G. S. Henderson, and D. E. Reichle. 1973. Patterns of trace-element distribution in a forested watershed. Proc. 1st Annu. NSF Trace Contaminants Conf., Oak Ridge National Laboratory, Oak Ridge, Tennessee, pp. 640-655.

Warren, J. L. 1973. Green Space for Air Pollution Control. Tech. Report No. 50, School of Forest Resources, North Carolina State Univ., Raleigh, North Carolina, 118 pp.

Zimdahl, R. L., and R. K. Skogerboe. 1977. Behavior of lead in soil. Environ. Sci. Technol. 11:1202-1207.

Zimmerman, P., and R. Rasmussen. 1975. Identification of soil denitrification peak as N_2O. Environ. Sci. Technol. 9:1077-1079.

Zunino, H., and J. P. Martin. 1977a. Metal-binding organic macromolecules in soil: 1 Hypothesis interpreting the role of soil organic matter in the translocation of metal ions from rocks to biological systems. Soil Sci. 123:65-76.

Zunino, H., and J. P. Martin. 1977b. Metal-binding organic macromolecules in soil: 2. Characterization of the maximum binding ability of the macromolecules. Soil Sci. 123:188-202.

5
Forests as Sinks for Air Contaminants: Vegetative Compartment

In addition to the soil compartment, the vegetative compartment of forest eco-systems functions as a sink for atmospheric contaminants. As in the case of soils a complex variety of biological, chemical, and physical processes are involved in the transfer of pollutants from the air to the surfaces of vegetation. For certain contaminants, for example, persistent heavy metal particles, the repository functions of vegetation and soils are intimately linked as a portion of the heavy metals input to the soil are derived from vegetative sources contributing litter to the forest floor. Interest in the ability of plants to remove pollutants from the air has grown considerably in recent years as individuals have become increasingly aware of the amenity functions (Heisler, 1975; Smith, 1970a) of woody plants, particularly in urban and suburban areas. The capability of plants to act as a sink for air contaminants has been addressed by a variety of recent reviews, for example, U.S. Environmental Protection Agency (1976a), Smith and Dochinger (1976), Bennett and Hill (1975), Hanson and Thorne (1972), Hill (1971), Environmental Health Science Center (1975), Keller (1978), and Warren (1973). These papers indicate that the surfaces of vegetation provide a major filtration and reaction surface to the atmosphere and importantly function to transfer pollutants from the atmosphere to the biosphere.

A. Forest Vegetation as a Sink for Particulate Contaminants

Much of the understanding of the mechanics of deposition of particles on natural surfaces has been gleaned from studies with particles in the size range 1-50 μm
and is reviewed in the excellent papers of Chamberlain (1967,1970,1975), Ingold
(1971), Gregory (1973), and Slinn (1976).

The physics and theory of interception and retention of fine particles by vegetation are well beyond the scope of this book. It is useful for the interpretation
of the evidence to follow, however, to have an introduction to some basic observations.

Particulates are deposited on plant surfaces by three processes: sedimentation
under the influence of gravity, impaction under the influence of eddy currents,
and deposition under the influence of precipitation. Sedimentation usually results
in the deposition of particles on the upper surfaces of plant parts and is most important with large particles. Sedimentation velocity varies with particle density,
shape, and other factors. Impaction occurs when air flows past an obstacle and
the airstream divides, but particles in the air tend to continue in a straight path
due to their momentum and strike the obstacle. The efficiency of collection via
impaction increases with decreasing diameter of the collection obstacle and
increasing diameter of the particle. Chamberlain (1967) suggested that impaction
is the principal means of deposition if (1) particle size is of the order of tens of
micrometers or greater, (2) obstacle size is of the order of centimeters or less,
(3) approach velocity is of the order of meters per second or more, and (4) the
collecting surface is wet, sticky, hairy, or otherwise retentive. Ingold (1971) presented data indicating that leaf petioles are considerably more efficient particulate impactors than either twigs (stems) or leaf lamina. For particles of dimensions 1-5 μm, impaction is not efficient and interception by fine hairs on
vegetation is possibly the most efficient retentive mechanism. The efficiency of
washout of particles by rain is high for particles approximately 20-30 μm in size.
The capturing efficiency of raindrops falls off very sharply for particles of 5 μm
or less.

Following deposition, particles may be retained on vegetative surfaces, they
may rebound from the surface, or they may be temporarily retained and subsequently removed. If either the particle or the tree surface is wet or sticky, deposited particles are generally retained. Surficial salt accumulation by plants in
marine or north-temperate roadside environments (where deicing chemicals are
employed) results as vegetation acts to trap salt particles. Fluorine, sulfate, and
nitrate molecules associated with moisture droplets (fog) in the atmosphere may
be distributed to vegetative surfaces with great efficiency (Chamberlain, 1975).

The transfer of particles from the atmosphere to natural surfaces is commonly expressed via deposition velocity which has been previously defined (Chapter
4). For small particles, for example, condensation aerosols less than 1 μm,
deposition velocities are much less than for large particles, for example, spores
and pollen 20-40 μm in diameter.

In addition to spores and pollen, particles in the atmosphere larger than 10 μm
are frequently the result of mechanical processes, for example, wind erosion,

grinding, or spraying. Soil particles, process dust, industrial combustion products, and marine salt particles are typically between 1 and 10 μm in diameter. Particles in the 0.1 to 1 μm range frequently represent gases that have condensed to form nonvolatile products.

The interaction of these variously sized particles with exceedingly diverse vegetative surfaces under conditions of extremely variably microclimate and particle source characteristics suggests an enormously complex relationship. Since this is the case, field evidence to quantify the amounts of natural or anthropogenic particles removed by trees is very sparse. Numerous investigations have studied detached plant parts, small plants, or seedlings under wind tunnel, growth chamber, or greenhouse conditions. This is an appropriate and necessary initial step and these studies have yielded considerable qualitative perspective on the capacity of plants to filter air. The studies reviewed in the following sections were typically not conceived nor conducted specifically to evaluate the role of plants as repositories for atmospheric contaminants. Nevertheless, the hypothesis that trees are important particulate sinks is supported by evidence obtained from studies dealing with diverse particulates including radioactive, trace element, pollen, spore, salt, precipitation, dust, and other unspecified particles.

1. Radioactive Particles

Because of the considerable interest in the distribution of radioactive material following the use of nuclear weapons or nuclear accidents and because of ease of counting, several investigations have examined the ability of aboveground plant parts to intercept radioactive aerosols (Chamberlain, 1970; Oak Ridge National Laboratory, 1969).

Witherspoon and Taylor (1969) treated potted, seedling white pine and red oak with 88-175 μm diameter quartz particles tagged with ^{134}Cs under field conditions. Initial particle retention by oak foliage was 35% while for pine only 24%. After 1 hr, however, the oak leaves lost 91% of their initial concentrations while pines lost only 10%. Most pine particles were trapped at the base of needle bundles around branch termini while oak particles were retained in small, hairy recesses along leaf veins. Particle half-lives were calculated at intervals of 0-1 day, 1-7 days, and 7-33 days. For pine the values were 0.25, 5, and 21 days, respectively. For oak they were 0.12, 1, and 25 days, respectively. Particulate loss was primarily attributed to the action of wind and rain.

Contamination of tree foliage with radioactive fallout, despite its obvious disconcerting implications, has provided especially valuable perspective because it has frequently been examined on large trees in natural environments. Romney et al. (1963) concluded from fallout evidence that Utah juniper foliage principally intercepted particles smaller than 44 μm. Interior canopy elm leaves were shown to be contaminated with elevated levels of radioactivity in selected New England and eastern New York sites following the 1957 atom bomb test series (Bormann et al., 1958). More recently Russell (1974) and Russell and Choquette (1974) have measured concentrations of fission product radionuclides, resulting

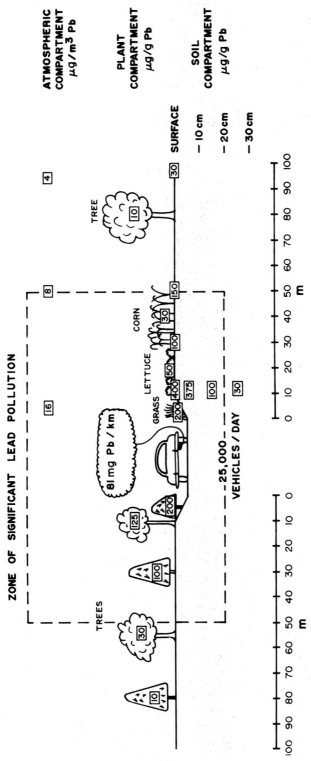

Figure 5-1. Distribution of particulate lead from motor vehicle exhaust in the atmospheric, vegetative, and soil components of a hypothetical roadside ecosystem bisected by a roadway averaging 24,000 vehicles per day. Surface contamination of roadside trees with particulates containing lead may elevate lead burdens to 200 times baseline levels. (From Smith, 1976.)

from megaton range Chinese nuclear explosions, in coniferous and deciduous trees in the New England area between 1968 and 1974. Peak contamination was determined to be reached 6-9 months following injection of the stratospheric source the preceding year. It was hypothesized that primary acquirement was due to attachment of rain droplets to leaf surfaces with subsequent diffusion of soluble radionuclides to leaf cuticles, where they were fixed or transported to leaf interiors. Dry deposition was concluded to be relatively unimportant.

2. Trace Metal Particles

Trace metals, especially heavy metals, are most commonly associated with fine particles in contaminated atmospheres. Trace element investigations conducted in roadside, industrial, and urban environments have dramatically demonstrated the impressive burdens of particulate heavy metals that can accumulate on vegetative surfaces.

In the case of lead in the roadside ecosystem, for example, the increased lead burden of plants, largely due to surface deposition, may be 5-20, 50-200, and 100-200 times baseline (nonroadside environment) lead levels for unwashed agricultural crops, grasses, and trees, respectively (Smith 1976) (Figure 5-1).

Eastern white pine is widely planted in the roadside environment in New England and its capacity to accumulate fine particles (~ 7 μm diameter; Heichel and Hankin, 1972) has been shown to be substantial (Smith, 1971). Heichel and Hankin (1976) have investigated the distribution of lead deposited on this species in roadsides and have advanced several important observations. The lead burden of older needles and twigs was consistently greater than that of younger organs, and was greater in samples taken adjacent to rather than far from the road. These are consistent with observations we made with the same species (Smith, 1971) and are important as the former indicates that lead accumulates over time on the trees while the latter argues against the importance of soil uptake as a mechanism of lead acquisition. Heichel and Hankin further concluded that twigs retained particles more effectively than needles throughout the season. This was judged to be due to the roughness of twigs relative to needles. The authors observed that a 12 m tall white pine growing in a dense planting would have about 15×10^4 cm^2 of woody surface and about 15×10^5 cm^2 of foliage surface. Although white pine exposed approximately 10-fold more foliage than woody surface, the woody surfaces retained about 20-fold the lead burden of foliage.

Like the roadside environment, urban atmospheres also have elevated amounts of particles containing trace metals. In New Haven, Connecticut, we have examined the surfaces of a variety of city trees and have found substantial accumulations of certain metals, particularly lead, zinc, and iron (Smith, 1973; Smith and Staskawicz, 1977). Observations of the leaves of mature London plane trees in New Haven throughout the growing season indicated nickel and zinc foliar surface amounts remained relatively constant. Aluminum, iron, manganese, and lead, on the other hand, appeared to accumulate through the spring and early summer and decrease during the late summer and fall on this species (Figure 5-2).

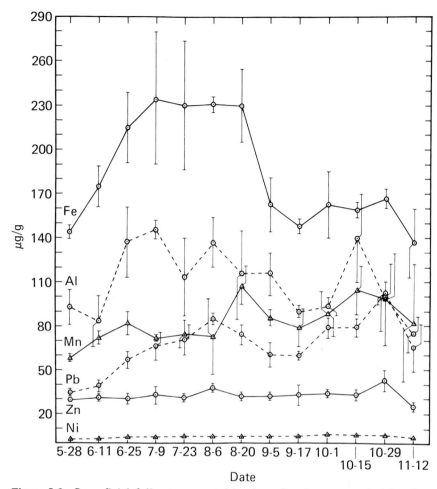

Figure 5-2. Superficial foliar trace metal burden of mature, unwashed London plane leaves collected throughout an entire growing season in New Haven, Connecticut. Concentrations are μg g^{-1} (dry weight basis), vertical bars represent 95% confidence intervals. (From Smith and Staskawicz, 1977.)

This latter decrease and late season decreased particle density indicated by observation with the scanning electron microscope suggest the particles on foliage are weathered or transported off the leaf. Precipitation, wind, insect activity, and other forces may cause particles to be lost from the leaves and transported by way of the petiole to twig tissue. Our leaf observations with the scanning electron microscope allowed a variety of additional observations:

1. In regard to spatial distribution throughout the growing season, particles were more prevalent on the adaxial (upper, facing twig) surface than the abaxial (lower, away from twig) surface. Peripheral leaf areas were always the cleanest. Midlaminar areas were generally only lightly contaminated. Most particu-

lates were located in the midvein, center portion of the leaves. The greatest particulate burden was located on the adaxial surface at the base of the blade just above the petiole junction. It is probable that precipitation washing plays an important role in this distribution pattern.

2. Late in the growing season, particulate loads could increase to very high levels.
3. Particle morphology was very variable (Figure 5-3). Carbonaceous and aggregate particles were especially common. Viable particles, including pollen grains, fungal spores, and mycelium (Figure 5-4), were particularly prevalent after early July.
4. Particle size was extremely variable. Most particles appeared to fall in the 5-50 μm range. Significant numbers outside this range were observed. Sub-

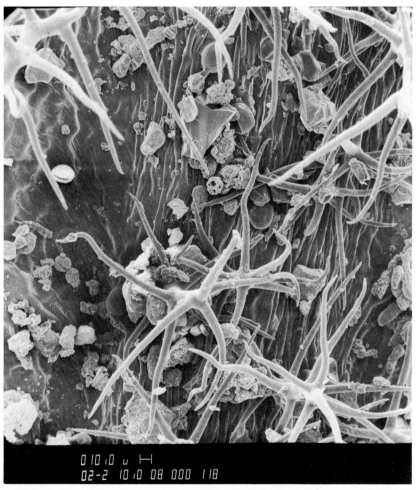

Figure 5-3. Scanning electron microscope micrograph of the adaxial surface of an 8-week-old London plane leaf. Spore, pollen, carbonaceous, angular, and aggregate particles are visible. Scale, 10 μm.

micron particles could easily be found, particularly on trichomes (leaf hairs). Large aggregates in excess of 100 μm could also be easily found.

5. On the lower leaf surface, complete stomatal blockage by particulates, partial blockage or contamination could be seen. It was evident, however, that particulate association with stomates was the exception, rather than the rule.
6. Trichomes, particularly abundant on the center portion of upper and lower leaf surfaces are particulate accumulators (Figure 5-5). As the season progresses the trichomes are reduced in size by "weathering" and occasionally completely broken off.
7. Fungal mycelium, which becomes particularly abundant on leaf surfaces as the growing season progresses, is in intimate association with particulate contaminants. The surficial trace metal burden of London plane leaves appeared

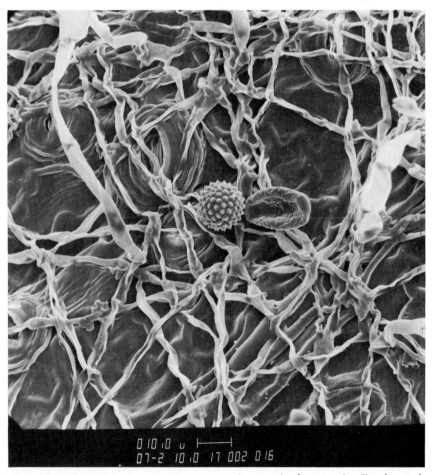

Figure 5-4. Scanning electron microscope micrograph of ragweed pollen (center), and collapsed fungus spore (right-center) and mycelium on the abaxial surface of a 15-week-old London plane leaf. Scale, 10 μm.

to be generally less than the average for several other New Haven deciduous species (Smith, 1973). This disparity may have resulted from a comparison of late season averages from the latter with the more representative "throughout the growing season" average of the former.

The literature is replete with studies demonstrating significant trace metal particle accumulation on trees in roadside, urban, and industrial situations. Industrial regions particularly those with metal smelters may excessively contaminate surrounding woody vegetation with particles containing trace metals. A representative study is that conducted by Little and Martin (1972) in the Avonmouth industrial complex, Severnside, England. Close to this complex elm leaves exhibited 8000, 5000, and 50 $\mu g\ g^{-1}$ zinc, lead, and cadmium, respectively. Page and Chang (1979) have reviewed the trace element contamination of vegetation in the vicinity of coal-fired power plants.

Figure 5-5. Scanning electron microscope micrograph of a trichome on the adaxial surface of a 17-week-old London plane leaf. The leaf hair has accumulated numerous particles. Scale, 10 μm.

Data provided, however, rarely permit a quantitative estimate of the sink capability of the woody plants. The required mass or vegetative surface area calculations are not presented as the purpose of the studies typically did not include sink function assessment. When quantitative estimates of sink capacity are made, the results can be impressive. During the course of interpretation of some of our urban lead data we combined sugar maple dimension analysis information from New Hampshire with contamination data from New Haven to speculate on the sink capacity of a single urban sugar maple (Table 5-1).

3. Pollen and Spores

Pollen studies have provided important evidence of vegetative interception of large particles. G. S. Raynor of the Brookhaven National Laboratory, Upton, Long Island, N.Y., has conducted a series of dispersion experiments employing ragweed pollen emitted from sources at various distances and heights upwind of a forest edge. Pollen loss from the plume occurred in two stages and by two mechanisms, impaction near the forest edge and deposition well within the forest. Pollen loss to the forest was considerably greater than over open terrain (Raynor, 1967; Raynor et al., 1966). Interception of ragweed pollen by a Pennsylvania forest canopy reduced pollen concentration in the forest atmosphere to only 70% of the concentration in a nearby open field (Elder and Hosler, 1954). Neuberger et al. (1967) measured ragweed pollen concentrations in and out of forests and found that 100 m inside a dense coniferous forest over 80% of the pollen had been substracted from the atmosphere. Data indicate that deciduous species are less effective than conifers for filtration of pollen (Neuberger et al., 1967; Steubing and Klee, 1970). For these large ($\sim$ 20 μm) particles the dominant transfer to vegetative surfaces is via sedimentation and not impaction (Aylor, 1975).

Fungal spore (size range 1.5-30 μm) interception studies have provided important evidence for understanding particulate capture (Gregory, 1971). Ingold (1971) has concluded that the most efficient plant parts for spore collection are petioles, twigs, and leaf lamina, respectively. Efficiency of spore collection increases with decreasing diameter of the collecting cylinder. Observations of basidiomycete spores in Washington Douglas fir forests have emphasized the

Table 5-1. Calculated Particulate Metal Sink Capacity for the Leaves and Current Twigs of a Single, 30 cm (12 inch) Diameter Urban Sugar Maple during the Course of a Growing Season

Metal contaminant	Growing season removal (mg tree^{-1})
Lead	5800
Nickel	820
Chromium	140
Cadmium	60

Source: Smith (1974).

extraordinary importance of microclimate and forest stand structure on the distribution and deposition of these particles in the forest. Wind speed, air temperature, inversions, cloud cover, and forest openings all influenced particle movement (Edmonds and Driver, 1974; Fritschen et al., 1970).

4. Salt Particles

Vegetative interception of saline aerosol (primarily NaCl) and nutrient particles has also contributed to our understanding of plant sink function. Particulate deposition of salt particles occurs in roadside environments where deicing salts are employed, in the vicinity of cooling towers, and in maritime regions (Eaton, 1979; Moser, 1979). Conifers planted close to roads receiving deicing salt applications frequently exhibit needle necrosis due to the accumulation of toxic levels of salt transported from the road to the leaves via the atmosphere (Constantini and Rich, 1973; Hofstra and Hall, 1971; Smith, 1970b). In coastal ecosystems subject to airborne marine salt, accumulation of salt particles by above ground plant parts, injures foliage and twigs (Boyce, 1954; Wells and Shunk, 1938; Oosting, 1945; Oosting and Billings, 1942) and may control species success or failure depending on tolerance to salt loading (Martin, 1959). Clayton (1972) described the trapping of particulate salts by *Baccharis* brushlands in coastal California. Woodcock (1953) provided evidence that the shape of plant leaves influences the amount of salt deposited. By employing plates of various shapes, he found that long narrow plates accumulated more salt per unit area than did circular plates. Edwards and Claxton (1964) found over four times the deposition of salt on the windward side of a hedgerow compared to the leeward side.

Where foliar capture of marine particulates is below the threshold of foliar injury, particle accumulation may be an important mechanism for nutrient acquisition (Art, 1971; Art et al., 1974). Numerous investigations, reviewed by White and Turner (1970), have indicated that nonmaritime trees also catch airborne nutrient particles. These authors found that a mixed deciduous forest was capable of annually removing 125 kg ha^{-1} sodium, 6 kg ha^{-1} potassium, 4 kg ha^{-1} calcium, 16 kg ha^{-1} magnesium, and 0.1 kg ha^{-1} phosphorus from the atmosphere. Degree of leaf hairiness was inversely correlated with particle retention. Apparently the small droplets employed had insufficient inertia to penetrate the stable boundary layer created by the hairy leaves. Small diameter branches were more efficient particle collectors than large diameter branches in all species examined.

In their examination of the impact of saline aerosols of cooling tower origin, McCune et al. (1977) emphasized the importance of particle wetness in causing damage to surrounding trees. Dry particles appeared less toxic than hydrated particles. This supports the contention that moist particles are more effectively retained by vegetative surfaces than dry ones.

5. Precipitation, Dust, and Other Particles

Foliar interception of precipitation has been intensively investigated (Zinke, 1967), but the relatively large size of the particles (range 50-700 μm) makes these data of limited application for considerations of fine particle retention. The enormous importance of rainout in transferring fine particles from the atmosphere to vegetation is recognized. Numerous precipitation studies support the general observation that conifers intercept more particles than deciduous species, for example, Helvey (1971) who reported canopy interception loss greatest in a spruce-fir-hemlock type, intermediate in pine, and least in broad-leaved deciduous forests.

Numerous additional studies employing dust, synthetic, or unspecified particles have contributed to our understanding of particulate capture by vegetation. Rosinki and Nagamoto (1965) investigated the deposition of 2 μm particles on Rocky Mountain juniper and Douglas fir. At low dosage particles preferentially accumulated on the windward leaf edge. Eventually a new layer was formed on the previously deposited layer. Thickness increased until an equilibrium was reached. Total deposition was increased when wind exposed different leaf areas for deposition. Langer (1965) concluded that dust deposition on coniferous leaves was not significantly influenced by electrostatic effects. Podgorow (1967) investigated the relative effectiveness of pine, birch, and aspen in filtering dust particulates. Pine proved most effective. Interior crown needles accumulated more and retained more dust than exterior needles. Bach (1972) also presented evidence supporting the superior collecting capacity of pines relative to deciduous species. In an Ohio study, Rochinger (1972) examined dustfall and suspended particulate matter in three areas—treeless, deciduous canopy, and conifer canopy—and concluded that trees have the capacity to reduce particulate pollutants in the ambient atmosphere.

Wedding et al. (1975) found, under controlled wind tunnel conditions, that particulate deposition on rough pubescent sunflower leaves was 10 times greater than on smooth, waxy tulip poplar leaves. In a unique study, Graustein (1978) employed the ratio of strontium isotopes in soil dust to determine strontium input to forested watersheds in New Mexico. Most of the atmospherically transported strontium entered the watershed by impaction of soluble particles on spruce foliage. Aspen, also present in the ecosystem was judged to trap little, if any, dust. The flux of dust-derived strontium to the forest floor was four times greater than the flux to an unforested area.

In an extremely informative set of experiments, Little (1977) exposed freshly collected leaves of several tree species in a wind tunnel to various sizes of polystyrene aerosols labeled with technetium. Particles sized 2.75, 5.0, and 8.5 μm were tested with leaves from European beech, white poplar, and nettle (*Urtica dioica*). Surface texture was critical in capture efficiency with the rough and hairy leaves of nettle more effective than the densely tomentose leaves of poplar or the smooth surfaces of beech. For each species there was a strong negative linear correlation between leaf area and deposition velocity, the latter being smallest for the largest leaves. Deposition was heaviest at the leaf tip and along leaf margins, where a turbulent boundary layer was present. Leaves with complex shapes and largest circumference to area ratio were the most efficient collectors. Increased wind speed and particle size both were reflected in increased deposition velocities. Deposition velocities to petioles and stems were many times greater than deposition velocities to leaf laminas, even though the majority of the total catch was intercepted by the leaf lamina (Table 5.2). This table reveals that nonlaminar catch is significant, however, and Little (1977) suggested that this may cause deposition of atmospheric particles to trees to be relatively high even during the winter when deciduous species are devoid of leaves.

a. Case Study of Street Tree Particulate Sink Capacity

The U.S. Environmental Protection Agency has developed a demonstration plan to explore the capability of urban woody vegetation to improve air quality (U.S. Environmental Protection Agency, 1976c). This plan, which utilized pollutant fluxes summarized and extrapolated from the literature and air quality and environmental conditions as they existed in the St. Louis, Missouri, area,

Table 5-2. Average Percentage of Total Catch Intercepted by Leaf Laminas, Petioles, and Stems of Freshly Cut European Beech, White Poplar, and Nettle Exposed to Polystyrene Particles in a Wind Tunnel

Wind speed (cm sec^{-1})	Particle size (μm)	Plant part	Beech	White poplar	Nettle
150	5.0	Leaf laminas	85.01	49.16	90.68
		Petioles	11.18	33.52	4.76
		Stems	3.80	17.32	4.56
250	2.75	Leaf laminas	63.08	73.71	68.30
		Petioles	30.19	11.08	15.56
		Stems	6.73	15.20	16.11
	5.0	Leaf laminas	70.94	57.10	78.59
		Petioles	23.08	26.78	6.66
		Stems	5.99	16.12	7.00
	8.5	Leaf laminas	62.68	45.39	68.30
		Petioles	17.17	29.12	11.29
		Stems	19.61	25.48	20.41
500	2.75	Leaf laminas	63.87	64.81	82.85
		Petioles	28.81	18.81	5.50
		Stems	7.32	16.36	11.06
	5.0	Leaf laminas	73.86	53.39	77.71
		Petioles	14.94	26.12	9.23
		Stems	11.19	20.49	13.06
	8.5	Leaf laminas	90.83	69.27	83.27
		Petioles	3.35	12.58	7.21
		Stems	5.81	18.15	9.82

Source: Little (1977).

included an assessment of the particulate removal capacity of selected and hypothetical street trees. It was proposed that trees be planted on both sides of the streets within the city boundaries of St. Louis. The trees would be planted 8.5 m (30 feet) apart. The total street length in St. Louis was determined to be 2316 linear km (6.6×10^6 feet) requiring a total of 440,000 trees for complete planting. The three tree species proposed for the street plantings included red oak, Norway maple, and linden. An average particulate flux rate of 2.5×10^3 μg m^{-2} hr^{-1} guesstimated from the literature was employed (U.S. Environmental Protection Agency, 1976b). Table 5.3 presents the dimensions of the tree species and the estimated quantity of particles that would be removed by the 440,000

Table 5-3. Estimation of the Amount of Particulates Absorbed by Hypothetical St. Louis Street Trees

1. Number of trees planted

Maple	146,666
Oak	146,667
Linden	146,667
Total	440,000

2. Dimensions of the maple
   Height = 6 m
   Diameter of canopy = 3 m
   Total surface area[a] tree^{-1} = 36.8 m^2
   Total surface area of 146,666 trees = 5.40×10^6 m^2

3. Dimensions of the oak
   Height = 6 m
   Diameter of canopy = 3 m
   Total surface area tree^{-1} = 36.1 m^2
   Total surface area of 146,667 trees = 5.30×10^6 m^2

4. Dimension of the linden
   Height = 5 m
   Diameter of canopy = 2.4 m
   Total surface area tree^{-1} (including undergrowth) = 23.0 m^2
   Total surface area for 146,667 trees = 3.40×10^6 m^2

5. Total surface area for the 440,000 trees = 1.4×10^7 m^2

6. Estimated particulate flux to vegetation = 2.5×10^3 μg^{-2} hr^{-1}

7. Calculation to determine the amount of particulates absorbed by street trees
   1.4×10^7 m$^2 \times 2.5 \times 10^3$ μg m^{-2} hr$^{-1} \times$ gm/10^6 μg $\times$ lb/453.59 gm $\times$ T/2000 lbs $\times$ 24 hr/day $\times$ 365 days/yr = 3.40×10^2 tons particulate yr^{-1}

Source: U.S. Environmental Protection Agency (1976a).
[a] Maple canopy diameter = 3 m
estimated ground area covered by maple canopy = 7.1 m^2
area index for maple = 5.18 (Raunder, 1976)
area index = surface area $\div$ ground area
$\therefore$ 5.18 = x/7.1 m^2
surface area of maple = 36.8 m^2.

street trees. The hypothetical transfer of particles from the atmosphere to the tree surfaces totaled 340 tons annually. The 1980 estimate for total particulate emission in the St. Louis area equaled 126,290 tons. The biological and medical significance of the transfer of the 340 tons from the atmosphere to the vegetation is unclear.

B. Forest Vegetation as a Sink for Gaseous Contaminants

Substantial evidence is available to support the potential that plants in general (Bennett and Hill, 1975; Hill, 1971; Rasmussen et al., 1975) and trees in particular (Smith, 1979; Smith and Dochinger, 1975,1976; Roberts, 1971; Warren, 1973) have to function as sinks for gaseous pollutants. The latter are transferred from the atmosphere to vegetation by the combined forces of diffusion and flowing air movement. Once in contact with plants gases may be bound or dissolved on exterior surfaces or taken up by the plants via stomata. If the surface of the plant is wet and if the gas is water soluble, the former process can be very important. When the plant is dry or in the case of gases with relatively low water solubilities the latter mechanism is assumed to be the most important.

1. Stomatal Uptake

Stomatal pores are small openings, typically approximately 10 μm in length and 2 to 7 μm in width, in the epidermal surface of leaves through which plants naturally exchange carbon dioxide, oxygen, and water vapor with the atmosphere (Figure 5-6). The waxy cuticle of leaf surfaces restricts diffusion so that essentially all gas exchange carried out by leaves is via stomatal openings. Even though these openings make up only approximately 1% of the leaf surface area, their orientation and mechanics prove to be nearly optimal for maximum gas diffusion in and out of the leaf (Salisbury and Ross, 1978). Stomates undergo diurnal opening and closing with the pores of most plants opened within an hour of sunrise and closed by dark. Gas diffusion to and from leaves and the timing and degree of opening of stomatal aperatures is strongly influenced by a number of complex environmental factors.

During daylight periods when plant leaves are releasing water vapor and taking up carbon dioxide, other gases, including trace pollutant gases, in the vicinity of the leaf will also be taken up through the stomates. Once inside the leaf these gases will diffuse into intercellular spaces and be absorbed on or in the surfaces of palisade or spongy parenchyma cell walls (Figure 5-7).

The rate of pollutant gas transfer from the atmosphere to interior leaf cells is regulated by a series of resistances conveniently thought of as atmospheric, stomatal, and mesophyllic. Factors controlling atmospheric resistance include wind speed, leaf size and geometry, and gas viscosity and diffusivity. Stomatal resistance is regulated by stomatal aperature which is influenced by water deficit, carbon dioxide concentration, and light intensity. Mesophyllic resistance is regulated by gas solubility in water, gas liquid diffusion, and leaf metabolism (Kabel

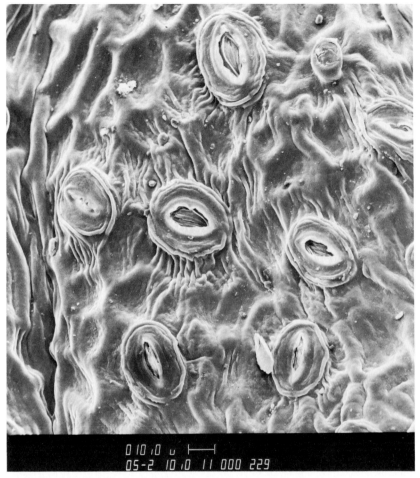

Figure 5-6. Scanning electron microscope micrograph of the abaxial surface of a 3-week-old London plane leaf showing stomates. Scale, 10 μm.

et al., 1976). Because the rate of pollutant uptake is regulated by numerous forces and conditions, the rate of removal under field conditions is highly variable. If leaf characteristics, wind speed, atmospheric moisture, temperature, and light intensity are quantified, however, the pollutant uptake rate can be estimated (Kabel et al., 1976; Bennett et al., 1973).

2. General Plant Uptake

The fundamental investigations of A. Clyde Hill and Jesse H. Bennett of the University of Utah allow several general conclusions concerning gaseous pollutant uptake (Hill, 1971; Bennett and Hill, 1973,1975). Their studies have concentrated on alfalfa, oats, barley, and grass. Standard alfalfa canopies removed gaseous pollutants from the atmosphere in rates of the following order: hydrogen fluoride

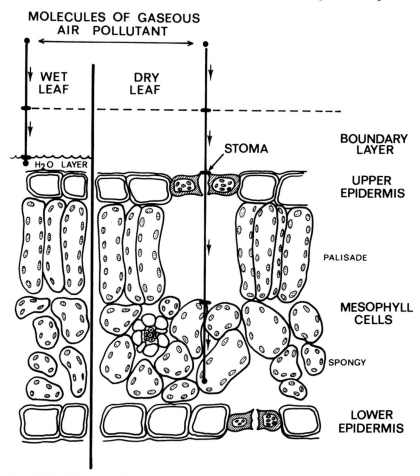

Figure 5-7. Schematic diagram portraying the interaction of molecules of gaseous air pollutants with leaves. When leaf surfaces are wet the gases may be dissolved in the moisture film on the leaf. When the leaves are dry, the gases are taken up through the stomates and ultimately may react on or in mesophyll cells of the leaf interior. Movement of gas molecules into the leaf involves several "resistances" imposed by the boundary layer (quiescent zone of retarded air flow surrounding the leaf), substomatal cavity (open space below the stoma), and the mesophyll cells themselves. Since these "resistances" can be measured or estimated, the rate of gas uptake can be modeled for a given set of gas, plant, and environmental circumstances.

> sulfur dioxide > chlorine > nitrogen dioxide > ozone > peroxyacetylnitrate > nitric oxide > carbon monoxide. In general plant uptake rates increased as the solubility of the pollutant in water increased. Hydrogen fluoride, sulfur dioxide, nitrogen dioxide, and ozone, which are soluble and reactive, were readily absorbed. Nitric oxide and carbon monoxide, which are very insoluble, were absorbed relatively slowly or not at all (Table 5-4). The rate of pollutant removal

Table 5-4. Solubility in Water and Uptake Rate of Pollutants by Alfalfa

Pollutant	Uptake rate by alfalfa at 1 pphm (liters min^{-1} m^{-2})	Equivalent deposition velocity (cm sec^{-1})	Solubility at 20°C (cm^3 gas cm^{-3} H$_2$O)
CO	0.0	0.00	0.02
NO	0.6	0.10	0.05
CO$_2$	2.0	0.33	0.88
PAN	3.8	0.63	–
O$_3$	10.0	1.67	0.26
NO$_2$	11.4	1.90	Decomposes
Cl$_2$	12.4	2.07	2.30
SO$_2$	17.0	2.83	39.40
HF	22.6	3.77	446

Source: Hill and Chamberlain (1974).

was found to increase linearly as the concentration of the pollutant was increased over the ranges of concentration that are encountered in ambient air and which were low enough not to cause stomatal closure.

Under growth chamber conditions, wind velocity, canopy height, and light intensity were shown to affect the rate of pollutant removal by vegetation. As previously stressed, light plays a critical role in determining physiological activities of the leaf and stomatal opening and as such exerts great influence on foliar removal of pollutants. Under conditions of adequate soil moisture, however, pollutant uptake by vegetation was judged almost constant throughout the day, as the stomata were fully open. Pollutants were absorbed most efficiently by plant foliage near the canopy surface where light-mediated metabolic and pollutant diffusivity rates were greatest. Sulfur and nitrogen dioxides were taken up by respiring leaves in the dark, but uptake rates were greatly reduced relative to rates in the light.

3. Tree Uptake

a. Sulfur Dioxide

Because of its high solubility in water, large amounts of sulfur dioxide are absorbed to external tree surfaces when they are wet. In the dry condition, sulfur dioxide is readily absorbed by tree leaves and rapidly oxidized to sulfate in mesophyll cells. At low uptake rates sulfur dioxide is presumed to be oxidized about as rapidly as it is absorbed (Bennett and Hill, 1975).

Roberts (1974) measured sulfur dioxide sorption by single leaves or shoots of several one-year-old seedlings of numerous woody species. All species examined were capable of reducing high ambient levels within his test chambers (Table 5-5). Because of the large dose employed, 1 ppm (2620 μg m^3) for 1 hr, Roberts reduced the concentration in subsequent trials and examined uptake at concentrations of 0.2 and 0.5 ppm (524 and 1310 μg m^3). At the lower concentration uptake by birch and firethorn was significantly less. It was speculated that higher concentrations of sulfur dioxide may maintain stomatal opening. Under controlled environmental conditions, comparable to those employed by Roberts, Jensen (1975) fumigated hybrid poplar cuttings with sulfur dioxide ranging in concentration from 0.1 to 5 ppm (262 to 13.1 × 10^3 μg m^3) for periods of 5 to 80 hr. Uptake was determined by measuring total sulfur content of the leaves. At low levels of fumigation [0.1 and 0.25 ppm (262 and 655 μg m^{-2})] leaf sulfur initially increased but then declined to unfumigated levels as fumigation continued. This reduction was judged by the author to be due to one or more of the following: reduction in absorption rate, translocation of sulfur out of the leaves, leaching of sulfate from the roots, or release of hydrogen sulfide by the leaves.

Jensen and Kozlowski (1975) have provided additional perspective on tree seedling uptake of sulfur dioxide in their experiments that exposed one-year-old sugar maple, bigtooth aspen, white ash, and yellow birch to 2.75 ppm (7205 μg m^{-3}) for 2 hr. Prefumigation with 0.75 ppm (1965 μg m^{-3}) sulfur dioxide for 20 hr or more reduced the rate of absorption in all species except white ash. The authors speculated that tolerance to sulfur dioxide injury following uptake may be related to the rate at which accumulated sulfur can be moved out of the leaves. Roberts and Krause (1976) monitored sulfur dioxide uptake of intact plants of rhododendron (three-month-old) and firethorn (12-month-old) and suggested that the greater uptake of the latter may have been partially due to abundant trichomes (leaf hairs).

These various controlled-environment studies are important as they qualify and caution our efforts to extrapolate the sink function to trees in natural environments. Uptake under ambient conditions may be less than under experimental conditions as the latter frequently employ unnaturally high concentrations of sulfur dioxide. Prefumigation, common in natural situations, may further reduce natural uptake. Uptake rates in the field may decline over time as pollution episodes continue. Even though mentioned by several investigators, the seedling studies have not addressed two very important questions concern-

Table 5-5. Foliar Sorption of Sulfur Dioxide by Selected Seedlings Fumigated at 1.0 ppm (2620 μg m^{-3}) for 1 hr in a Controlled Environment Chamber (27 ± 1°C, 51 ± 7% RH, 1300 ft-c)

Species	SO$_2$ uptake	
	mg SO$_2$ dm^{-2} hr^{-1}	mg SO$_2$ g^{-1} hr^{-1}
Red maple	0.088	0.260
White birch	0.086	0.268
Sweetgum	0.074	0.267
Firethorn	0.072	0.213
Privet	0.068	0.134
Rhododendron	0.056	0.079
White ash	0.046	0.118
Azalea	0.044	0.072

Source: Roberts (1974).

ing uptake, namely the relative importance of stomatal uptake versus adsorption to the surface of dry plant parts and the relative uptake of dry plants versus plants with moisture films on their surfaces. Garland and Branson (1977), however, have recently provided evidence supporting the importance of stomates and wet surfaces in uptake. These investigators determined the rates of water vapor and sulfur dioxide conductance in detached Scotch pine needles collected from a 45-year-old stand and in intact trees in a 10-year-old plantation. The similarity of the conductances observed for these two processes led the authors to conclude that they were both controlled by diffusion through the stomata. By analogy with transpiration, the authors further estimated that the deposition velocity of sulfur dioxide to a dry pine canopy would vary from 0.2 to 0.6 cm sec^{-1} during daytime (0.05-0.1 cm sec^{-1} at night) but might be 10 times this rate if the canopy was wet from precipitation or dew. This deposition velocity appears slightly conservative when compared with the dry deposition rates provided by other investigators (Sheih, 1977; Sheih et al., 1979). Martin and Barber (1971) monitored the sulfur dioxide loss in the immediate vicinity of a large hawthorne hedge (approximately 4 m high $\times$ 3 m wide) subject to effluent from an electric generating facility. With ambient concentrations generally less than 10 pphm (262 μg m^{-3}), the authors observed significant loss of sulfur dioxide near ($\approx$ 150 mm) the foliage. The greatest loss was during rain or dew periods when the hedge foliage was wet.

b. Oxidants

Ozone is relatively insoluble in water (0.052 g 100 g^{-1} H_2O at 20°C) but readily diffuses into stomatal cavities (Rich and Turner, 1972; Rich et al., 1970; Thorne and Hanson, 1972; Wood and Davis, 1969). The very reactive nature of this gas undoubtedly causes it to rapidly react on the surface of leaf mesophyll cells.

Under controlled environmental conditions, Townsend (1974) has monitored ozone uptake by a variety of seedling tree species (Table 5-6). Ozone sorption exhibited a linear increase up to 0.5 ppm (980 μg m^{-3}) for both white birch and

Table 5-6. Foliar Sorption of Ozone by Selected Seedlings Fumigated at 0.20 ppm (392 μg m^{-3}) for a Few Hours in a Controlled Environment Chamber (26 ± 1.5°C, 45 ± 5% RH, 2100 ft-c)

Species	O_3 uptake	
	mg O_3 dm^{-2} hr^{-1}	mg $O_3 g^{-1}$ hr^{-1}
White oak	0.635	1.318
White birch	0.536	2.347
Coliseum maple	0.502	0.991
Sugar maple	0.371	0.863
Ohio buckeye	0.362	0.927
Redvein maple	0.285	0.911
Sweetgum	0.278	0.854
Red maple	0.272	0.555
White ash	0.239	0.562

Source: Townsend (1974).

red maple. These two species were also capable of reducing ambient ozone throughout a prolonged 8-hr exposure.

While tree removal of atmospheric peroxyacetylnitrate has not been reported, Garland and Penkett (1976) have suggested that the deposition velocity of this gas to grass was approximately 0.25 cm sec^{-1} which is lower than the value for ozone (0.8 cm sec^{-1}) or sulfur dioxide (1 cm sec^{-1}).

For the other gases judged to be significantly removed from the atmosphere by vegetation, hydrogen fluoride and nitrogen dioxide, relatively little work has been reported on trees as sinks. Hydrogen fluoride is very water soluble and very reactive and can be adsorbed onto plant surfaces and absorbed through stomates. Leaves exposed to hydrogen fluoride may accumulate fluoride to one million times the ambient concentration. Nitrogen dioxide dissolved in water yields nitrite and nitrate ions in solution. The latter can be reduced to ammonia in leaf cells (Bennett and Hill, 1975). Rogers et al. (1979) have provided nitrogen dioxide uptake rates for loblolly pine and white oak.

4. Models of Forest Gas Sink Function

Efforts to estimate the sink capability of forest vegetation under natural conditions must consider a complex set of variables including pollutant concentration and deposition velocity, meteorological parameters, and dimensions (leaf or canopy area, dry weight) and conditions of the trees. The systematic approach of model development is desirable and necessary despite the obvious deficiencies in field data.

Murphy et al. (1977) have modeled the sulfur dioxide uptake of a simulated loblolly pine forest exposed to 50 ppb (131 μg m^3) sulfur dioxide on two clear days in January and June using climate data from a station near Aiken, South Carolina. The simulated uptake compared favorably, but was smaller than, the seedling uptake rates reported by Roberts (1974) (Table 5.7). Murphy et al. (1977) also applied their model to regions where forest vegetation was dominant and where actual frequency distributions of sulfur dioxide concentrations were known. At a site on the Savannah River Laboratory with an average sulfur dioxide concentration of 8 ppb (21 μg m^{-3}) during the spring the model predicted an uptake of 11 metric tons day^{-1} over the 778 km^2 area of the southern pine forest site. For Long Island, New York, over an area of 1723 km^2 in June, the model predicted a sulfur dioxide uptake of 103 metric tons day^{-1} (SO$_2$ 32 ppb, 84 μg m^{-3}) for a west wind condition. According to the authors the New York estimate was larger due to the larger land area, higher ambient sulfur dioxide, and greater leaf area employed.

Kabel et al. (1976) have argued, and appropriately so, that the uptake rate of sulfur dioxide on a leaf area basis must be extrapolated to a ground area basis in order to predict large area pollutant removal. In addition deposition velocities, or mass transfer coefficients as these authors prefer (compare Kabel, 1976), must be given for uptake of stems and branches as well as leaves. Fortunately, Whittaker and Woodwell (1967) have provided generalized area ratios for temper-

Table 5-7. Comparison of Simulated Sulfur Dioxide Uptake for a South Carolina Loblolly Pine Forest with Experimental Uptake by Deciduous Seedlings

Species	SO$_2$ uptake kg ha^{-1} hr^{-1} ppb^{-1} by volume
Loblolly pine (simulated)	
January	
day 1	1.3×10^{-3}
day 2	2.2×10^{-4}
June	
day 1	2.0×10^{-3}
day 2	2.0×10^{-4}
Seedlings (experimental)	
Red maple	8.8×10^{-5}
White birch	8.6×10^{-6}
Sweetgum	7.4×10^{-5}
White ash	4.6×10^{-5}

Source: Murphy et al. (1977).

ate forest communities (Table 5-8). Kabel et al. (1976) calculated the following deposition velocities using the appropriate ratios: 0.015 m sec^{-1} for dry condition (stomatal plus soil) and 0.21 m sec^{-1} for damp canopy condition, yielding a total deposition velocity of 0.23 m sec^{-1} for a moist forest canopy. Uptake rates were calculated for a model forest (dry condition) downwind from a sulfur dioxide source. Gas concentration profile and uptake rates are presented in Figure 5-8.

The radioactive measurements of Garland (1977) propose that the deposition velocity for a dry forest canopy varies from 0.001 to 0.006 m sec^{-1}. The Kabel et al. (1976) figure is presumably higher than this due to the inclusion of soil uptake in the latter.

Unfortunately models describing sink capabilities of forests for other gaseous contaminants are not readily available. Waggoner (1971,1975) has made some preliminary observations with ozone, but no formal model has been proposed.

Table 5-8. Area Ratios for Forest Communities

Stem (bark) to ground surface
    0.5-0.7 for mature, closed forests
    0.2-0.4 for small, open forests
    0.7-1.0 for dense, young stands

Branch (bark) to ground surface
    1.5-1.6 for mature, deciduous forests

Leaf to ground surface
    4.0-6.0 for closed, deciduous forests
    6.0-7.0 for dense, evergreen forests

Source: Whittaker and Woodwell (1967).

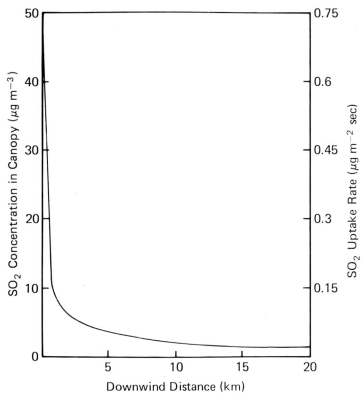

Figure 5-8. Ground level SO_2 concentration and uptake rate over a model forest with a deposition velocity of 0.015 m sec^{-1} and eddy diffusivity of 7 m^2 sec^{-1}. (From Kabel et al., 1976.)

a. Case Study of Forest Gas Sink Capability

As part of the U.S. Environmental Protection Agency's demonstration plan to explore the capability of urban trees to improve air quality previously described in this chapter (U.S. Environmental Protection Agency, 1976c), a model forest hectare was developed and employed to estimate gas uptake. The model forest consisted of six species including red oak, Norway maple, linden, poplar, birch, and eastern white pine. The arrangement and spacing of the proposed forest is presented in Figure 5-9. The estimations of total tree surface area (canopy plus woody) at five years after planting were as follows:

maple	(6 m ht)	36.8 m^2
oak	(6 m ht)	36.1 m^2
poplar	(6 m ht)	52.5 m^2
linden	(5 m ht)	23.0 m^2
birch	(5 m ht)	27.2 m^2
pine	(3 m ht)	4.2 m^2

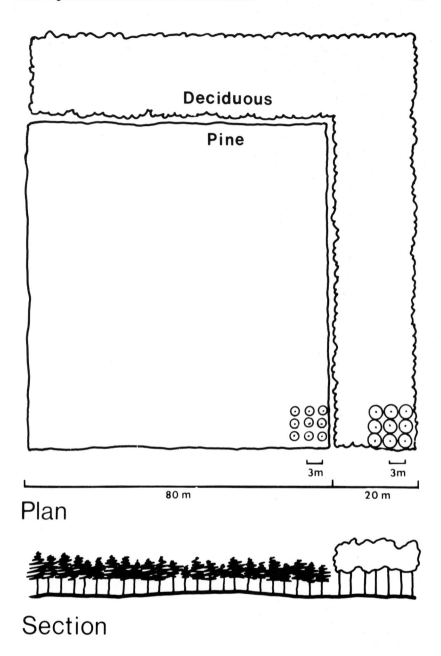

Plan

Section

Figure 5-9. Model forest hectare developed by the U.S. Environmental Protection Agency to examine the potential of open space to serve as an air quality management strategy. (From U.S. Environmental Protection Agency, 1976c.)

The total number of each species planted in the model forest and total vegetative area was as follows:

69 maple	2.54	$\times$ 10^3 m^2
69 oak	2.50	$\times$ 10^3 m^2
69 poplar	3.63	$\times$ 10^3 m^2
68 linden	1.56	$\times$ 10^3 m^2
69 birch	1.88	$\times$ 10^3 m^2
700 pine	2.90	$\times$ 10^3 m^2
total	15.0	$\times$ 10^3 m^2

The soil area of the model forest, not covered by tree trunks, was estimated to be 9.98×10^3 m^2. Pollutant flux rates guesstimated from the literature and used in the calculations are presented in Table 5-9. Table 5-10 lists the estimated sink capability. The U.S. Environmental Protection Agency concluded that if 122,517 ha (473 mi^2) of the model forest were in place in the St. Louis air quality region studied, that this forest could remove 80.5×10^6 tons of sulfur dioxide per year and function to maintain the air quality standard for this pollutant.

C. Summary

Our examination of the literature permits us to conclude that abundant field data are available to support the suggestion that tree surfaces accumulate a variety of natural and anthropogenic particles from the atmosphere and that controlled environment and wind tunnel studies allow the following generalizations:

1. The interception and retention of atmospheric particles by plants is highly variable and primarily dependent on:
   a. size, shape, wetness, and surface texture of the particles
   b. size, shape, wetness, and surface texture of the intercepting plant part
   c. micro- and ultramicroclimatic conditions surrounding the plant
2. More is known concerning the physical-mechanical aspects of particle deposition under controlled conditions than is known about the relative capture

Table 5-9. Guesstimated Gaseous Pollutant Flux Rates for Dry Soil and Vegetative Surfaces

	μg m^{-2} hr^{-1}	
Pollutant	Soil surface	Vegetative surface
Carbon monoxide	1.9×10^4	2.6×10^3
Nitrogen oxides	2.0×10^2	2.3×10^3
Ozone	1.0×10^9	6.2×10^4
Peroxyacetylnitrate	—	1.2×10^3
Sulfur dioxide	7.7×10^6	4.1×10^4

Source: U.S. Environmental Protection Agency (1976b).

and retention efficiencies of various plants and different species under natural conditions.

3. Generally greater leaf surface roughness increases particle capture efficiency for particles approximately 5 μm (and less) in diameter. Smooth leaved species (for example, horse chestnut and yellow poplar) are less efficient than rough leaved species (for example, elm and hazel).
4. Surface roughness acts to decrease the stability of the boundary layer (region of retarded air flow) surrounding the leaf and thus acts to increase particle impaction. Leaf hairs and leaf veins are principal contributors to surface roughness.
6. Smaller leaves are generally more efficient particle collectors than larger leaves.
7. Particle deposition (but probably not retention) is heaviest at the leaf tip and along leaf margins where a turbulent boundary layer is present. Leaves with complex shape and large circumference area ratio collect particles most efficiently.
8. Increased wind speeds and increased particle size typically increases particulate deposition velocities.
9. Deposition velocities to petioles and stems are generally many times greater than deposition velocities to leaf laminas. Collection of atmospheric particles by leafless deciduous species in the winter may remain quite high due to twig and shoot impaction.
10. Conifers are generally more effective particulate sinks than deciduous species.
11. Mechanisms by which particles are resuspended or otherwise removed from tree surfaces must be investiged more thoroughly.

Unfortunately only a modest portion of all experiments conducted with artificially generated particles has been conducted in natural forest environments and very few models have been developed to estimate the removal capacity of groups of trees. Hosker (1973) applied a standard plume diffusion model to hypothetical sources located at several heights above homogeneous stretches of grassland and forest. He concluded that the amount of effluent physically deposited on the foliage would be significantly larger for the forest than for the field. Slinn (1975) has provided the best analysis of the problems associated with

Table 5-10. Guesstimated Gaseous Pollutant Removal for Model Forest Hectare[a]

Pollutant	tons yr^{-1}
Ozone	9.6 $\times$ 10^4
Sulfur dioxide	748
Carbon monoxide	2.2
Nitrogen oxides	0.38
Peroxyacetylnitrate	0.17

Source: U.S. Environmental Protection Agency (1976c).
[a] Total includes both soil and tree removal, dry condition.

developing a model for dry deposition in a plant canopy. He stresses the importance of considering "resuspension" of particles from vegetation in model design. Bache (1979) stresses the importance of considering all trapping mechanisms, for example, sedimentation and impaction, when designing particle trapping models.

It is tempting to conclude that trees may be especially efficient filters of airborne particles because of their large size, high surface to volume ratio of foliage, petioles and twigs, and frequently hairy or rough leaf, twig, or bark surfaces. Because the interior portions of forest stands act to still the air, mean wind speeds are reduced and particle sedimentation will be augmented. We are unable to quantify the filtration capacity of forests at the present time, however, because of deficient field data. Much of the data on particle loading of trees is expressed on a μg g^{-1} (ppm) dry weight basis. Little (1977) has very appropriately indicated that judgments regarding particle collection efficiencies which are expressed as a function of plant dry weight must be used with caution and consider differences in area:dry weight ratios if they are to be meaningful. Further unless accurate dimension analysis data (leaf, twig dry weight, leaf, and trunk surface area) are available along with ambient microclimatological information, tree or stand loading can only be speculated. Much of the information on deposition velocities available for particle transfer to natural surfaces was not accumulated from experiments designed to access sink capability, and did not employ particles small enough (< 1 μm) to be particularly relevant to individuals primarily interested in air quality.

We conclude that there is also substantial evidence that trees remove *gaseous* contaminants from the atmosphere. Experiments, again largely performed under controlled environmental circumstances and with seedling or young plants, allow the following generalizations:

1. Plant uptake rates increase as the solubility of the pollutant in water increases. Hydrogen fluoride, sulfur dioxide, nitrogen dioxide, and ozone, which are soluble and reactive, are readily sorbed pollutants. Nitric oxide and carbon monoxide, which are very insoluble, are absorbed relatively slowly or not at all by vegetation.
2. When vegetative surfaces are wet (damp) the pollutant removal rate may increase up to 10-fold. Under damp conditions, the entire plant surface—leaves, twigs, branches, stems—is available for uptake.
3. Light plays a critical role in determining physiological activities of the leaf and stomatal opening and as such exerts great influence on foliar removal of pollutants. Under conditions of adequate soil moisture pollutant uptake by vegetation is almost constant throughout the day, as the stomata are fully open. Moisture stress sufficient to limit stomatal opening, and relatively common in various urban environments, would severely restrict gaseous pollutant uptake.
4. Pollutants are absorbed most efficiently by plant foliage near the canopy surface where light-mediated metabolic and pollutant diffusivity rates are greatest.

5. Sulfur and nitrogen dioxides are taken up by respiring leaves in the dark, but uptake rates are greatly reduced relative to rates in the light.

Models of gas pollutant uptake by forests based on flux rates derived from controlled environment investigations, for example, the U.S. Environmental Protection Agency's model forest hectare, probably overestimate the sink capacity as they employ flux rates determined in environments with unnaturally high pollutant concentrations, with limited or no prefumigation treatments and with moisture, light, and temperature conditions optimal for stomatal function. Models based on pollutant concentrations actually monitored in field situations, for example, Murphy et al. (1977) and Kabel et al. (1976) are more informative.

Under certain environmental circumstances, especially when tree surfaces are wet and when leaves are metabolically active, biologically and medically significant reductions in ambient levels of sulfur dioxide, nitrogen dioxide, ozone, and hydrogen fluoride may be realized by stands of trees for extended periods as long as the atmospheric loading of the contaminant gases is not excessive.

References

Art, H. W. 1971. Atmospheric salts in the functioning of a maritime forest ecosystem. Unpublished Ph.D. Thesis, Yale University, School of Forestry and Environmental Studies, New Haven, Connecticut, 135 pp.

Art, H. W., F. H. Bormann, G. K. Voigt, and G. M. Woodwell. 1974. Barrier island forest ecosystem: Role of meteorologic nutrient inputs. Science 184: 60-62.

Aylor, D. E. 1975. Deposition of particles of ragweed pollen in a plant canopy. J. Appl. Meteorol. 14:52-57.

Bach, W. 1972. Atmospheric Pollution. McGraw-Hill, New York, 144 pp.

Bache, D. H. 1979. Particle transport within plant canopies—I. A framework for analysis. Atmos. Environ. 13:1257-1262.

Bennett, J. H., and A. C. Hill. 1973. Absorption of gaseous air pollutants by a standardized plant canopy. J. Air Pollut. Control Assoc. 23:203-206.

Bennett, J. H., and A. C. Hill. 1975. Interactions of air pollutants with canopies of vegetation. In: J. B. Mudd and T. T. Kozlowski (Eds.), Responses of Plants to Air Pollution. Academic Press, New York, pp. 273-306.

Bennett, J. H., A. C. Hill, and D. M. Gates. 1973. A model for gaseous pollutant sorption by leaves. J. Air Pollut. Control Assoc. 23:957-962.

Bormann, F. H., P. R. Shafer, and D. Mulcahy. 1958. Fallout on the vegetation of New England during the 1957 atom bomb test series. Ecology 39:376-378.

Boyce, S. G. 1954. The salt spray community. Ecol. Monogr. 24:29-67.

Chamberlain, A. C. 1967. Deposition of particles to natural surfaces. In: P. H. Gregory and J. L. Monteith (Eds.), Airborne Microbes. 17th Symp. Soc. Gen. Microbiol., Cambridge Univ. Press, London, pp. 138-164.

Chamberlain, A. C. 1970. Interception and retention of radioactive aerosols by vegetation. Atmos. Environ. 4:57-78.

Chamberlain, A. C. 1975. The movement of particles in plant communities. *In*: J. L. Monteith (Ed.), Vegetation and the Atmosphere, Vol. I. Academic Press, New York, pp. 155-203.

Clayton, J. L. 1972. Salt spray and mineral cycling in two California coastal ecosystems. Ecology 53:74-81.

Costantini, A., and A. E. Rich. 1973. Comparison of salt injury to four species of coniferous tree seedlings when salt was applied to the potting medium and to the needles with or without an anti-transpirant. Phytopathology 63:200.

Dochinger, L. S. 1972. Can trees cleanse the air of particulate pollutants? Int. Shade Tree Conf. Proc. 48:45-48.

Eaton, T. E. 1979. Natural and artificially altered patterns of salt spray across a forested barrier island. Atmos. Environ. 13:705-709.

Edmonds, R. L., and C. H. Driver. 1974. Dispersion and deposition of spores of *Fomes annosus* and fluorescent particles. Phytopathology 64:1313-1321.

Edwards, R. S., and S. M. Claxton. 1964. The distribution of airborne salt of marine origin in the Aberystwyth area. J. Appl. Ecol. 1:253-263.

Elder, F., and C. Hosler. 1954. Ragweed pollen in the atmosphere. Report, Dept. of Meteorology, Pennsylvania State Univ., University Park, Pennsylvania.

Environmental Health Science Center. 1975. The Role of Plants in Environmental Purification. Environ. Health Sci. Ctr., Oregon State Univ., Corvallis, Oregon, 34 pp.

Fritschen, L. J., C. H. Driver, C. Avery, J. Buffo, R. Edmonds, R. Kinerson, and P. Schiess. 1970. Dispersion of air traces into and within a forested area (3). Report No. OSD01366, College of Forest Resources, Washington Univ., Seattle, Washington, 53 pp.

Garland, J. A. 1977. The dry deposition of sulfur dioxide to land and water surfaces. Proc. Royal Soc. London 354:245-268.

Garland, J. A., and J. R. Branson. 1977. The deposition of sulfur dioxide to pine forest assessed by a radioactive tracer method. Tellus 29:445-454.

Garland, J. A., and S. A. Penkett. 1976. Absorption of peroxyacetylnitrate and ozone by natural surfaces. Atmos. Environ. 10:1127-1131.

Graustein, W. D. 1978. Measurement of dust input to a forested watershed using $^{87}Sr/^{86}Sr$ ratios. Geol. Soc. Am. Abst. 10:411.

Gregory, P. H. 1971. The leaf as a spore trap. *In*: T. F. Preece and C. H. Dickinson (Eds.), Ecology of Leaf Surface Microorganisms. Academic Press, New York, 640 pp.

Gregory, P. H. 1973. The Microbiology of the Atmosphere. Wiley, New York, 377 pp.

Hanson, G. P., and L. Thorne. 1972. Vegetation to reduce air pollution. Lasca Leaves 20:60-65.

Heichel, G. H., and L. Hankin. 1972. Particles containing lead, chlorine and bromine detected on trees with an electron microprobe. Environ. Sci. Technol. 6:1121-1122.

Heichel, G. H., and L. Hankin. 1976. Roadside coniferous windbreaks as sinks for vehicular lead emissions. J. Air. Pollut. Control Assoc. 26:767-770.

Heisler, G. M. 1975. How trees modify metropolitan climate and noise. *In*: Forestry Issues in Urban America. Proc. 1974 National Convention Society of American Foresters, New York, pp. 103-112.

Helvey, J. D. 1971. A summary of rainfall interception by certain conifers of North America. *In*: Proc. Biological Effects in the Hydrological Cycle. U.S.D.A. Forest Service, Washington, D.C., pp. 103-113.

Hill, A. C. 1971. Vegetation: A sink for atmospheric pollutants. J. Air Pollut. Control Assoc. 21:341-346.

Hill, A. C., and E. M. Chamberlain, Jr. 1974. The removal of water soluble gases from the atmosphere by vegetation. Atmospheric-Surface Exchange of Particulate and Gaseous Pollutants Symp. Richland, Washington, Sept. 4-6, 1974, 12 pp.

Hofstra, G., and R. Hall. 1971. Injury on roadside trees: Leaf injury on pine and white cedar in relation to foliar levels of sodium chloride. Can. J. Bot. 49: 613-622.

Hosker, R. P., Jr. 1973. Estimates of dry deposition and plume depletion over forests and grassland. Proc. IAEA Symposium on the Physical Behavior of Radioactive Contaminants in the Atmosphere, Vienna, Austria, Nov. 12-16, 1973.

Ingold, C. T. 1971. Fungal Spores. Clarendon Press, Oxford, 302 pp.

Jensen, K. F. 1975. Sulfur content of hybrid poplar cuttings fumigated with sulfur dioxide. U.S.D.A. Forest Service, Res. Note No. NE-209, Upper Darby, Pennsylvania, 4 pp.

Jensen, K. F., and T. T. Kozlowski. 1975. Absorption and translocation of sulfur dioxide by seedlings of four forest tree species. J. Environ. Qual. 4:379-382.

Kabel, R. L. 1976. Natural removal of gaseous pollutants. 3rd Symp. Atmospheric Turbulence, Diffusion and Air Quality. Amer. Meteorological Soc., Oct. 19-22, 1976, Raleigh, North Carolina.

Kabel, R. L., R. A. O'Dell, M. Taheri, and D. D. Davis. 1976. A preliminary model of gaseous pollutant uptake by vegetation. Center for Air Environment Studies, Publ. No. 455-76, Pennsylvania State Univ., University Park, Pennsylvania, 96 pp.

Keller, T. 1978. How effective are forests in improving air quality? Eighth World Forestry Conference, Jakarta, Indonesia, Oct. 16-28, 1978, 9 pp.

Langer, G. 1965. Particle deposition and re-entrainment from coniferous trees. Part II. Experiments with individual leaves. Kolloid Z. Z. Polym. 204:119-124.

Little, P. 1977. Deposition of 2.75, 5.0 and 8.5 μm particles to plant and soil surfaces. Environ. Pollut. 12:293-305.

Little, P., and M. H. Martin. 1972. A survey of zinc, lead and cadmium in soil and natural vegetation around a smelting complex. Environ. Pollut. 3:241-254.

Martin, A., and F. R. Barber. 1971. Some measurements of loss of atmospheric sulfur dioxide near foliage. Atmos. Environ. 5:345-352.

Martin, W. E. 1959. The vegetation of Island Beach State Park, New Jersey. Ecol. Monogr. 29:1-46.

McCure, D. C., D. H. Silberman, R. H. Mandl, L. H. Weinstein, P. C. Freudenthal, and P. A. Giardina. 1977. Studies on the effects of saline aerosols of cooling tower origin on plants. J. Air Pollut. Control Assoc. 27:319-324.

Moser, B. C. 1979. Airborne salt spray techniques for experimentation and its effects on vegetation. Phytopathology 69:1002-1006.

Murphy, C. E., Jr., T. R. Sinclair, and K. R. Knoerr. 1977. An assessment of the use of forests as sinks for the removal of atmospheric sulfur dioxide. J. Environ. Qual. 6:388-396.

Neuberger, H., C. C. Hosler, and C. Koemond. 1967. Vegetation as an aerosol filter. *In*: S. W. Tromp and W. H. Weihe (Eds.), Biometeorology 2. Pergamon Press, New York, pp. 693-702.

Oak Ridge National Laboratory. 1969. Progress report in postattack ecology. Interim Progress Report NO. ORNL-TM-2466. Oak Ridge, Tennessee, 60 pp.

Oosting, H. J. 1945. Tolerance to salt spray of plants of coastal dunes. Ecology 26:85-89.

Oosting, H. J., and W. D. Billings. 1942. Factors affecting vegetational zonation on coastal dunes. Ecology 23: 131-142.

Page, A. L., and A. C. Chang. 1979. Contamination of soil and vegetation by atmospheric deposition of trace elements. Phytopathology 69:1007-1011.

Podgorow, N. W. 1967. Plantings as dust filters. Les. Khoz. 20:39-40.

Rasmussen, K. H., M. Taheri, and R. L. Kabel. 1975. Global emissions and natural processes for removal of gaseous pollutants. Water, Air, Soil Pollut. 4:33-64.

Rauner, J. L. 1976. Deciduous forests. *In*: J. L. Monteith (Ed.), Vegetation and the Atmosphere. Vol. 2. Academic Press, New York, pp. 241-264.

Raynor, G. S. 1967. Effects of a forest on particulate dispersion. *In*: C. A. Mawson (Ed.), Proc. USAEC Meteorological Information Meeting, Chalk River Nuclear Laboratories, Chalk River, Ontario, Canada, Sept. 11-14, 1967, pp. 581-586.

Raynor, G. S., M. E. Smith, I. A. Singer, L. A. Cohen, and J. V. Hayes. 1966. The dispersion of ragweed pollen into a forest. Proc. 7th National Conf. Agricultural Meteorology, Aug. 29-Sept. 1, 1966. Rutgers Univ., New Brunswick, New Jersey.

Rich, S., and N. C. Turner. 1972. Importance of moisture on stomatal behavior of plants subjected to ozone. J. Air Pollut. Control Assoc. 22:369-371.

Rich, S., P. E. Waggoner, and H. Tomlinson. 1970. Ozone uptake by bean leaves. Science 169:79-80.

Roberts, B. R. 1971. Foliar absorption of gaseous air pollutants. Am. Nursery 133:44-45.

Roberts, B. R. 1974. Foliar sorption of atmospheric sulfur dioxide by woody plants. Environ. Pollut. 7:133-140.

Roberts, B. R., and C. R. Krause. 1976. Changes in ambient SO_2 by rhododendron and pyracantha. Hort. Sci. 11:111-112.

Rogers, H. H., H. E. Jeffries, and A. M. Witherspoon. 1979. Measuring air pollutant uptake by plants: Nitrogen dioxide. J. Environ. Qual. 8:551-557.

Romney, E. M., R. G. Lindberg, H. A. Hawthorne, B. G. Bystrom, and K. H. Larson. 1963. Contamination of plant foliage with radioactive fallout. Ecology 44:343-349.

Rosinki, J., and C. T. Nagamoto. 1965. Particle deposition on and re-entrainment from coniferous trees. Part I. Experiments with trees. Kolloid Z.Z. Polym. 204:111-119.

Russell, I. J. 1974. Some factors affecting beta particle dose to tree populations in the eastern New England area from stratospheric fallout to 1974. Report No. CH-3015-8, Atomic Energy Commission, Chicago, Illinois, 58 pp.

Russell, I. J., and C. E. Choquette. 1974. Scale factors for foliar contamination by stratospheric sources of fission products in the New England area. Report NO. CH-3015-13, Atomic Energy Commission, Chicago, Illinois, 47 pp.

Salisbury, F. B., and C. W. Ross. 1978. Plant Physiology. Wadsworth, Belmont, California, 422 pp.

Sheih, C. M. 1977. Application of a statistical trajectory model to the simulation of sulfur pollution over northeastern United States. Atmos. Environ. 11:173-178.

Sheih, C. M., M. L. Wesely, and B. B. Hicks. 1979. A guide for estimating dry deposition velocities of sulfur over the eastern United States and surrounding regions. Argonne National Laboratory Report NO. ANL-RER-79-2, 55 pp.

Slinn, W. G. N. 1975. Dry deposition and resuspension of aerosol particles—A new look at some old problems. Proc. Conf. Atmosphere—Surface Exchange of Particles and Gases, ERDA Conf. Series, No. CONF-740921, Washington, D.C., pp. 1-40.

Slinn, W. G. N. 1976. Some approximations for the wet and dry removal of particles and gases from the atmosphere. Atmos. Sciences Dept., Battelle Memorial Institute, Pacific Northwest Laboratory, Richland, Washington.

Smith, W. H. 1970a. Technical review: Trees in the city. J. Am. Inst. Planners 36:429-436.

Smith, W. H. 1970b. Salt contamination of white pine planted adjacent to an interstate highway. Plant Dis. Reptr. 54:1021-1025.

Smith, W. H. 1971. Lead contamination of roadside white pine. For. Sci. 17: 195-198.

Smith, W. H. 1973. Metal contamination of urban woody plants. Environ. Sci. Technol. 7:631-636.

Smith, W. H. 1974. Air pollution—Effects on the structure and function of the temperate forest ecosystem. Environ. Pollut. 6:111-129.

Smith, W. H. 1976. Lead contamination of the roadside ecosystem. J. Air Pollut. Control Assoc. 26:753-766.

Smith, W. H. 1979. Urban vegetation and air quality. In: Proc. National Urban Forestry Conference, Washington, D.C., Nov. 13-16, 1978, U.S.D.A. Forest Service, Washington, D.C. and State Univ. of New York, Publica. No. 80-003, Syracuse, New York, pp. 284-305.

Smith, W. H., and L. S. Dochinger. 1975. Air Pollution and Metropolitan Woody Vegetation. Pinchot Institute, Consortium for Environmental Forestry Research, Publica. No. PIEFR-PA-1. U.S.D.A. Forest Service, Upper Darby, Pennsylvania, 74 pp.

Smith, W. H., and L. S. Dochinger. 1976. Capability of metropolitan trees to reduce atmospheric contaminants. In: H. Gerhold, F. Santamor, and S. Little (Eds.), Proc. Better Trees for Metropolitan Landscapes, U.S.D.A. Forest Service, Gen. Tech. Report No. NE-22, Upper Darby, Pennsylvania, pp. 49-59.

Smith, W. H., and B. J. Staskowicz. 1977. Removal of atmospheric particles by leaves and twigs of urban trees: Some preliminary observations and assessment of research needs. Environ. Manage. 1:317-328.

Steubing, L., and R. Klee. 1970. Comparative investigations into the dust filtering effects of broad leaved and coniferous woody vegetation. Angew. Bot. 4:73-85.

Thorne, L., and G. P. Hansen. 1972. Species differences in rates of vegetal ozone absorption. Environ. Pollut. 3:303-312.

116   5: Forests as Sinks for Air Contaminants: Vegetative Compartment

Townsend, A. M. 1974. Sorption of ozone by nine shade tree species. J. Am. Soc. Hort. Sci. 99:206-208.

U.S. Environmental Protection Agency. 1976a. Open Space as an Air Resource Management Measure. Vol. I. Sink Factors. U.S.E.P.A. Publica. No. EPA-450/3/76-028a, Research Triangle Park, North Carolina.

U.S. Environmental Protection Agency. 1976b. Open Space as an Air Resource Management Measure. Vol. II. Design Criteria. U.S.E.P.A. Publica. No. EPA-450/3/76-028b, Research Triangle Park, North Carolina.

U.S. Environmental Protection Agency. 1976c. Open Space as an Air Resource Management Measure. Vol. III. Demonstration Plan (St. Louis, Mo.). U.S. E.P.A. Publica. No. EPA-450/3-76/028c, Research Triangle Park, North Carolina.

Waggoner, P. E. 1971. Plants and polluted air. Bioscience 21:455-459.

Waggoner, P. E. 1975. Micrometeorological models. *In*: J. L. Monteith (Ed.), Vegetation and the Atmosphere, Vol. I. Academic Press, New York, pp. 205-228.

Warren, J. L. 1973. Green space for air pollution control. School of Forest Resources, Tech. Rep. No. 50, North Carolina State Univ., Raleigh, North Carolina, 118 pp.

Wedding, J. B., R. W. Carlson, J. J. Stukel, and F. A. Bazzaz. 1975. Aerosol deposition on plant leaves. Environ. Sci. Tech. 9:151-153.

Wells, B. W., and I. V. Shunk. 1938. Salt spray: An important factor in coastal ecology. Torr. Bot. Club Bull. 65:485-492.

White, E. J., and F. Turner. 1970. Method of estimating income of nutrients in catch of airborne particles by a woodland canopy. J. Appl. Ecol. 7:441-461.

Whittaker, R. H., and G. M. Woodwell. 1967. Surface area relations of woody plants and forest communities. Am. J. Bot. 8:931-939.

Witherspoon, J. P., and F. G. Taylor, Jr. 1969. Retention of a fallout simulant containing [134]Cs by pine and oak trees. Health Phys. 17:825-829.

Wood, F. A., and D. D. Davis. 1969. Sensitivity to ozone determined for trees. Pennsylvania State Univ., Sci. Agr. 17:4-5.

Woodcock, A. H. 1953. Salt nuclei in marine air as a function of altitude and wind force. J. Meteorol. 10:362-371.

Zinke, P. J. 1967. Forest interception studies in the United States. *In*: Forest Hydrology, Pergamon Press, Oxford, England, pp. 137-160.

6

Class I Summary: Relative Importance of Forest Source and Sink Strength and Some Potential Consequences of These Functions

The Class I relationship between forest ecosystems and air pollution is of primary importance when the atmospheric load of air contaminants from anthropogenic sources is relatively low. This situation exists locally and regionally when the sources of air pollutants produced by the activities of human beings are not operating or operating at low level or when meteorological conditions are not conducive to atmospheric accumulation. On a global scale, the Class I relationship may be extensive throughout those regions relatively remote from the activities of people. The specific concentration of air contaminants under "low" conditions is variable depending on the pollutant, but in general is meant to approximate "background," clean-air concentrations as, for example, presented by Rasmussen et al. (1975) for the major trace gases in μg m^{-3}: sulfur dioxide (1-4), hydrogen sulfide (0.3), dinitrogen oxide (460-490), nitric oxide (0.3-2.5), nitrogen dioxide (2-2.5), ammonia (4), carbon monoxide (100), ozone (20-60), and reactive hydrocarbons ($<$ 1). Since the majority of air contaminants of greatest significance to vegetative and human health (Table 1-1) originate from, and are removed by, both anthropogenic and natural agents it is essential to evaluate the importance of forest ecosystems within the latter group.

The evidence supporting the hypotheses that forest systems are important sources and sinks of air contaminants (Table 6-1) has been presented in Chapters 2 through 5.

Table 6-1. Interaction of Air Pollution and Temperate Forest Ecosystems under Conditions of Low Air Contaminant Load, Designated Class I Interaction

Forest soil and vegetation: Activity and response	Ecosystem consequence and impact
1. Forest soils and vegetation release particulate and gaseous contaminants to the atmosphere	1. Atmospheric burden of contaminants from anthropogenic sources supplemented by forest additions—scale may be local, regional, or global
2. Forest soils and vegetation remove particulate and gaseous contaminants from the atmosphere	2. Air contaminants transferred from the atmosphere to the biosphere, forest ecosystems supplement natural removal mechanisms
3. No or minimal alteration of structure or metabolism of forest soils or vegetation	3. No adverse ecosystem change discernible, slight fertilization possible

A. Forests as Sources of Air Pollutants

Efforts to assess the global importance of forest ecosystems as direct and indirect sources of air pollutants are frustrated by the very great imprecision of global estimates. On a global scale, air monitoring is absent and calculation based on approximation and intuition is abundant. Table 6-2 contains, at best, approximate order of magnitude guesstimations of air pollution materials released from anthropogenic relative to forest sources. For the purpose of these estimates the size of the global forest as estimated by Woodwell (1978) (5700×10^6 ha), Persson (1974) (4030×10^6 ha), and Ingersoll et al. (1974) (5010×10^6 ha) were averaged to yield 5000×10^6 ha global forest or very approximately one-third of global terrestrial ecosystems. While little absolute faith should be placed in any of the figures of Table 6-2, the table does suggest that the hypothesis that global forests are important sources of air contaminants is supported. In the case of carbon dioxide, trace gases containing sulfur and nitrogen, and reactive hydrocarbons, the quantities released directly or indirectly by forests may exceed those produced by people on a global scale. Only in the cases of carbon monoxide and particulates do anthropogenic sources potentially exceed forest sources.

A comparison of air pollutants produced from the activities of people in the United States relative to production by American forests (Table 6-3) also suggests that nationwide, forests are important sources. The size of this forest is estimated to approximate 300×10^6 ha (U.S. Department of Agriculture Forest Service, 1978,1979) which is roughly one-third of United States terrestrial ecosystems. Total forest generated carbon dioxide, sulfur and nitrogen containing gases, and reactive hydrocarbons may exceed amounts resulting from human activity.

On a global basis, the forest-related production of carbon dioxide and hydrocarbons may be of particular significance. The latter may importantly influence regional atmospheric oxidant patterns. The consequences to forest ecosystems of air contaminants of the global atmosphere are generally not considered adverse as the concentrations are generally well below the thresholds of subtle (Class II) or acute (Class III) effects. A primary exception, however, may be the continually increasing carbon dioxide concentration of the global atmosphere.

1. Elevated Carbon Dioxide, Global Warming, and Forest Health

It is presumed that a primary result of an elevated global atmospheric concentration of carbon dioxide will be warming (Schneider, 1975). While incoming solar radiation is uninfluenced by atmospheric carbon dioxide, portions of outgoing infrared radiation returned from earth to space are adsorbed by carbon dioxide. Over time, the net influence of reduced loss of infrared radiation from the earth should act to warm the planet (Baes et al., 1977). While the forces controlling global temperature change are varied and complex, the increase of 0.5°C since the mid-1800s is generally agreed to be at least partially caused by increased carbon dioxide. By 2000 it may increase an additional 0.5°C (McLean, 1978). Numerous models have been advanced to estimate the amount of the average global increase in temperature per doubling of carbon dioxide and these projections are in the range of 0.7 to 9.6°C (Schneider, 1975). Despite considerable uncertainty, it is reasonable to conclude that in the early part of the next century the average temperature of the earth will be beyond the limits experienced during the last 1000 years (Broecker, 1975).

The consequences of a warmer global climate, with even a very modest temperature increase, on the development of forest ecosystems could be profound. Even a summary discussion of the complex relationship between forest growth and health as related to climate is beyond the scope of this volume. Elevation of temperature, particularly with increased carbon dioxide available in the atmosphere, might enhance forest growth. This suggestion, however, is largely based on observations in greenhouses and other "controlled" growth facilities where numerous factors influencing plant development are regulated. In nature, especially with long-lived plants, the interaction of vegetation with climate is very complex and sensitive to subtle alteration over several years. It has been suggested that the numerous die-back and decline diseases that periodically influence a large number of forest species throughout the temperate zone are at least partially a reflection of species "out-of-phase" with optimal climatic conditions.

Physiological processes of plants, perhaps most importantly photosynthesis, transpiration, respiration, and reproduction, are extremely sensitive to temperature change. With general warming respiration and decomposition may increase faster than photosynthetic production. Transpiration and evaporation increases may impose enhanced stress on drier sites. Reproductive strategies may be altered due to changes in population dynamics of pollinating insect species or seedling survival. The geographic or host ranges of significant exotic microbial pathogens or insect pests may be caused to expand. Previously innocuous endemic microbes or insects may be elevated to important pest status following climatic warming.

Table 6-2. Guesstimated Annual Emission Strength for Global Forest Ecosystems and Anthropogenic Sources [a]

Pollutant	Forest ecosystems		Anthropogenic sources	
	Mechanism	kg	kg	Mechanism
Carbon dioxide	Forest destruction, humus oxidation forest burning	78×10^{12} [b]	3.5×10^{12} [l]	Fossil fuel combustion, cement production
Carbon monoxide	Total of a and b below a. Metabolism and chlorophyll loss b. Forest burning	36×10^{10} 30×10^{10} [c] 6×10^{10} [d]	60×10^{10} [m]	Motor vehicle exhaust, other combustion
Sulfur (inorganic and organic volatiles)	Forest soil release	36×10^{12} [e]	0.08×10^{12} [n]	Coal combustion, petroleum refining, smelting, other industrial
Nitrogen (in NO_x and NH_3)	Forest soil release	17×10^{10} [f]	2.3×10^{10} [o]	Coal and other combustion, fertilizer, waste treatment
Hydrocarbons (reactive)	Tree foliage release	175×10^9 [g]	27×10^9 [p]	Motor vehicle exhaust, oil combustion
Hydrocarbons (nonreactive)	Forest soil CH_4 release	?	70×10^9	Motor vehicle exhaust, oil combustion
Particulates	Total of a-d below a. Pollen shed b. Gas-particle conversion c. Photochemical (terpenes) d. Forest burning	191×10^9 31×10^9 [h] 91×10^9 [i] 66×10^9 [j] 3×10^9 [k]	296×10^9 [k]	Combustion, industrial operations, gas-particle conversion, photochemical

[a] Calculations based on estimates provided in references reviewed in Chapters 2 and 3.

[b] [(Woodwell, 1978; and Wong, 1978) + (Bolin, 1977)] ÷ 2 (for mean of gross release).

[c] (Nozhevnikova and Yurganov, 1978) x 0.33 (forest contribution).

[d] Seiler, 1974.

[e] (Adams et al., 1979) x 5000 x 10^6 (forest area ha).

[f] [(Kim, 1973) + (Focht and Verstraete, 1977)] ÷ 2 (for mean) x 5000 x 10^6 (forest area ha).

[g] Rasmussen, 1972.

[h] (0.1 kg tree^{-1}) x (62 trees ha^{-1}) x 5000 x 10^6 (forest area ha).

[i] (Robinson and Robbins, 1971) x 0.1 (forest contribution).

[j] (Robinson and Robbins, 1971) x 0.33 (forest contribution).

[k] Robinson and Robbins, 1971.

[l] Bolin, 1977.

[m] Seiler, 1974.

[n] Bremner and Steele, 1978.

[o] Rasmussen et al., 1975 (NH_3, NO_x converted to N, NO_x = ½ NO_2, ½ NO).

[p] Rasmussen et al., 1975.

Table 6-3. Guesstimated Annual Emission Strength for United States Forest Ecosystems and Anthropogenic Sources[a]

Pollutant	Forest ecosystems		Anthropogenic sources	
	Mechanism	kg	kg	Mechanism
Carbon dioxide	Forest destruction humus oxidation, forest burning	5.2×10^{12} [b]	1.75×10^{12} [n]	Fossil fuel combustion
Carbon monoxide	Total of a and b below	2.2×10^{10}	8.7×10^{10} [o]	Motor vehicle exhaust, other combustion
	a. Metabolism and chlorophyll loss	1.8×10^{10} [c]		
	b. Forest burning	0.4×10^{10} [d]		
Sulfur (in inorganic and organic volatiles)	Forest soil release	2.2×10^{12} [e]	$.014 \times 10^{12}$ [p]	Coal combustion, petroleum refining, smelting, other industrial
Nitrogen (in NO_x and NH_3)	Forest soil release	9.9×10^{9} [f]	8.9×10^{9} [q]	Coal and other combustion, motor vehicle exhaust
Hydrocarbons (reactive)	Total of a and b below	10.6×10^{9}	7.8×10^{9} [r]	Motor vehicle exhaust, refining, industrial solvents
	a. Tree foliage	10.5×10^{9} [g]		
	b. Forest burning	0.09×10^{9} [h]		
Hydrocarbons (nonreactive)	Total of a and b below	?	20×10^{9} [s]	Motor vehicle exhaust, refining, industrial solvents
	a. Forest soil CH_4	?		
	b. Forest burning	0.21×10^{9} [i]		
Particulates	Total of a–d below	13×10^{9}	12.8×10^{9} [t]	Motor vehicle exhaust, refining, industrial solvents
	a. Pollen shed	3×10^{9} [j]		
	b. Gas-particle conversion	5.5×10^{9} [k]		
	c. Photochemical (terpenes)	4×10^{9} [l]		
	d. Forest burning	0.45×10^{9} [m]		

a Calculations based on estimates provided in references reviewed in Chapters 2 and 3.

b (Woodwell, 1978) (net global temperate forest) × 0.25 (for net U.S. temperate forest) × 4 (net to gross conversion).

c Table 6-1 (global forest CO metabolism + chlorophyll loss) × 0.06 (for U.S. forest contribution).

d 71 kg CO ton^{-1} fuel burned × 56 × 10^6 tons (U.S. total forest fuel burned).

e (Adams, 1979) × 300 × 10^6 (forest area ha).

f [(Kim, 1973) + (Focht and Verstraete, 1977)] ÷ 2 (for mean) × 1300 × 10^6 (forest area ha).

g Table 6-1 (global forest reactive HC tree foliage) × 0.06 (for U.S. forest contribution).

h 5.9 kg HC ton^{-1} fuel burned × 56 × 10^6 tons (U.S. total forest fuel burned) × 0.3 (for portion reactive).

i 5.9 kg HC ton^{-1} fuel burned × 56 × 10^6 tons (U.S. total forest fuel burned) × 0.7 (for portion nonreactive).

j (0.1 kg tree^{-1}) × (100 trees ha^{-1}) × 300 × 10^6 (forest area ha).

k (Robinson and Robbins, 1971) × 0.1 (global forest contribution) × 0.06 (U.S. forest contribution).

l (Robinson and Robbins, 1971) × 0.33 (global forest contribution) × 0.06 (U.S. forest contribution).

m [5.15 kg particulates ton^{-1} fuel burned × 56 × 10^6 tons (U.S. total forest fuel burned) + 0.6 (U.S. Environmental Protection Agency, 1977)] ÷ 2 (for mean).

n Table 6-1 (global CO$_2$ production) × 0.5 (U.S. contribution).

o U.S. Environmental Protection Agency (1977).

p 26.9 × 10^9 kg SO$_2$ (U.S. Environmental Protection Agency, 1977) × 0.5 (for S).

q 23.0 × 10^9 kg NO$_x$ (U.S. Environmental Protection Agency, 1977) × 0.385 (for N).

r 27.9 × 10^9 kg HC (U.S. Environmental Protection Agency, 1977) × 0.28 (for reactive HC).

s 27.9 × 10^9 kg HC (U.S. Environmental Protection Agency, 1977) × 0.72 (for unreactive HC).

t 13.4 × 10^9 kg particulates (U.S. Environmental Protection Agency, 1977) − 0.6 × 10^9 kg forest fire particulates (U.S. Environmental Protection Agency, 1977).

2. Regional Importance of Forest Emissions: Forests as Sources of Oxidant Precursors

On a regional scale, primary forest emissions may be converted to secondary pollutants. The hypothesis has been advanced that hydrocarbons released from vegetation may play a role in ozone synthesis in the atmosphere (Middleton, 1967). While the chemistry of oxidant synthesis is exceedingly complex in detail, it is straightforward in summary outline (Figure 6-1).

In the presence of sunlight nitrogen dioxide is dissociated and forms equal numbers of nitric oxide molecules and oxygen atoms. The oxygen atoms rapidly combine with molecular oxygen to form ozone. This ozone then reacts with the nitric oxide, on a one-to-one basis, to reform nitrogen dioxide. The steady-state concentrations of ozone that is produced by this cycle is very small. When hydrocarbons, aldehydes, or other reactive atmospheric constituents are present, how-

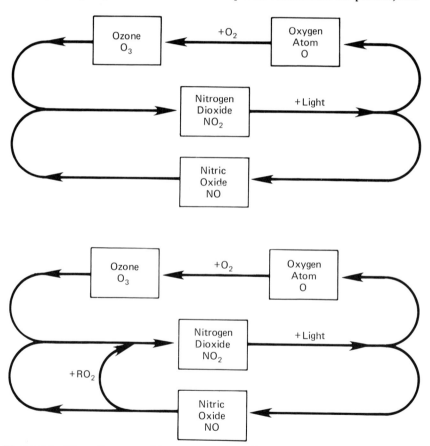

Figure 6-1. The nitrogen oxide-ozone cycles. The upper cycle portrays the atmosphere containing only nitrogen oxides and oxygen. The lower cycle portrays the situation in the presence of reactive contaminants (for example, hydrocarbons). (From NAS, 1977.)

ever, they can form peroxy radicals that oxidize the nitric oxide back to nitrogen dioxide. With reduced nitric oxide available to react with the ozone, the latter may accumulate to relatively high concentrations (Maugh, 1975; Spedding, 1974; National Academy of Sciences, 1977).

a. Laboratory Evidence

In 1962, Stephens and Scott reported that terpenes could react photochemically with oxides of nitrogen to form ozone and peroxyacetylnitrates. They demonstrated that the monoterpenes pinene and α-phellandrene both showed the high reactivity predicted by their olefinic structure. Isoprene oxidation of nitric oxide to nitrogen dioxide was found by Glasson and Tuesday (1970) to be at a rate intermediate between the rates for ethylene and $trans$-2-butene. Ripperton and Lillian (1971) reported the formation of oxidants, ozone, condensation nuclei, and nitric oxide, following irradiation of systems containing 0.1 ppm (188 μg m^{-3}) nitrogen dioxide and 0.5 ppm (2780 μg m^{-3}) α-pinene. Japar et al. (1974) have provided rate constants for α-pinene and terpinolene.

Some of the initial experiments on terpene reactivity were conducted with gas concentrations quite dissimilar from ambient forest conditions. Westberg and Rasmussen (1972) and Grimsrud et al. (1975) employed realistic concentrations of 10 ppb (56 μg m^{-3}) hydrocarbons and approximately 7 ppb (12 μg m^{-3}) nitrogen oxides in their determinations of monoterpene reactivity. The data from these studies suggest that olefinic terpenes react with ozone at rates similar with those of the most reactive alkenes (Bufalini et al., 1976) and that structure plays a minor role in reactivity (Table 6-4).

Table 6-4. Photooxidation Rates of Monoterpene Hydrocarbons

Hydrocarbon	Photooxidation rate at 7 ppb NO$_x$ and 10 ppb HC (sec^{-1} $\times$ 10^4)
p-Methane	0.11
p-Cymene	0.25
Isobutene	0.84
β-Pinene	1.1
Isoprene	1.3
α-Pinene	1.3
3-Carene	1.4
β-Phellandrene	1.9
Carvomethene	2.1
Limonene	2.4
Dihydromyrcene	3.0
Mycene	3.8
Oximene	5.3
Terpinolene	11
α-Phellandrene	12
α-Terpinene	55-110

Source: Grimsrud et al. (1975).

This brief review of laboratory evidence clearly suggests that photochemical reactions of terpenes and nitric oxide can occur and therefore support the hypothesis that forests may be involved in ozone synthesis in nature.

b. Field Evidence

If forest hydrocarbons do participate in the generation of ozone under ambient conditions, it should be possible to find elevated ozone concentrations in forest (non-urban) regions. High rural ozone has been detected in numerous situations: Maryland and West Virginia (Research Triangle Institute, 1973), West Virginia (Richter, 1970), California (Miller et al., 1972), and New York (Stasiuk and Coffey, 1974). The latter study simultaneously monitored ozone concentrations at eight rural and urban sites throughout New York State. Ozone concentrations at rural sites were found to be comparable or in excess of ozone amounts in urban areas. The authors considered three potential sources for the high rural ozone they observed: (1) transport from the stratosphere, (2) generation from anthropogenic precursors, and (3) generation from hydrocarbons released from vegetation. The latter explanation appeared most plausible to Stasiuk and Coffey.

Sandberg et al. (1978) examined correlations between ozone levels in the San Francisco air basin and winter precipitation. These authors concluded that the good correlation between precipitation and ozone they found supported the hypothesis that vegetative precursors played a major role in the high oxidant levels of San Francisco.

Analyses of daily ozone concentrations in New Jersey and New York throughout the week have revealed high ozone levels on summer weekends relative to weekdays despite considerably attenuated motor vehicle use during the former periods (Bruntz et al., 1974; Cleveland et al., 1974). Graedel et al. (1977) have employed detailed chemical kinetic computations representing workdays and Sundays in Hudson County, New Jersey, to evaluate this so-called "weekend or Sunday effect." These authors conclude that weekday ozone is subject to greater scavenging by oxides of nitrogen and thereby infer that while greater ozone may be formed during the week relative to weekends, it does not accumulate. They also, however, suggest that "weekend ozone" may have an enhanced component of oxidant advected from less urban areas. Cleveland and McRae (1978) have expanded the observations of daily ozone levels beyond northern New Jersey and New York City and have included Connecticut and Massachusetts. By examining a larger region it was apparent that while a reduction of primary emissions (weekend situation) may not lead to a reduction in ozone concentration in the metropolitan region itself, it may result in reduced ozone concentrations downwind from the primary source area. The authors showed that sites downwind of the New York City region with respect to summer prevailing winds (Connecticut and Massachusetts) generally show a reduction of daily maximum ozone concentrations of approximately 10-25% on weekends. As both Connecticut and Massachusetts are very approximately 75% forested, it is possible that widespread production of oxidant precursors occurs in these woodlands.

One of the greatest difficulties in defining the specific origin of rural and urban ozone is the realization that oxidants or their precursors may be transported considerable distances from site of origin (Table 6-5). Ancora, a small nonindustrial, low-traffic community in southern New Jersey frequently had daily maximum ozone concentrations in excess of the 0.08 ppm (157 $\mu g\ m^{-3}$) federal standard during summer 1973. Cleveland and Kleiner (1975) investigated this enigma and concluded that Ancora's ozone was transported 37 km from the Camden-Philadelphia urban complex. With appropriate wind direction, ozone from this urban complex was moved to sites as far as 49 km away. During this same year, Coffey and Stasiuk (1975) were commonly recording concentrations of ozone in excess of 0.08 ppm in rural areas of New York State. These authors argued that this high rural ozone was transported *into* urban areas via advection and vertical mixing and contributed to the total oxidant level in the urban areas of New York, Syracuse, and Buffalo.

Chameides and Stedman (1976) presented theoretical model evidence supporting the hypothesis that it was the interaction of anthropogenic nitrogen oxides generated over urban areas with photochemical methane oxidation chain products generated over natural areas that caused elevated rural ozone levels. Isaksen et al. (1978), on the other hand, argued that anthropogenic or natural hydrocarbons, *other* than methane, must be emitted to raise ambient ozone above 0.08 ppm in rural areas. When these nonmethane hydrocarbons are present and during anticyclonic weather situations, Isaksen et al. concluded that ozone concentrations in excess of 0.1 ppm (196 $\mu g\ m^{-3}$) may form in rural regions.

The data of Wolff et al. (1977) support the obvious conclusion that long-range ozone transport can be significant. Their observations of ozone in the urban corridor from Washington, D.C., to Boston, Massachusetts, suggest that an additive effect of sequential urban emissions on ambient ozone concentrations may in fact occur in the Northeast. Cleveland and Graedel (1979) have pointed out that areas as far as 45 to 60 km downwind of major sources of primary emissions, and which themselves have lesser emissions, can have the highest regional ozone concentrations.

Clearly, resolution of the origin of photochemical oxidant precursors in forested or urban regions within several hundred kilometers of each other must await the development of even greater sophistication in our understanding of atmospheric chemical and transport processes. An additional effort that would

Table 6-5. Maximum Downwind Movement of Urban Generated Oxidants

Area	Downwind distance (km)	Reference
Camden-Philadelphia	49	Cleveland and Kleiner (1975)
St. Louis	160	White et al. (1976)
Los Angeles	161	Altshuller (1975)
New York	300	Cleveland et al. (1976)

greatly facilitate our estimate of the importance of plant communities in ozone generation would be the direct monitoring of plant hydrocarbon release under field conditions. While little information of this kind exists for forest tree species, Lonneman et al. (1978) have monitored natural hydrocarbon emissions from citrus trees in Florida. Isoprene, which has been detected in other rural samples and is generally associated with deciduous vegetation, was the only hydrocarbon of natural origin detected in the vicinity of the leaves. The authors concluded that citrus production of hydrocarbons was low and that they could not possibly contribute to the production of significant levels of ambient ozone.

B. Forests as Sinks for Air Pollutants

The sink capability of forest ecosystems is guesstimated in Table 6-6. The figures of this table are every bit as approximate and imprecise as those of Tables 6-2 and 6-3, but nevertheless do encourage some reasoned speculation. The role of forests in the carbon dioxide global cycle must be refined. While the efficiency of the photosynthetic sink for carbon dioxide removal may balance the liber-

Table 6-6. Guesstimated Annual Sink Strength for Global and United States Forest Ecosystems[a]

Pollutant	Mechanism	Global (kg)		United States (kg)	
Carbon dioxide	Photosynthesis	70	$\times 10^{12}$ [b]	4.2	$\times 10^{12}$ [c]
Carbon monoxide	Total of a and b below	55	$\times 10^{10}$	2.3	$\times 10^{10}$
	a. Vegetation	5	$\times 10^{10}$ [d]	0.3	$\times 10^{10}$ [e]
	b. Soils	50	$\times 10^{10}$ [f]	2	$\times 10^{10}$ [f]
Sulfur dioxide	Total of a and b below	201	$\times 10^{12}$	13	$\times 10^{12}$
	a. Vegetation	0.675	$\times 10^{12}$ [g]	0.041	$\times 10^{12}$ [h]
	b. Soils	200	$\times 10^{12}$ [i]	13	$\times 10^{12}$ [j]
Nitrogen oxides	Total of a and b below	36	$\times 10^{11}$	2.1	$\times 10^{11}$
	a. Vegetation	1	$\times 10^{11}$ [k]	0.06	$\times 10^{11}$ [l]
	b. Soils	35	$\times 10^{11}$ [m]	2	$\times 10^{11}$ [n]
Hydrocarbons (ethylene only)	Soil	40	$\times 10^{9}$ [o]	2.3	$\times 10^{9}$ [p]
Ozone	Total of a and b below	45	$\times 10^{13}$	2.7	$\times 10^{13}$
	a. Vegetation	.003	$\times 10^{13}$ [q]	.0002	$\times 10^{13}$ [r]
	b. Soils	45	$\times 10^{13}$ [s]	2.7	$\times 10^{13}$ [t]
Particulates (lead only)	Soil	100	$\times 10^{6}$ [u]	6	$\times 10^{6}$ [v]

[a] Calculations based on estimates provided in references reviewed in Chapters 4 and 5.
[b] 50×10^{15} g C terrestrial plant fixation (Woodwell, 1978) x 0.38 (forest contribution) ÷ 0.27 (conversion C to CO_2).
[c] Footnote b x 0.006 (U.S. forest contribution).
[d] Bidwell and Fraser (1972) mean CO vegetation uptake x 0.1 (controlled environment to nature extrapolation) x 0.38 (forest contribution).
[e] Footnote d x 0.06 (U.S. forest contribution).

^{f}Ingersoll et al. (1974).

^{g}Murphy et al. (1977) mean of Savanna River and N.Y. predictions extrapolated to 1 yr = 0.135 tons SO_2 ha^{-1} x 5000 x 10^6 ha.

^{h}Murphy et al. (1977) mean of Savanna River and N.Y. predictions extrapolated to 1 yr = 0.135 tons SO_2 ha^{-1} x 300 x 10^6 ha.

^{i}Abeles et al. (1971) (footnote j) ÷ 300 x 10^6 ha x 5000 x 10^6 ha (conversion to global forest).

^{j}Abeles et al. (1971) x 0.33 (forest contribution).

^{k}NO_x vegetation uptake average (U.S. Environmental Protection Agency, 1976) extrapolated to 1 year and 1 ha x 5000 x 10^6 ha (global forest) x 0.1 (controlled environment to nature extrapolation).

^{l}NO_x vegetation uptake average (U.S. Environmental Protection Agency, 1976) extrapolated to 1 year and 1 ha x 300 x 10^6 ha (U.S. forest) x 0.1 (controlled environment to nature extrapolation).

^{m}Abeles et al. (1971) (footnote n) ÷ 300 x 10^6 ha (conversion to global forest).

^{n}Abeles et al. (1971) x 0.33 (forest contribution).

^{o}Abeles et al. (1971) (footnote q) ÷ 300 x 10^6 ha x 5000 x 10^6 ha (conversion to global forest).

^{p}Abeles et al. (1971) x 0.33 (forest contribution).

^{q}O_3 vegetation uptake average (U.S. Environmental Protection Agency, 1976) extrapolated to 1 year and 1 ha x 5000 x 10^6 ha (global forest) x 0.001 (controlled environment to nature extrapolation).

^{r}O_3 vegetation uptake average (U.S. Environmental Protection Agency, 1976) extrapolated to 1 year and 1 ha x 300 x 10^6 ha (U..S. forest) x 0.001 (controlled environment to nature extrapolation).

^{s}O_3 soil uptake average (U.S. Environmental Protection Agency, 1976) extrapolated to 1 year and 1 ha x 5000 x 10^6 ha (global forest) x 0.001 (controlled environment to nature extrapolation).

^{t}O_3 soil uptake average (U.S. Environmental Protection Agency, 1976) extrapolated to 1 year and 1 ha x 300 x 10^6 ha (U.S. forest) x 0.001 (controlled environment to nature extrapolation).

^{u}0.2 kg ha^{-1} yr^{-1} (Chap. 4) x 5000 x 10^6 ha (global forest) x 0.1 (eastern U.S. forest to global forest conversion).

^{v}0.2 kg ha^{-1} yr^{-1} (Chap. 4) x 300 x 10^6 ha (U.S. forest) x 0.1 (eastern U.S. forest to U.S. forest conversion).

ation of this gas via humus oxidation and burning in temperate forest zones, the extensive destruction and burning of tropical forests may be contributing excess carbon dioxide to the atmosphere. The Woodwell hypothesis must be pursued and additional evidence supporting or rejecting this critically important possibility must be developed.

The most striking feature of the forest ecosystem sink review is the apparent importance of the soil compartment of forest ecosystems as a sink for air contaminants. The quality and quantity of the evidence provided are strongest in the case of particles containing lead, and a few additional trace metal elements, and carbon monoxide. It is suggested by at least some evidence that the sink capabilities of forest soils may be very substantial, and well in excess of the capability of forest vegetation, in the case of sulfur dioxide, nitrogen oxides, and ozone.

Forest ecosystems on a global basis appear to have the capacity to transfer air contaminants from the atmosphere, in amounts approximately equal to or in excess of current anthropogenic production levels, in the instance of carbon monoxide, sulfur dioxide, nitrogen oxides, and hydrocarbons.

Balance sheets of global source and sink strength are academically interesting but practically of limited value as air contaminants are not uniformly produced nor distributed around the globe. Review of United States forest sink and anthropogenic production are more interesting. United States forests may have the capability to remove roughly 25% of the annual carbon monoxide production and most of the sulfur and nitrogen oxide production. Our uncertainty concerning particulate and hydrocarbon removal is very great as the data are grossly incomplete. Again because air contaminant distribution over forest ecosystems is not uniform throughout the country, generalizations concerning efficiency may be misleading. Excessive local generation of air contaminants in urban and industrial regions results in atmospheric accumulation as natural sinks do not exist, are saturated, or are inefficient.

1. Forest Fertilization as a Consequence of Sink Function

Forest trees require sixteen essential elements for growth. Ten of these elements, capitalized below may be added to forest ecosystems as air pollutants in the particulate or gaseous form:

Macroelements required	*Microelements required*
CARBON	BORON
hydrogen	CHLORINE
oxygen	COPPER
NITROGEN	IRON
SULFUR	MANGANESE
phosphorus	MOLYBDENUM
potassium	ZINC
magnesium	
calcium	

The indicated elements are potentially introduced to forest systems in the following states: carbon as gaseous carbon monoxide and dioxide; nitrogen as gaseous oxides or ammonia and particulate nitrates; sulfur as sulfur dioxide, hydrogen sulfide, or organic gases or particulate sulfates and the microelements generally in association with particles. It has been hypothesized that the addition of essential elements to forests via air pollution, at levels below those causing Class II relationships, may act to stimulate the growth of forests by fertilization. What is the evidence?

a. Carbon

A great amount of data shows that plants grow better in atmospheres enriched in carbon dioxide. Most of this information has been gleaned from experiments with various herbaceous agricultural and ornamental plants. Wright (1974), however, examined the photosynthetic response of several California woody plants, under controlled environmental conditions, to a range of carbon dioxide concentrations spanning past and likely future ambient concentrations. Small increases in atmospheric carbon dioxide generally produced significant increases in the rate of net photosynthesis. Wright observed considerable difference in response to carbon dioxide enrichment for his various San Bernardino Mountain tree species according to experimental run and conditioning temperatures. He concluded that rising carbon dioxide would tend to favor angiosperms over gymnosperms. Wright felt that the carbon dioxide "fertilization" he documented in the laboratory would be equally great under field conditions and did in fact, show that enhancement of carbon dioxide up to about 100 ppm (18×10^4 μg m^{-3}) above ambient levels produced a sharp linear rise in photosynthesis of leafy twigs in a Long Island, New York, oak-pine forest (Wright and Woodwell, 1970).

Botkin (1977) very correctly pointed out the considerable risks associated with extrapolating carbon dioxide enrichment data from controlled environment to natural situations. In the former, environmental factors, most importantly temperature, moisture, and light, are not limiting and the rate of photosynthesis is controlled by the rate at which carbon dioxide can diffuse into leaves. In natural situations, moisture and temperature may impose restrictions on photosynthetic rates as well as carbon dioxide availability. Also natural forest systems are frequently comprised of a diverse number of species that may exhibit differential response to carbon dioxide enrichment. Botkin applied his computer model of northern hardwood forest growth to an assessment of carbon dioxide fertilization and presented several interesting conclusions. Over time, shade tolerant species, such as sugar maple and red spruce, were favored over shade intolerant species, such as white birch. Further, unless the effect of the carbon dioxide increase was large enough to stimulate the annual increment of each tree approximately 50% above normal, changes were not observed in the productivity or biomass of the forest.

In summary, the modest information currently available suggests that the fertilizing effect of increased atmospheric carbon dioxide may be less important in natural forest ecosystems in increasing biomass than in altering the capability of tree species to compete with one another. The demonstration that soil and meteorological factors are not limiting tree photosynthesis in natural forests would be necessary to establish that increased atmospheric carbon dioxide was acting as a "fertilizer."

The potential forest ecosystems have to convert carbon monoxide to carbon dioxide remains unspecified. Current information regarding the possibility that elevated carbon monoxide atmospheric burdens supplement the carbon in tree biomass is unresolved (Nozhevnikorg and Hurganov, 1978).

b. Nitrogen

Soils very infrequently contain enough nitrogen for maximum plant growth. Coniferous forest soils in the north temperate zone are commonly deficient in tree-available nitrogen. Forest trees respond more favorably to nitrogen than to any other fertilizer element. Can the atmospheric input of ammonia, nitrogen dioxide, nitric acid, potassium nitrate, sodium nitrate, or calcium nitrate result in significant forest fertilization?

Approximately 99% of the combined nitrogen in soil is contained in organic matter (Thompson and Troeh, 1978). The latter supplies exchangeable nitrogen to plants but does not increase the total nitrogen content of the ecosystem. Biological fixation and atmospheric deposition are the sole external sources of nitrogen to forest ecosystems (Noggle et al., 1978). The anticipated experiment dealing with nitrogen dioxide uptake by an agricultural species under controlled environmental conditions can be found. Troiano and Leone (1977) have demonstrated increased nitrogen in tomato plant tissue following exposure to "realistic" concentrations of atmospheric nitrogen dioxide. Rogers et al. (1979) have shown that nitrogen dioxide taken up by snap beans is rapidly metabolized and incorporated into organic nitrogen compounds.

Bormann and Likens (1979) have attempted to determine the relative importance of biological fixation versus atmospheric deposition in their study of the northern hardwood forest in central New Hampshire. They concluded that 70% of the nitrogen added to the ecosystem was added via biological fixation and that 30% was added via precipitation. Noggle et al. (1978) observed that the estimated 14.1 kg ha^{-1} yr^{-1} entering the northern hardwood forest in the form of nitrogenous gas or aerosol presumably would exert a fertilizing effect as it compared with levels of forest fertilization employed in studies that have demonstrated positive forest growth responses.

c. Sulfur

Sulfur is similar to nitrogen in many respects; it is held in soil in association with organic compounds, most available to plants in its most oxidized state, and readily lost via leaching in the soil solution (Thompson and Troeh, 1978). It differs from nitrogen, however, in that it is required by trees in substantially lesser amounts and its presence in soils is usually adequate to supply the needs of tree growth. Regionally, however, soils can be deficient in sulfur. Many forest soils of the Pacific Northwest (Will and Youngberg, 1978) and in the Southeast (U.S. Department of Agriculture, 1964), for example, are deficient in this element. Sulfur deficiencies in agricultural soils may become more prevalent in the future as sulfur percentages in fertilizers have declined steadily in recent years (Thompson and Troeh, 1978).

Plants absorb sulfur from their leaves or from the soil as the sulfate ion or directly from the atmosphere as sulfur dioxide. A variety of controlled environment studies has demonstrated the efficiency of forest tree seedling sulfur dioxide uptake (for example, Jensen, 1975; Jensen and Kozlowski, 1975). Wainwright (1978) made field collections of sycamore leaves and litter in polluted and unpolluted sites in England. Leaves collected from the vicinity of a chemical coking plant had 118 μg g^{-1} elemental sulfur, 213 μg g^{-1} tetrathionate, and 3.7 mg g^{-1} sulfate while leaves from an unpolluted location exhibited no elemental sulfur nor tetrathionate and only 0.2 mg g^{-1} sulfate.

The Tennessee Valley Authority has conducted studies to measure the amount of sulfur that is contributed by the atmosphere to Tennessee Valley agricultural ecosystems (Noggle and Jones, 1979; Maugh, 1979). Cotton grown 4 and 3 km from coal-fired power plants accumulated 125 and 245 mg sulfur 100 g^{-1}, while fescue accumulated 65 and 58 mg 100 g^{-1} at the same locations. As a result of sulfur fertilization, cotton produced more biomass near power plants than that grown at locations remote from sulfur dioxide sources.

Fortunately the northern hardwood forest studies of Likens et al. (1977) provide us with one of the few quantitative, field estimates of atmospheric input of sulfur to a forested ecosystem. These investigators suggested that 6.1 kg ha^{-1} yr^{-1} of sulfur are provided to the northern hardwood forest via particulate deposition and direct absorption of gaseous sulfur. Throughfall and stemflow data suggested that sulfur aerosols may be impacted on tree surfaces in large amounts during the summer! The net gaseous sulfur input was judged to be about double that from aerosol deposition on an annual basis.

Fertilization trials with pole-size ponderosa pine on the Deschutes National Forest in Oregon failed to demonstrate a stimulatory effect from artificial applications of sulfur containing fertilizer (Cochran, 1978).

It is possible to conclude that sulfur containing air contaminants are introduced abundantly into some forest ecosystems and that these sulfur compounds can be metabolized and utilized in sulfur pathways by forest trees. It is not possible to generalize about the ability of this atmospheric sulfur to stimulate tree development, however, except to suggest that it is probably important only in those limited forest regions that have soils deficient in sulfur.

d. Micronutrients

All micronutrients are required by plants in relatively small amounts. As a result, deficiencies are the exception rather than the rule. Since all the micronutrients are potential components of anthropogenic effluents, it is possible that atmospheric input of these elements might function to supply a portion of the micro-element requirement in areas deficient in these elements and subject to urban or industrial contamination.

Boron deficiencies are widespread in humid regions. In the United States they are prevalent in the uplands of the Atlantic coastal plain, the Great Lakes region, and the coastal region of the Pacific Northwest. Copper deficiencies are most common in organic soils and have been identified in the United States in Washington, California, Florida, South Carolina, and in the Great Lakes region. Iron is one of the most commonly deficient micronutrients because it is frequently unavailable to plants even though present in the soil. Manganese deficiencies generally occur in humid areas and in acid sandy soils. In the United States deficiencies have been documented in California, along the Atlantic coastal plain, and in the muck soils of the Great Lakes region. Molybdenum plant requirements are so small that deficiencies are probably infrequent, but do occur, especially in acid soils. Zinc deficiencies may also occur in acid soils and occur in small, widely scattered areas throughout the United States. Chlorine is so abundant that deficiencies are thought to be extremely rare (Thompson and Troeh, 1978).

The importance of acid soils in the incidence of the deficiencies cited above suggests that forest vegetation might be especially benefited by input of micronutrients from the atmosphere. There is, however, no direct evidence to support this suggestion.

References

Abeles, F. B., L. E. Craker, L. E. Forrence, and G. R. Leather. 1971. Fate of air pollutants: Removal of ethylene, sulfur dioxide, and nitrogen dioxide by soil. Science 175:914-916.

Adams, D. F., S. O. Farwell, M. R. Pack, and W. L. Banesberger. 1979. Preliminary measurements of biogenic sulfur-containing gas emissions from soils. J. Air Pollut. Control Assoc. 29:380-383.

Altshuller, A. P. 1975. Evaluation of oxidant results at CAMP sites in the United States. J. Air Pollut. Control Assoc. 25:19-24.

Baes, C. F., Jr., H. E. Goeller, J. S. Olson, and R. M. Rotty. 1977. Carbon dioxide and climate: The uncontrolled experiment. Am. Scien. 65:310-320.

Bidwell, R. G. S., and D. E. Fraser. 1972. Carbon monoxide uptake and metabolism by leaves. Can. J. Bot. 50:1435-1439.

Bolin, B. 1977. Changes of land biota and their importance for the carbon cycle. Science 196:613-615.

Bormann, F. H., and G. E. Likens. 1979. Pattern and Process in a Forested Ecosystem. Springer-Verlag, New York, 253 pp.

Botkin, D. B. 1977. Forests, lakes, and the anthropogenic production of carbon dioxide. Bio. Sci. 27:325-331.

Bremner, J. M., and C. G. Steele. 1978. Role of microorganisms in the atmospheric sulfur cycle. Adv. Microb. Ecol. 2:155-201.

Broecker, W. S. 1975. Climatic change: Are we on the brink of a pronounced global warming? Science 189:460-463.

Bruntz, S. M., W. S. Cleveland, T. E. Graedel, B. Kleiner, and J. L. Warner. 1974. Ozone concentrations in New Jersey and New York: Statistical association with related variables. Science 186:257-258.

Bufalini, J. J., T. A. Walter, and M. M. Bufalini. 1976. Ozone formation potential of organic compounds. Environ. Sci. Technol. 10:908-912.

Chameides, W. L., and D. H. Stedman. 1976. Ozone formation from NO_x in "clean air." Environ. Sci. Technol. 10:150-153.

Cleveland, W. S., and T. E. Graedel. 1979. Photochemical air pollution in the northeast United States. Science 204:1273-1278.

Cleveland, W. S., and B. Kleiner. 1975. Transport of photochemical air pollution from Camden-Philadelphia urban complex. Environ. Sci. Technol. 9:869-872.

Cleveland, W. S., and J. E. McRae. 1978. Weekday-weekend ozone concentrations in the northeast United States. Environ. Sci. Technol. 12:558-563.

Cleveland, W. S., T. E. Graedel, B. Kleiner, and J. L. Warner. 1974. Sunday and workday variations in photochemical air pollutants in New Jersey and New York. Science 186:1037-1038.

Cleveland, W. S., B. Kleiner, J. E. McRae, and J. L. Warner. 1976. Photochemical air pollution: Transport from the New York City area into Connecticut and Massachusetts. Science 191:179-181.

Cochran, P. H. 1978. Response of a pole-size ponderosa pine stand to nitrogen, phosphorus and sulfur. U.S.D.A. Forest Service, Research Note No. PNW-319, Portland, Oregon, 8 pp.

Coffey, P. E., and W. N. Stasiuk. 1975. Evidence of atmospheric transport of ozone into urban areas. Environ. Sci. Technol. 9:59-62.

Focht, D. D., and W. Verstraete. 1977. Biochemical ecology of nitrification and denitrification. Adv. Microb. Ecol. 1:135-214.

Glasson, W. A., and C. S. Tuesday. 1970. Hydrocarbon reactivities in the atmospheric photooxidation of nitric oxide. Environ. Sci. Technol. 4:916-924.

Graedel, T. E., L. A. Farrow, and T. A. Weber. 1977. Photochemistry of the "Sunday Effect." Environ. Sci. Technol. 7:690-694.

Grimsrud, E. P., H. H. Westberg, and R. A. Rasmussen. 1975. Atmospheric reactivity of monoterpene hydrocarbons. NO_x photooxidation and ozonolysis. Int. J. Chem. Kinet. Symp. 1:183-195.

Ingersoll, R. B., R. E. Inman, and W. R. Fisher. 1974. Soils potential as a sink for atmospheric carbon monoxide. Tellus 26:151-158.

Isaksen, I. S. A., Ø. Hov, and E. Hesstvedt. 1978. Ozone generation over rural areas. Environ. Sci. Technol. 12:1279-1284.

Japar, S. M., C. H. Wu, and H. Niki. 1974. Rate constants for the gas phase reaction of ozone with α-pinene and terpinolene. Environ. Lett. 7:245-249.

Jensen, K. F. 1975. Sulfur content of hybrid poplar cuttings fumigated with sulfur dioxide. U.S.D.A. Forest Service Research Note No. NE-209, Broomall, Pennsylvania, 4 pp.

Jensen, K. F., and T. T. Kozlowski. 1975. Absorption and translocation of sulfur dioxide by seedlings of four forest tree species. J. Environ. Qual. 4:379-382.

Kim, C. M. 1973. Influence of vegetation types on the intensity of ammonia and nitrogen dioxide liberation from soil. Soil Biol. Biochem. 5:163-166.

Likens, G. E., F. H. Bormann, R. S. Pierce, J. S. Eaton, and N. M. Johnson. 1977. Biochemistry of a Forested Ecosystem. Springer-Verlag, New York, 146 pp.

Lonneman, W. A., R. L. Seila, and J. J. Bufalini. 1978. Ambient air hydrocarbon concentrations in Florida. Environ. Sci. Technol. 12:459-463.

Maugh, T. H. 1975. Air pollution: Where do hydrocarbons come from? Science 189:277-278.

Maugh, T. H. 1979. SO_2 pollution may be good for plants. Science 205:383.

McLean, D. M. 1978. A terminal Mesozoic "greenhouse": Lessons from the past. Science 201:401-406.

Middleton, J. T. 1967. Air—An essential resource for agriculture. In: N. C. Brady (Ed.), Agriculture and the Quality of Our Environment. Amer. Assoc. Advan. Sci., Publica. No. 85, AAAS, Washington, D.C., pp. 3-9.

Miller, P. R., M. H. McCutchan, and H. D. Milligan. 1972. Oxidant air pollution in the Central Valley, Sierra Nevada Foothills and Mineral King Valley of California. Atmos. Environ. 6:623.

Murphy, C. E., Jr., T. R. Sinclair, and K. R. Knoerr. 1977. An assessment of the use of forests as sinks for the removal of atmospheric sulfur dioxide. J. Environ. Qual. 6:388-396.

National Academy of Sciences. 1977. Ozone and Other Photochemical Oxidants. National Academy of Sciences, Washington, D.C., 717 pp.

Noggle, J. C., and H. C. Jones. 1979. Accumulation of Atmospheric Sulfur by Plants and Sulfur-Supplying Capacity of Soils. U.S. Environmental Protection Agency, Publica. No. 600/7-79-109, Washington, D.C., 37 pp.

Noggle, J. C., H. C. Jones, and J. M. Kelly. 1978. Effects of accumulated nitrates on plants. Paper presented at the 71st Annual Meeting, Air Pollution Control Association, June 25-30, 1978. Houston, Texas, 16 pp.

Nozhevnikova, A. N., and L. N. Yurganov. 1978. Microbiological aspects of regulating the carbon monoxide content in the earth's atmosphere. Adv. Microb. Ecol. 2:203-244.

Persson, R. 1974. World Forest Resources. Review of the World's Forest Resources in the Early 1970's. Research Note No. 17, Royal College of Forestry, Stockholm, Sweden.

Rasmussen, R. A. 1972. What do the hydrocarbons from trees contribute to air pollution? J. Air Pollut. Control Assoc. 22:537-543.

Rasmussen, K. H., M. Taheri, and R. L. Kabel. 1975. Global emissions and natural processes for removal of gaseous pollutants. Water, Air, Soil Pollut. 4:33-64.

Research Triangle Institute. 1973. Investigation of high ozone concentration in the vicinity of Garrett County, Maryland and Preston County, West Virginia. RTI Publication NTIS No. PB-218540, Research Triangle Park, North Carolina.

Richter, H. G. 1970. Special ozone and oxidant measurements in vicinity of Mount Storm, West Virginia. Research Triangle Institute, Research Triangle Park, North Carolina.

Ripperton, L. A., and D. Lillian. 1971. The effect of water vapor on ozone synthesis in the photo-oxidation of alpha-pinene. J. Air Pollut. Control Assoc. 21:629-635.

Robinson, E., and R. C. Robbins. 1971. Emission Concentration and Fate of Particulate Atmospheric Pollutants. Final Report, SRI Project SCC-8507. Stanford Research Institute, Menlo Park, California.

Rogers, H. H., J. C. Campbell, and R. J. Volk. 1979. Nitrogen-15 dioxide uptake and incorporation by *Phaseolus vulgaris* (L.). Science 206:333-335.

Sandberg, J. S., M. J. Basso, and B. A. Okin. 1978. Winter rain and summer ozone: A predictive relationship. Science 200:1051-1054.

Schneider, S. H. 1975. On the carbon dioxide-climate confusion. J. Atmos. Sci. 32:2060-2066.

Seiler, W. 1974. The cycle of atmospheric CO. Tellus 26:116-135.

Spedding, D. J. 1974. Air Pollution. Oxford Chemistry Series. Clarendon Press, Oxford, 76 pp.

Stasiuk, W. N., Jr., and P. E. Coffey. 1974. Rural and urban ozone relationships in New York State. J. Air Pollut. Control Assoc. 24:564-568.

Stephens, E. R., and W. E. Scott. 1962. Relative reactivity of various hydrocarbons in polluted atmospheres. Am. Petroleum Institute Proc. 42:665-670.

Thompson, L. M., and F. R. Troeh. 1978. Soils and Soil Fertility. McGraw-Hill, New York, 516 pp.

Troiano, J. J., and I. A. Leone. 1977. Changes in growth rate and nitrogen content of tomato plants after exposure to NO_2. Phytopathology 67:1130-1133.

U.S. Department of Agriculture. 1964. Sulfur as a Plant Nutrient in the Southern United States. Tech. Bull. No. 1297, Washington, D.C., 45 pp.

U.S. Department of Agriculture, Forest Service. 1978. Forest Statistics of the U.S. 1977. U.S.D.A., Washington, D.C., 133 pp.

U.S. Department of Agriculture, Forest Service. 1979. A Report to Congress on the Nation's Renewable Resources. RPA Assessment and Alternative Program Directions. U.S.D.A., Washington, D.C., 209 pp.

U.S. Environmental Protection Agency. 1976. Open Space as an Air Resource Management Measure. Vol. II. Design Criteria. Publica. No. EPA-450/3-76-028b. U.S.E.P.A., Research Triangle Park, North Carolina.

U.S. Environmental Protection Agency. 1977. National Air Quality and Emissions Trends Report, 1976. U.S.E.P.A. Publica. No. EPA-450/1-77-002, Research Triangle Park, North Carolina.

Wainwright, M. 1978. Distribution of sulphur oxidation products in soils and on *Acer pseudoplatanus* L. growing close to sources of atmospheric pollution. Environ. Pollut. 17:153-160.

Westberg, H. H., and R. A. Rasmussen. 1972. Atmospheric photochemical reactivity of monoterpene hydrocarbons. Chemosphere 4:163-168.

White, W. H., J. A. Anderson, D. L. Blumenthal, R. B. Husar, N. V. Gillani, J. D. Huser, and W. E. Wilson, Jr. 1976. Formation and transport of secondary air pollutants: Ozone and aerosols in the St. Louis urban plume. Science 194: 187-189.

Will, G. M., and C. T. Youngberg. 1978. Sulfur status of some central Oregon pumice soils. Soil Sci. Soc. Am. 42:132-134.

Wolff, G. T., P. J. Lioy, R. E. Meyers. R. T. Cederwall, G. D. Wight, R. E. Pasceri, and R. S. Taylor. 1977. Anatomy of two ozone transport episodes in the Washington, D.C. to Boston, Mass. corridor. Environ. Sci. Technol. 11:506-510.

Wong, C. S. 1978. Atmospheric input of carbon dioxide from burning wood. Science 200:197-200.

Woodwell, G. M. 1978. The carbon dioxide question. Sci. Am. 238:34-43.

Woodwell, G. M., R. H. Whittaker, W. A. Reiners, G. E. Likens, C. C. Delwiche, and D. B. Botkin. 1978. The biota and the world carbon budget. Science 199: 141-146.

Wright, R. D. 1974. Rising atmospheric CO_2 and photosynthesis of San Bernardino Mountain plants. Am. Mid. Natural 91:360-370.

Wright, R. D., and G. M. Woodwell. 1970. Effect of increased CO_2 on carbon fixation by a forest. Brookhaven National Laboratory, Publica. No. 1444, Upton, New York, 7 pp.

SECTION II

FORESTS ARE INFLUENCED BY AIR CONTAMINANTS IN A SUBTLE MANNER—CLASS II INTERACTIONS

7

Forest Tree Reproduction: Influence of Air Pollutants

Sexual reproduction of forest trees is critically important for maintenance of genetic flexibility and the persistence of most species in natural forest communities. Reproductive strategies, however, are typically beset by a variety of "weak points" and reproductive growth of many forest trees is, at best, irregular and quite unpredictable. Generally there is a very good correlation between tree vigor and the capacity for flowering and fruiting (Kozlowski, 1971). A variety of environmental constraints imposes restrictions on tree reproductive processes. Because air contaminants may reduce tree vigor and in view of the fact that numerous potential points of interaction have been identified between air pollutants and reproductive elements (Figure 7-1), it has been hypothesized that air contaminants may impact forest ecosystems by influencing reproductive processes.

A. Pollen Production and Function

Copious quantities of pollen must be produced by trees in order to accomplish the infrequent occurrence of successful gamete transfer. Pollen grains must be transferred in a viable form from stamens to pistils in angiosperms and from staminate to ovulate strobili in gymnosperms. Adequate pollen germination and tube growth must occur in order to realize fertilization of ovaries.

Unfortunately we have little information on the influence of air pollution on the quantity of pollen produced and on pollen distribution under either artificial or natural conditions. As summarized in Chapter 3, pollen production by trees is naturally quite irregular and subject to a variety of environmental influences. This results in considerable variation in the specific time of pollen release from

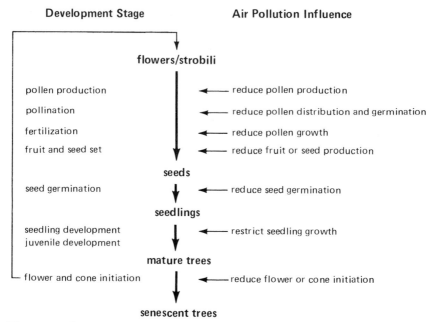

Figure 7-1. Potential points of interaction between air pollutants and sexual reproduction of forest trees.

year to year. Once initiated, however, the process of pollen shed is relatively rapid, ranging from a few hours to a few days (Kozlowski, 1971), and it should be possible to design experiments that would attempt to correlate air quality with pollen yield. Experimental design, however, would have to carefully account for variations in pollen production due to environmental factors others than air pollution.

We are equally uncertain about the influence of air quality on pollen distribution as well as production. In the instance of wind pollinated species, for example gymnosperms, *Populus, Quercus, Fraxinus, Ulmus, Carya,* and *Platanus* species, it is presumed that air contaminants are unimportant in pollen movement. In the instance of tree species depending on animal pollen transfer, however, it has been speculated that air pollutants may impact pollination if they exert an adverse influence on pollinating insects, birds, or other animal vectors. This consideration would apply to the important forest species of the *Acer, Salix,* and *Tilia* genera and, of course, fruit species such as *Malus, Pyrus, Ficus,* and others. Since forest ecosystems with a preponderance of animal-pollinated species are our most species diverse ecosystems (Ostler and Harper, 1978), an adverse impact on animal pollinators might potentially result in competitive weakening and reduction in species diversity.

Bees, by virtue of their ecology and anatomy, constitute the most important group of insect pollinators. Flies and midges follow bees in importance and are in turn followed by a variety of other insect orders including butterflies, moths, thrips, and beetles. Elevated fluoride concentrations in pollinators including bumblebees, honeybees, sphinx moths, and wood nymph butterflies, have been reported in the vicinity of an aluminum reduction plant (Carlson and Dewey, 1971; Dewey, 1973). We are deficient, however, in our understanding of the relationships between air pollutants and these various pollinating insects. Some evidence for toxicity of arsenic, fluoride, sulfur dioxide, and ozone to bees will be presented (Chapter 12). We are also deficient in our appreciation of which specific insect species are most important for forest tree fertilization. Future research must clarify these relationships as well as explore the impact of various pollutants on pollinating insect physiology and ecology (Bromenshenk and Carlson, 1975).

1. Pollen Germination and Tube Elongation

The greatest research effort regarding air contaminants and pollen physiology has dealt with in vitro and some in vivo efforts to assess pollutant influence on germination percentage and pollen tube elongation. Once deposited on the pistil or ovulate cone, the pollen grain must germinate and produce a tube several millimeters in length in order to reach the ovule. These complex processes are adequately reviewed in Heslop-Harrison (1973) and Street and Öpik (1976).

A variety of air contaminants have suppressed pollen function as indicated in Table 7-1. Sulzbach and Pack (1972) recorded reduced pollen grains retained on stigma, pollen germination and pollen tubes reaching ovules when tomato plants were grown in growth chambers and exposed to hydrogen fluoride doses of 4.2 μg m^{-3} for several weeks. This is, of course, an excessive dose, and restricts extrapolation to natural environments. Working with corn in a greenhouse environment, Mumford et al. (1972) employed realistic doses of ozone: 3, 6, and 12 pphm (95, 118, 235 μg m^{-3}) for 5.5 hr day^{-1} for 60 days, to examine various biochemical changes induced in pollen. They concluded that ozone exposure caused autolysis of structural glycoproteins and stimulated amino acid synthesis. Free amino acids of the corn pollen increased 50% following exposure to ozone at 3 pphm (59 μg m^{-3}). The higher doses further enhanced amino acid and peptide accumulation and inhibited germination 40-90%. Ultrastructural examination of petunia pollen, exhibiting reduced germination percentage following ozone exposure to 50 pphm (980 μg m^{-3}) for 3 hr, demonstrated a movement of organelles "away from" the plasma membrane in pollen tubes (Harrison and Feder, 1974). The authors felt this change may have been associated with impaired germination and pollen tube growth. Masaru et al. (1976) collected pollen from greenhouse lilies and exposed the grains to various doses of several air contaminants singly and in combination. Synergistic interaction between various combinations of sulfur dioxide, nitrogen dioxide, ozone, and aldehydes was evident in inhibition of tube elongation. Doses employed were variable but generally related to typical urban ambient levels. Flückiger et al. (1978) monitored tobacco pollen germination and tube growth under ambient atmospheric conditions at the edge, 30 and 200 m from a highway. After 8 hr exposure, germination was inhibited 98% and tube growth 89% at the road edge. At 200 m tube elongation was still reduced by 22%.

Table 7-1. Response of Pollen to Air Pollution under Laboratory and Field Conditions

Plant	Pollutant	Pollen parameter suppressed	Reference
Petunia	Ozone	Germination, tube elongation	Feder (1975)
Tobacco	Ozone	Germination, tube elongation	Feder (1975)
Corn	Ozone	Germination, tube elongation	Feder (1975)
Tomato	Hydrogen fluoride	Germination, tube elongation	Sulzbach and Pack (1972)
Lily	Sulfur dioxide, acrolein, nitrogen dioxide, formaldehyde	Tube elongation	Masaru et al. (1976)
Tobacco	Motor vehicle exhaust (ambient)	Germination, tube elongation	Flückiger et al. (1978)
Cottonwood, red pine, Austrian pine, blue spruce	Sulfur dioxide	Germination, tube elongation	Karnosky and Stairs (1974)
Red pine, white pine	Sulfur dioxide (ambient)	Germination, tube elongation (red pine only)	Houston and Dochinger (1977)
Scotch pine	Sulfur dioxide	Grain size	Mamajev and Shkarlet (1972)
White fir	Sulfur dioxide	Germination	Keller (1976)

Karnosky and Stairs (1974) collected pollen from several forest tree species and observed germination and tube elongation on agar following exposure to sulfur dioxide. Moist quaking aspen pollen germination was reduced at sulfur dioxide concentrations of 0.75 ppm (1965 μg m^{-3}) and above for 4 hr. Highly significant decreases in tube length occurred at 0.30 ppm (786 μg m^{-3}) for 4 hr. A four-hr exposure to 1.4 ppm (3668 μg m^{-3}) sulfur dioxide severely restricted moist pollen germination and tube elongation of red pine, Austrian pine, and blue spruce. The authors cautioned, however, that much of the inhibition of germination and pollen tube elongation may have been due to absorption of sulfur dioxide by the agar media and resulting acidification from pH 7 to 5. The authors speculate that if similar acidification of stigmatic or micropylar tissues occur in nature, low levels of sulfur dioxide may effectively limit seed production. Houston and Dochinger (1977) collected pollen from stands of red and white pines growing in areas of high and low air pollution incidence in central Ohio. The germination percentage of white pine pollen was higher in the material collected from the "cleaner" site. In the case of red pine, both germination percentage and average pollen tube length was greater in pollen gathered from the low pollution incidence region.

B. Flower, Cone, and Seed Production

Working with various nonwoody species, including duckweed, carnation, geranium, and petunia, W. A. Feder and co-workers, University of Massachusetts, Waltham, Massachusetts (Feder and Campbell, 1968; Feder and Sullivan, 1969; Feder, 1970) have shown that ozone at low doses, 10 pphm (196 μg m^{-3}) for 5-7 hr for 1-3 months, can reduce flower production.

One of the earliest observations of decreased seed production by forest tree species in response to air contamination was provided by G. G. Hedgcock in 1912. He recorded a few or no seed borne on conifers close to the Washoe smelter at Anaconda, Montana (Hedgcock 1912). Observations of reduced cone production by ponderosa pine presumably caused by field exposure to ozone (Miller, 1973) and sulfur dioxide (Scheffer and Hedgcock, 1955) have been reported. The latter observations were made near ore smelters at Trail, British Columbia, and Anaconda, Montana. Forest Service data collected in the late 1920s and early 1930s from the Colville National Forest close to smelter effluent revealed sparse cone production by western larch, lodgepole pine, and Douglas fir.

Mamajev and Shkarlet (1972) have reviewed the European literature on the impact of air pollution on cone production and presented their own data concerning Scotch pine response to sulfur dioxide from smelting operations in the Urals. In an area of high, approximately 1-3 μg m^{-3} sulfur dioxide concentration for decades the following reductions in Scotch pine cone dimensions were measured: 16-19% in mature female cone length and 37-50% in weight, 12-15% in seed weight, and 23-32% in staminate cone weights (Table 7-2). Houston and

Dochinger (1977) made similar measurements of cone parameters in their investigation of white and red pine in Ohio (Table 7-3).

Additional evidence for reductions in fruit production by a wide variety of agricultural crops and fruit trees in response to controlled environmental and ambient exposure to hydrogen fluoride and oxidants has been presented (Pack, 1972; Pack and Sulzback, 1976; Thompson and Taylor, 1969; Thompson and Kats, 1975). Thompson and his colleagues of the Air Pollution Research Center in Riverside, California, have demonstrated substantial citrus fruit yield reductions occasioned by photochemical oxidants but reported little effect from ambient fluorine compounds at concentrations varying from 0 to 1.2 μg m^{-3}. Brewer et al. (1960, 1967), on the other hand, obtained reduced yields of orange fruit from trees exposed continuously to 1-5 ppb (0.8-4 μg m^{-3}) hydrogen fluoride gas for a 2-year period and from trees sprayed periodically with hydrogen fluoride or sodium fluoride solutions to achieve foliar concentrations of 75-150 ppm fluoride. The most common response of fruiting of several agricultural species to hydrogen fluoride was the development of fewer seeds (Pack and Sulzbach, 1976). Pepper and corn plants exhibited flower development in this latter study.

Even if seeds are produced and distributed their chances for successful germination and initial development are extremely limited by a wide variety of natural constraints. Paramount among these are water and oxygen supply, temperature, salt, microbial infection, and insect infestation or consumption by rodents, birds, or other animals. Maguire (1972) suggested that air pollution should also be added to this list. Specific evidence relating seed germination and air quality is largely nonexistent. A notable exception is the impact of trace elements, especially heavy metals, on seed germination. Jordan (1975) has thoroughly studied the Lehigh Gap area of Pennsylvania in the vicinity of a smelter complex at Palmerton, Pennsylvania, where zinc ores have been smelted since 1898. Within 2 km of the primary smelter, up to 8% zinc, 1500 ppm cadmium, 1200 ppm copper, and 1100 ppm lead were found at the surface of the A1 soil horizon. As expected, very few tree seedlings were found near the smelters and the author hypothesized that inhibition of seed germination or seedling growth by high

Table 7-2. Reduction in Scotch Pine cone Dimensions and Seed Weight in the Vicinity of Russian Copper Smelters

Parameter	High SO$_2$ incidence	Low SO$_2$ incidence
Ripe average cone length (mm)	32.1-35.6	39.5-42.4
Ripe average cone weight (g)	2.39-3.03	3.81-5.91
1000 seed average weight (g)	4.85-5.72	5.47-6.67
Staminate average cone length (mm)	17-21	22-31

Source: Mamajev and Shkarlet (1972).

levels of soil zinc was responsible. Solution concentrations of up to 100 ppm zinc and 10 ppm cadmium did not affect seed germination of red oak, gray birch, and quaking aspen but these concentrations did preclude radicle elongation. Radicle elongation occurred, but was significantly reduced, at $\geq$ 1 ppm zinc or $\geq$ 5 ppm cadmium in solution culture. Jordan concluded that fire and zinc have interacted to decimate the forest in the smelter vicinity as high soil zinc levels have inhibited sexual reproduction and prevented normal succession of vegetative cover following burning.

C. Seedling Development

Forest trees may be especially vulnerable to air pollution stress in the seedling stage. During this period growth is rapid and gaseous air contaminants may be rapidly absorbed. Above ground organs lack complete protective coatings and are fragile. Epigeous germinating species (most gymnosperms and *Acer, Fagus, Cornus, Robinia*) that push their cotyledons above ground by elongation of the hypocotyl may be more severely injured than hypogeous germinating species (*Quercus, Juglans, Aesculus*) whose cotyledons remain underground while the epicotyl grows upward and develops leaves. All seedlings remain vulnerable, however, over the first 100 days as foliar expansion and development remains rapid. Leaf production by ponderosa pine seedlings during this period may average 1-2 leaves per day (Berlyn, 1972).

Table 7-3. Reduction of White and Red Pine Cone Dimensions and Seed Number, Weight, and Germination in Ohio Areas of High and Low Ambient Air Pollution

Parameter	High pollution incidence		Low pollution incidence	
	White pine	Red pine	White pine	Red pine
Aver. cone length (mm)	122	44	124	47^a
Aver. cone width (mm)	20	23	21	23
Aver. no. seeds cone^{-1}	55		67^a	
Aver. 100 seed wt (g)	1532	0.666	1.850^b	0.805^a
% filled seed	85	50	84	68^a
% seed germination	70	50	70	66^a

Source: Houston and Dochinger (1977).
[a] significant 0.01 level.
[b] significant 0.05 level.

While we do not have abundant data concerning young seedling response to air pollution, we do have several studies that have examined young trees several months or years old. Townsend and Dochinger (1974) studied the relationship between red maple seed source and susceptibility to ozone damage and recorded significant foliar damage during early developmental stages (Table 7-4). Overall leaf injury was greatest in the youngest seedling stages. Fumigation of 2-week-old red pine seedlings in the cotyledon stage at four sulfur dioxide concentrations, 0.5, 1, 3, or 4 ppm (1310, 2620, 7860, or 10,480 μg m^{-3}), and four exposure times (15, 30, 60, or 120 minutes) adversely impacted seedling development (Constantinidou et al., 1976). Sulfur dioxide decreased chlorophyll content and dry weight of both cotyledons and primary needles. The authors observed that the pollutant concentrations employed were high, but indicated that the exposure times, compared with field situations, were short and that they found increasing the time of sulfur dioxide exposure was much more harmful than increasing gas concentration. Because of their results the authors judged that even at low dosages continuous exposure to sulfur dioxide in the field may impact seedling development and restrict regeneration of pine communities. Davis et al. (1977) exposed 2- to 3-year-old black cherry seedlings biweekly to either 0.9 ppm (1764 μg m^{-3}) or 0.10 ppm (196 μg m^{-3}) ozone for 2, 4, 6, and 8 hr. Greatest ozone sensitivity occurred when the foliage was between 4 and 8 weeks old.

Field studies with tree seedlings have yielded more ambiguous results than controlled environment studies. One-year old seedlings of white birch (sulfur dioxide sensitive) and pin oak (sulfur dioxide resistant) were potted and placed in a field location in Akron, Ohio (average annual ambient sulfur dioxide high, 71 μg m^{-3}), and in Delaware, Ohio (average annual ambient sulfur dioxide low, < 10 μg m^{-3}) (Roberts, 1975). After 3 months the overall growth of white birch was greater in the higher, but subphytotoxic environment, and the growth of pin oak was greater at the low sulfur dioxide site. The author suggested that oak stomatal closure in the high sulfur dioxide location may have restricted growth. Two-year old white pine seedlings were field-grown for 5 months in either high or low ambient sulfur dioxide environments in Cleveland and Delaware, Ohio, respectively (Roberts, 1976). The seedlings consisted of susceptible (chlorotic dwarf syndrome) and tolerant clones. Foliar injury and reduced leaf growth were observed on susceptible clones growing in the high (annual average 92 μg m^{-3}) sulfur dioxide environment.

Table 7-4. Average Foliar Damage of Red Maple Seedlings from Four Seed Sources following Ozone Fumigation with 75 pphm (1470 μg m^{-3}) for 7 hr day^{-1} for 3 days

Seed source	% of leaf damaged	Seedling height (cm)
1	6	4
2	34	11
3	18	24
4	22	32

Source: Townsend and Dochinger (1974).

D. Summary

Reproduction of forest trees may be adversely impacted by a variety of air contaminants at numerous points in the reproductive cycle. We do not have sufficient evidence to judge impact on pollen production and distribution. We do, however, have substantial information suggesting inimical changes in the biochemistry and morphology of pollen that may result in reduced pollen germination or reduced pollen tube elongation under field conditions. The net result of the latter would be to reduce seed production.

We have reviewed several papers suggesting reduced cone and fruit production under field conditions. These data must be interpreted with caution, however, as most are observational in nature and in several instances did not appear to account for non-air pollution phenomena that may have also reduced fruit production. In any case, few of these studies evaluated the specific cause of cone reduction, that is whether it was a direct or indirect consequence of air pollution exposure.

Solid evidence has been presented to indicate that selected heavy metals in the forest floor may reduce seed germination. It is probable, however, that this only occurs in those limited environments subject to excessive input from smelters or other major sources. Controlled environment fumigations with ambient and above doses of numerous gaseous contaminants strongly indicate a potential for field damage of very young tree seedlings.

If one or more of these various reproductive stress mechanisms is operative in natural forest ecosystems, it is possible that changes in species composition may ultimately occur. Brandt and Rhoades (1972) investigated the effects of accumulated limestone dust on forest community structure in southwestern Virginia. The species composition and structure of control and dusty sites were intensively studied. The control site had an abundance of shrubs, good reproductive efficiency in all strata, and was assumed to be undergoing normal succession. The dusty site, however, evidenced disruptions in structure and composition resulting from dust accumulation. The authors judged that reproductive efficiency of some species had been decreased and the course of succession altered. Dominance in the tree stratum of the control site by chestnut oak, red oak, and red maple was being altered in the dusty site to ultimately include yellow poplar, sugar maple, and possibly chinkapin oak. In their study of ozone impact on the understory vegetation of an aspen ecosystem, Harward and Treshow (1975) concluded that only 1 or 2 years of ozone exposure might be sufficient to cause shifts in community composition because of seed production responses to ozone exposure.

References

Berlyn, G. P. 1972. Seed germination and morphogenesis. *In*: T. T. Kozlowski (Ed.), Seed Biology. Academic Press, New York, pp. 223-312.

Brandt, C. J., and R. W. Rhoades. 1972. Effects of limestone dust accumulation on composition of a forest community. Environ. Pollut. 3:217-225.

Brewer, R. F., F. H. Sutherland, F. B. Guillemet, and R. K. Creveling. 1960. Some effects of hydrogen fluoride gas on bearing navel orange trees. Proc. Am. Soc. Hort. Sci. 76:208-214.

Brewer, R. F., M. J. Garber, F. B. Guillemet, and F. H. Sutherland. 1967. The effects of accumulated fluoride on yields and fruit quality of "Washington" navel oranges. Proc. Am. Soc. Hort. Sci. 91:150-156.

Bromenshenk, J. J., and C. E. Carlson. 1975. Reduced reproduction—impact on insect pollinators. *In*: W. H. Smith and L. S. Dochinger (Eds.), Air Pollution and Metropolitan Woody Vegetation, U.S.D.A. Forest Service, Publica. No. PIEFR-PA-1, Upper Darby, Pennsylvania, pp. 26-28.

Carlson, C. E., and J. E. Dewey. 1971. Environmental pollution by fluorides in Flathead National Forest and Glacier National Park. U.S.D.A. Forest Service, Div. State and Private Forestry, Missoula, Montana, 57 pp.

Constantinidou, H., T. T. Kozlowski, and K. Jensen. 1976. Effects of sulfur dioxide on *Pinus resinosa* seedlings in the cotyledon stage. J. Environ. Qual. 5:141-144.

Davis, D. D., C. A. Miller, and J. B. Coppolino. 1977. Foliar response of eleven woody species to ozone (O_3) with emphasis on black cherry. Proc. Am. Phytopath. Soc. 4:185.

Dewey, J. E. 1973. Accumulation of fluorides by insects near an emission source in western Montana. Environ. Entomol. 2:179-182.

Feder, W. A. 1970. Plant response to chronic exposure of low levels of oxidant type air pollution. Environ. Pollut. 1:73-79.

Feder, W. A. 1975. Abnormal pollen, flower or seed development. *In*: W. H. Smith and L. S. Dochinger (Eds.), Air Pollution and Metropolitan Woody Vegetation, Pub. No. PIEFR-PA-1, U.S.D.A. Forest Service, Upper Darby, Pennsylvania, pp. 28-30.

Feder, W. A., and F. J. Campbell. 1968. Influence of low levels of ozone on flowering of carnations. Phytopathology 58:1038-1039.

Feder, W. A., and F. Sullivan. 1969. Ozone: Depression of frond multiplication and floral production in duckweed. Science 165:1373-1374.

Flückiger, W., S. Braun, and J. J. Oertli. 1978. Effect of air pollution caused by traffic on germination and tube growth of pollen by *Nicotiana sylvestris*. Environ. Pollut. 16:73-80.

Harrison, B. H., and W. A. Feder. 1974. Ultrastructural changes in pollen exposed to ozone. Phytopathology 64:257-258.

Harward, M., and M. Treshow. 1975. Impact of ozone on the growth and reproduction of understory plants in the aspen zone of western USA. Environ. Conserva. 2:17-23.

Hedgcock, G. G. 1912. Winter-killing and smelter-injury in the forests of Montana. Torrya 12:25-30.

Heslop-Harrison, J. (Ed.). 1973. Pollen: Development and Physiology. Butterworth, London, 338 pp.

Houston, D. B., and L. S. Dochinger. 1977. Effects of ambient air pollution on cone, seed, and pollen characteristics in eastern white and red pines. Environ. Pollut. 12:1-5.

Jordan, M. J. 1975. Effects of zinc smelter emissions and fire on a chestnut-oak woodland. Ecology 56:78-91.

Karnosky, D. F., and G. R. Stairs. 1974. The effects of SO₂ on *in vitro* forest tree pollen germination and tube elongation. J. Environ. Qual. 3:406-409.

Keller, T. 1976. Personal communication and preprint. Internat. Union For. Res. Organ. World Congress, Oslo, Norway, 12 pp.

Kozlowski, T. T. 1971. Growth and Development of Trees. Vol. II. Cambial Growth, Root Growth, and Reproductive Growth. Academic Press, New York, 514 pp.

Maguire, J. D. 1972. Physiological disorders in germinating seeds induced by the environment. *In*: W. Heydecker (Ed.), Seed Ecology, Pennsylvania State Univ. Press, University Park, Pennsylvania, pp. 289-310.

Mamajev, S. A., and O. D. Shkarlet. 1972. Effects of air and soil pollution by industrial waste on the fructification of Scotch pine in the Urals. Mitteil. Forst. Bundes-Versuch. Wien 97:443-450.

Masaru, N., F. Syozo, and K. Saburo. 1976. Effects of exposure to various injurious gases on germination of lily pollen. Environ. Pollut. 11:181-187.

Miller, P. L. 1973. Oxidant-induced community change in a mixed conifer forest. *In*: J. A. Naegele (Ed.), Air Pollution Damage to Vegetation. Adv. Chem. Series No. 122, Amer. Chem. Soc., Washington, D.C., pp. 101-117.

Mumford, R. A., H. Lipke, D. A. Laufer, and W. A. Feder. 1972. Ozone-induced changes in corn pollen. Environ. Sci. Technol. 6:427-430.

Ostler, W. K., and K. T. Harper. 1978. Floral ecology in relation to plant species diversity in the Wasatch Mountains of Utah and Idaho. Ecology 59:848-861.

Pack, M. R. 1972. Response of strawberry fruiting to hydrogen fluoride fumigation. J. Air Pollut. Control Assoc. 22:714-717.

Pack, M. R., and C. W. Sulzbach. 1976. Response of plant fruiting to hydrogen fluoride fumigation. Atmos. Environ. 10:73-81.

Roberts, B. R. 1975. The influence of sulfur dioxide concentration on growth of potted white birch and pin oak seedlings in the field. J. Am. Soc. Hort. Sci. 100:640-642.

Roberts, B. R. 1976. The response of field-grown white pine seedlings to different sulphur dioxide environments. Environ. Pollut. 11:175-180.

Scheffer, T. C., and G. G. Hedgcock. 1955. Injury to Northwestern Forest Trees by Sulfur Dioxide from Smelters. U.S.D.A. Forest Service, Tech. Bull. No. 1117, Washington, D.C., 49 pp.

Street, H. E., and H. Öpik. 1976. The Physiology of Flowering Plants: Their Growth and Development. Elsevier, New York, 280 pp.

Sulzbach, C. W., and M. R. Pack. 1972. Effects of fluoride on pollen germination, pollen tube growth and fruit development in tomato and cucumber. Phytopathology 62:1247-1253.

Thompson, C. R., and G. Kats. 1975. Effects of ambient concentrations of peroxyacetylnitrate on navel orange trees. Environ. Sci. Technol. 9:35-38.

Thompson, C. R., and O. C. Taylor. 1969. Effects of air pollutants on growth, leaf drop, fruit drop, and yield of citrus trees. Environ. Sci. Technol. 3:934-940.

Townsend, A. M., and L. S. Dochinger. 1974. Relationship of seed source and developmental stage to the ozone tolerance of *Acer rubrum* seedlings. Atmos. Environ. 8:957-964.

8
Forest Nutrient Cycling: Influence of Trace Metal Pollutants

Nutrients must move into, within, and out of forest ecosystems in appropriate amounts, at appropriate rates and along established pathways for normal forest growth to occur. The two major sources of nutrients for temperate forest ecosystems are (1) meteorologic input of dissolved, particulate, and gaseous chemicals from outside the ecosystem; and (2) release by weathering of nutrients from primary and secondary minerals stored within the ecosystem (Bormann and Likens, 1979). Healthy forest ecosystems conserve these nutrients and continually recycle them through the system via an elaborate litterfall-decomposition-uptake intrasystem cycle.

The high productivity of forest ecosystems is achieved and maintained through efficient nutrient recycling. For most forest ecosystems essential elements required to maintain productivity cannot be sustained by annual increments from precipitation and mineral substrates alone. Decomposition, mineralization, and reuptake are a must. Root uptake requirements will be met by release from detritus in those temperate forest ecosystems completing litter decay in one year. In cooler temperate regions litter input will exceed decay and humus will accumulate. In warmer temperate forests litter decay may exceed detrital input (Witkamp and Ausmus, 1976).

In the nutrient cycle, primary produced organic matter in the form of litterfall, leachates, root exudates, and sloughage is remineralized by a variety of soil macro- and microorganisms through a series of fractionation and solubilization decomposition steps (Bond et al., 1976; Mason, 1977). Once nutrients are transferred from organic matter to the available nutrient compartment they may again be taken up by forest vegetation.

Air pollutants may influence nutrient dynamics at several points in the cycle. Air contaminants may stimulate the forest by providing additional nutrients via meteorologic input to the system (Chapter 6). Air contaminants may adversely impact the system and reduce growth by altering decomposition rates (Chapter 8), leaching or weathering rates (Chapter 9), and by interfering with symbiotic microbes (Chapter 10). Figure 8-1 provides a summary of temperate forest nutrient cycling and an indication of potential points of air pollution impact.

Forest Soil Organic Matter Decomposition

The release of inorganic nutrients during decomposition of forest soil organic matter is of fundamental and profound importance in the maintenance of intra-system nutrient cycling. It has been estimated that 80-90% of net primary production in terrestrial ecosystems is eventually converted by decomposer organisms (Odum, 1971; Witkamp, 1971; Whittaker, 1975).

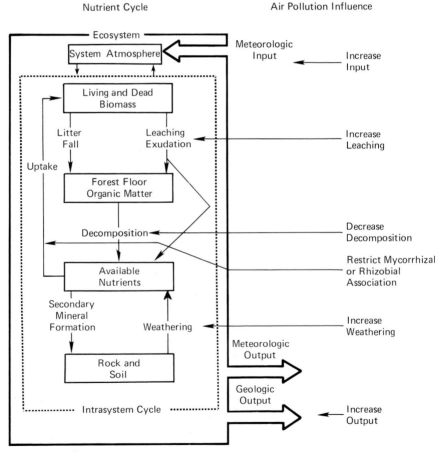

Figure 8-1. Potential points of interaction between nutrient cycling in forest ecosystems and air pollution. (Modified from Bormann and Likens, 1979.)

The primary input to the forest-soil organic-matter compartment is foliar litterfall which is concentrated in the autumn in temperate regions. Cromack and Monk (1975) investigated litter production in a mixed deciduous and white pine forest in North Carolina. The total annual litter production in the hardwood forest was 4369 kg ha^{-1} yr^{-1} of which 2773 kg ha^{-1} yr^{-1} (64%) was leaf litter. Total annual litter production in the young white pine forest was 3253 kg ha^{-1} yr^{-1}, 98% of which was needle litter.

A most informative comparison of nutrient elements stored in forest floor organic matter of eastern United States forests and respective turnover times has been presented by Lang and Forman (1978). Likens et al. (1977) have provided detailed information on the chemistry and amounts of various nutrient inputs to the forest floor for the northern hardwood forest. Mader et al. (1977) have characterized the forest floor for a variety of forests throughout the northeastern United States.

Decreasing turnover rates for element release are correlated with increasing latitude and can be related to general climatic change (Table 8-1). In cold temperate environments the activities of decomposing agents are slow and, as a result, forest floor organic matter accumulation is relatively large. In the warm, humid climates of the lower latitudes, decomposing agent activity is high, decomposition and nutrient element release are fast, and organic matter accumulation is relatively small. Temperate forest soils in regions of moderate rainfall and with well-drained, but nutrient-deficient parent material, undergo podsolization and produce "mor" type organic matter. Deciduous forests developing on more nutrient-rich parent material develop brown forest soils with "mull" type organic matter (Etherington, 1975). Bormann and Likens (1979) have judged that the mor-type forest floor of the New Hampshire northern hardwood forest exerts a major regulating influence on forest development through its capacity for net storage of nutrients and energy. It is not clear if the forest floor of mull-type forest soils exerts a similar regulatory influence of the same magnitude.

It is clear, however, that decomposition and mineralization rate is a critical component of all intrasystem forest nutrient cycles and it has been hypothesized that any suppression or interruption of the decomposition cycle might lead to a decrease in supply of available plant nutrients that will ultimately limit tree growth rates (Bond et al., 1976).

Table 8-1. Annual Production of Total Litter in Relation to Latitude

Latitude (N or S) (degrees)	tons ha^{-1} yr^{-1}	
	Mean	Max
0-10	9-11	13.5
30-40	3.5-8	
40-50	2-4	6
50-60	1.5-3.5	4.5
60-65	0.5-1	

Source: Kühnelt and Walker (1976).

A. Organisms Involved in the Decomposition Process

Decomposition of forest organic matter proceeds in stages and involves a complex variety of soil organisms. Microfloral components with primary roles include bacteria, actinomycetes, and fungi. Animal components with important roles include the following micro- and macrofauna: protozoa, nematodes, earthworms, Enchytraeidae, mollusks, Acari, Collembola, Dipteria, and other arthropods.

Bacteria are the most abundant soil organism with 10^6-10^9 viable cells cm^{-3} but due to their small size (approximately 1 μm) are generally not the major components of soil biomass (Lynch, 1979). They have primary roles in the soil transformations of carbon, nitrogen, phosphorus, iron, and sulfur. Goodfellow and Dawson (1978) have investigated the bacteria colonizing the litter of a mature Sitka spruce stand in Hamsterley Forest, England. Most bacteria were found in the H layer and fewest in the F layer. Over 90% of the 525 isolates identified belonged to the *Arthrobacter, Bacillus, Micrococcus,* and *Streptomyces* genera. Other studies have suggested the importance of the *Achromobacter, Flavobacterium,* and *Pseudomonas* genera.

Fungi are at least as important, perhaps more important in acid forest soils, in the decomposition process as bacteria. A very large number of species are involved with members of the *Aspergillus, Chaetomium, Fusarium, Gliomastix, Memnoniella, Penicillium, Stachybotrys,* and *Trichoderma* genera especially abundant in a variety of ecosystems (Garrett, 1963; Griffin, 1972). The microbial biomass in the forest floor of a black spruce ecosystem averaged 5.7 g m^{-2} in the L and F layers and was composed of 85% fungi and 15% bacteria (weight basis; Flanagan and Van Cleve, 1977).

A high proportion of the protozoan soil population is usually encysted, especially if the soil is dry. Our understanding of the ecology of soil protozoa is extremely incomplete, but a prime importance of these organisms is presumed to be as predators of bacteria.

The micro- and macrofauna are important in the decomposition of organic matter as they reduce the size of litter, improve soil structure, and graze the microflora (Reichle, 1977). Soil animals fragment litter from large to small pieces. The animals also create pore spaces of various dimensions in the soil that allow circulation of air and water. The walls of these same pore spaces provide habitats for bacteria and fungi (Kühnelt and Walker, 1976).

Nematodes are most abundant in well drained and well aerated forest soils. Enchytraeid worms reach maximum numbers in wet, cool, acid soils of high organic matter content. Lumbricid worms (earthworms), because of their large size, are able to move all but the largest soil particles. Mor-type forest floors have extreme horizon differentiation due to minimal mixing by soil animals, particularly earthworms. Mull-type forest floors, on the other hand, generally undergo continual horizon mixing due to the activity of a prolific earthworm fauna.

Termites, springtails (*Collembola*), and oribatid mites (*Acarina*) are commonly abundant in numerous forest soils. Ausmus (1977) has pointed out that one of the most important functions of invertebrates in temperate forest ecosystems is in the regulation of wood mineralization.

Carbon is the energy substrate for decomposition. Heterotrophic metabolism is a very important component of ecosystem metabolism and may range in value from 34 to 57% of the total respiratory carbon flux in the system. Heterotrophic community and microbial respiration may account for approximately 90% (excluding roots) of the total respiration from the soil (Reichle, 1977). Reichle has presented a comprehensive calculation of annual carbon dioxide release by soil decomposers for a yellow poplar (mesic, deciduous) forest (Table 8-2).

B. Measurement of Organic Matter Decomposition Rate and Microbial Biomass

Assessment of impacts on the soil biota commonly involve measurement of metabolic activity (for example, respiration rate), microbial biomass, or both. Unfortunately a prediction of one from the other is difficult as the correlation between the two may not be good under field conditions (Nannipieri et al., 1978). Generally respiration is a better measure of soil metabolic activity than is biomass alone (Reichle, 1977).

The most widely used measure of soil activity is soil respiration either as oxygen "uptake" or carbon dioxide "evolution" (Nannipieri et al., 1978). Other determinations of soil activity may be made by monitoring: mineralization rates of common biopolymers, carbohydrates, or other organic materials; sulfur oxidation; phosphorus solubilization; and activities of certain soil enzymes (Atlas et al., 1978; Lewis et al., 1978).

New strategies for estimating soil biomass include new microscopic observation and counting techniques, soil ATP content, agar-film technique, hexosamine (chitin) assay, and enzyme activity (Frankland and Lindley, 1978; Nannipieri et al., 1978; Todd et al., 1973). All these recent studies support the conclusion of Ausmus (1973) that it is too simplistic to attempt to use only one or two indices as a general means of estimating biomass or metabolism in the soil system.

Under natural conditions, and in the absence of anthropogenic pollutants, the two primary controls of litter decomposition rate are presumed to be the prevailing climatic environment and susceptibility of the substrate to attack by specialized decomposers, that is, substrate quality (Meentemeyer, 1978). Meentemeyer concluded, after examining litter decay rates from locations ranging in climate from subpolar to warm temperate, that actual evapotranspiration and lignin content are useful predictors of litter decomposition rates in unpolluted environments.

Considerable evidence that trace metal pollution of forest floors may reduce litter decomposition rates has been presented.

Table 8-2. Calculation of Annual Carbon Dioxide Respiration by Soil and Litter Invertebrate Decomposers for a Mesic Deciduous Forest (Yellow Poplar)

Decomposer taxon	Mean body weight (mg dw ind.$^{-1}$)	Mean annual biomass (mg dw m^{-2})	O_2 uptake rate at 15 C° (cm^3 O_2 g^{-1} dw day^{-1})	Annual CO_2 efflux population (g CO_2 m^{-2} yr^{-1})	Totals (g CO_2 m^{-2} yr^{-1})
MICROFLORA		124,000		2291	2291
NEMATODA	0.0003	950	90.7	49.43	49.43
PULMONATA	37.98	222.46	9.55	1.22	1.22
ARTHROPODA					139.76
Phalangida	0.30	5.79	24.1	0.080	
Pseudoscorpionida	0.413	9.3	22.6	0.120	
Chilopoda	0.69	32.32	20.5	0.380	
Diplopoda	8.757	249.6	12.6	1.80	
Araneae	0.199	115.5	26.1	1.73	
Acarina					
Gamasina	0.280	2,836.3	24.4	39.7	
Uropodina	0.182	237.7	26.5	3.61	
Oribatei	0.099	3,678.2	29.8	62.9	
Prostigmata	0.004	59.3	55.2	1.87	
Pauropoda	0.006	10.3	51.0	0.30	
Symphyla	0.089	104.1	30.4	1.82	
Protura	0.003	8.3	58.3	0.28	
Diplura	0.081	58.3	31.0	1.03	
Insecta					
Collembola					
Onychiuridae	0.005	37.6	52.9	1.14	
Poduridae	0.008	12.4	48.3	0.34	
Isotomidae	0.017	150.4	41.8	3.60	
Entomobryidae	0.045	295.6	34.7	5.88	

Sminthuridae	0.008	6.3	48.3	0.17
Orthoptera	7.69	26.91	12.9	0.20
Psocoptera	0.068	2.3	23.0	0.04
Coleoptera (larvae)	0.209	134.4	25.8	1.99
Hymenoptera	0.110	16.92	29.2	0.28
Lepidoptera (larvae)	0.103	5.73	29.6	0.10
Diptera (larvae)	0.564	849.6	21.3	10.4	55.6
ANNELIDA				
Enchytraeidae	0.080	500	31.0	8.90
Lumbricidae	118	10,640	7.65	46.7

Source: Reichle (1977).

C. Trace Metal Influence on Litter Dec

The strongest evidence available to support the importance of air pollu to forest nutrient cycling comes from studies that have been concerned with the impact of trace metals on components of the soil biota and the processes they perform. Investigators active in this research area have enjoyed the considerable benefit of an enormous literature concerning the relationship between trace metals and soil organisms, especially microbes. This literature has developed because of the importance certain trace metals have in the nutrition of microorganisms (Perlman, 1949; Weinberg, 1970; Zajic, 1969), the importance of trace metals as pesticidal components for the control of microorganisms (Horsfall, 1956; Somers, 1961), and the relatively recently recognized significance of microorganisms in the environmental metabolism (transformation) of metals (Jernelöv and Martin, 1975; Saxena and Howard, 1977; Summers, 1978).

The extraordinary accumulation of trace metals, particularly lead, cadmium, zinc, and copper, in the organic horizon of the forest floor (Jackson et al., 1978a) as reviewed in Chapter 4, has led to the hypothesis that heavy metals depress decomposition rates. Tyler (1972) has proposed that decomposition of forest litter and remobilization of nutrients will be slower or less complete as heavy metal ions bind with colloidal organic matter and increase resistance to decomposition or exert a toxic effect directly on decomposing microbes or the enzymes they produce.

1. Microbes

Many studies have been conducted that have examined the influence of lead on soil microbes. Doelman (1978) has provided a recent review. He concluded that lead impacts appear greater in sandy soils relative to clay or peat soils, that lead influences are restricted to short periods of time, and that soil functions resist alteration and tend to return to a steady state. He cautioned, however, that return to the steady-state condition may be accomplished with a different microbiota in a changed microhabitat in response to lead amendment. Doelman reviewed several studies that artificially applied lead salts to soils and then monitored impact on fungal or bacterial species. Generally these studies employed lead concentrations ranging from 500 to 5000 ppm. The results were very variable and unfortunately somewhat ambiguous. In polluted soils, bacteria were less sensitive to lead than bacteria from non-lead-polluted soils. Gram-negative bacteria appeared to be less sensitive to lead than Coryneform bacteria. While Doelman did report that the growth rate of some strains of *Arthrobacter globiformis* might be decreased by lead concentrations as low as 1-5 ppm, the more general bacterial threshold of growth impact appeared to be in the range of 2000-5000 ppm lead. Almost no information is available on the in vivo or in vitro influence of lead on actinomycetes. Limited research on fungal reaction to soil lead show variable response but generally support the suggestion that fungi are more resistant to lead influence than bacteria. Jensen (1976) reported that

th 5000 ppm lead nitrate reduced bacterial counts but
.. ~ ~ounts.

.. n...mium, because it is one of the most toxic contaminants introduced into
soil, has also received considerable attention by soil microbiologists and others.
Babich and Stotzky (1978) have provided a most comprehensive review of the
cadmium impact on microbiota. Their summary presents several pertinent generalizations. Cadmium toxicity to microbes appears potentiated at elevated soil
pH which may suggest reduced significance in forest soils. Toxicity of cadmium
to yeasts appears greater in aerobic rather than anaerobic conditions. Evidence
has been presented to show that cadmium can decrease and prolong the logarithmic growth rate of microbes, reduce microbial respiration, inhibit formation
of fungal spores, induce abnormal microbial morphologies, inhibit bacterial transformation, and reduce fungal spore germination. Bond et al. (1976) have employed microcosms to examine the influence of cadmium on the biota of forest
litter. In experiments with Oregon Douglas fir litter amended with up to 10 ppm
cadmium and observed for 4 weeks, they detected no change in bacterial and
fungal densities.

In addition to lead and cadmium, zinc and copper have been implicated in
altered decomposition rates. Despite the fact that these elements are required
nutrients for normal fungal metabolism (Devi, 1962), their extraordinarily high
concentrations in soils adjacent to metal smelters has proved to be inimical to
microbial mineralization.

Jackson and Watson (1977) have studied the influence of lead, cadmium,
zinc, and copper added to hardwood forest soil in the Clark National Forest
from a lead smelter in southeastern Missouri. Annual deposition rates within
0.4 km from the smelter for lead, cadmium, zinc, and copper were 103, 0.72,
6.4, and 2.1 g m^{-2}, respectively. Soil samples were taken at 0.4, 0.8, 1.2, and
2.0 km intervals in line with the prevailing wind direction. Effects on accumulation of litter and the soil biota were not apparent at the 1.2- and 2.0-km
sampling stations. At 0.4 and 0.8 km, however, there was indication of depleted
soil and litter nutrient pools and evidence of depressed decomposer communities and nutrient translocation (Table 8-3). Significant accumulation of 02 litter
(01 litter has original conformation of plant material while 02 litter is fragmented
and unrecognizable to species; after Lutz and Chandler, 1946) was measured at
all sites along the transect relative to the control site (Figure 8-2).

Jordan and Lechavalier (1975) also recorded a significantly greater 02 horizon weight nearby, relative to some distance from, a zinc smelter in Palmerton,
Pennsylvania. Within 2 km of the smelter 13.5% zinc was measured in the 02
horizon and 8% zinc, 1500 ppm cadmium, 1200 ppm copper, and 1100 ppm
lead were found in the A1 soil horizon. The total number of bacteria, actinomycetes, and fungi were counted by dilution plate technique and found to be greatly reduced in the most severely zinc contaminated soils compared with control
soils. The authors judged that reduction of microbial populations may be a partial cause of the decreased rate of litter decomposition observed in the vicinity
of the smelter. The thresholds of zinc toxicity for various microbial groups iso-

Table 8-3. Concentration of Trace Metals, Fungal Biomass, and Standing Oak Tree Foliar Calcium Concentration in the Clark National Forest at Various Distances from a Lead Smelter

| Distance from smelter (km) | Forest floor litter (μg g^{-1}) | | | | Fungal biomass (mg m^{-2}) | Oak leaf calcium concentration (mg g^{-1}) |
	Pb	Cd	Zn	Cu		
0.4	88,349[a]	129[a]	2,189[a]	1,315[a]	0.8[b]	7.6[a]
0.8	30,420[a]	59[a]	917[a]	448[a]	2.4[b]	9.5[a]
1.2	11,872[a]	36[a]	522[a]	183[a]	5.8	11.6
2.0	6,856[a]	21[a]	351[a]	113[a]	16.1	10.4
21.0 (control)	398	2	111	26	22.1	13.0

Source: Jackson and Watson (1977).
[a] Significantly different from control ($P \leq 0.05$).
[b] Significantly different from control ($P \leq 0.1$).

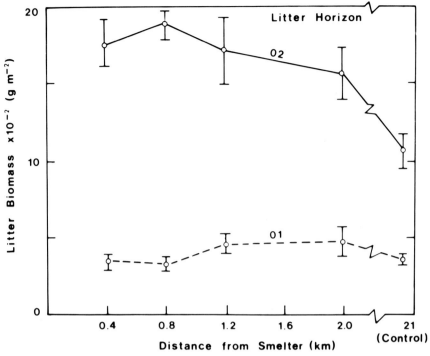

Figure 8-2. Biomass of forest floor litter of the Clark National Forest at various distances from a lead smelter. (From Jackson and Watson, 1977.)

lated from smelter influenced and noninfluenced soils as determined by Jordan and Lechavalier are presented in Table 8-4.

    Williams et al. (1977) compared the litter and soil microbiota from the site of an abandoned Wales metal mine and a control site 500 m distant. Retarded litter decomposition was observed at the mine site. Dilution plate count revealed depressed microbial populations in the soil, but not in the litter (Table 8-5).

Table 8-4. Minimum Zinc Concentration Toxic to Various Microorganisms as Determined on Pablum Extract Agar

Microbe	Isolates from non-Zn-contaminated soil (ppm)	Isolates from Zn-contaminated soil (ppm)
Bacillus spp.	200	100-200
Non-spore-forming bacteria	100-200	600-2000
Actinomycetes	100	600-2000
Fungi	100-1000	100-2000

Source: Jordan and Lechavalier (1975).

Table 8-5. Lead and Zinc Concentrations and Dilution Plate Counts of Litter and Soil from the Site of an Abandoned Mine and a Control Pasture in Wales

	Soil			Litter	
	Mine waste	Control		Mine waste	Control
			$\mu g\ g^{-1}$		
Lead	21,320	274[a]		14,207	172[a]
Zinc	1,273	79[a]		406	84[a]
			number g^{-1} x 10^5		
Bacteria	1.64	169[a]		504	247
Actinomycetes	0.42	70[a]		466	55
Fungi	0.10	69[a]		46	117

Source: Williams et al. (1977).

[a] Means of two sites significantly different ($P \leqslant 0.05$).

2. Soil Animals

Micro- and macrosoil animals are presumed not to play primary roles in the chemical decomposition of soil organic matter. As summarized earlier, however, they perform critically important changes in physical properties in the sequence of organic matter destruction. Fortunately, several investigators have included some of these organisms in their field assessments of trace metal impact.

Arthropod biomass in the 02 litter of the Clark National Forest was significantly reduced in those areas subject to excessive trace metal contamination by a lead smelter (Watson et al., 1976; Jackson and Watson, 1977). Significant reduction of arthropod predators, detritivores, and fungivores were recorded within 0.8 km of the smelter (Table 8-6). Williams et al. (1977) observed that the commonest animals in the litter of an abandoned mine site grossly contaminated with trace metals and a nearby control site were mites (*Acarina*) and springtails (*Collembola*) (Table 8-7). They observed fewer animals in the litter on the mine waste, primarily due to a considerable reduction in the mite population. Collembola numbers were greater on the mine waste litter. Joosse and Buker (1979) also presented evidence supporting the notion that Collembola may be tolerant of trace metal pollution. These investigators fed *Orchesella cincta* green algae contaminated with more than 10,000 ppm lead and analyzed Collembola collected from a roadside environment and concluded that high levels of lead in the food of Collembola had no apparent harmful short-term effect. While they did not specify the contaminants involved, Singh and Tripathi (1978) have recorded fewer soil microarthropods ($28,107\ m^{-2}$) close to as opposed to away from ($39,217\ m^{-2}$) a chemical plant in Varanasi, India. Acarina were the most prevalent group in both the control and polluted sites and comprised approximately one-half the total fauna in both cases.

Table 8-6. Average 02 Litter Biomass of Arthropod Predators, Detritivores, and Fungivores in the Clark National Forest at Various Distances from a Lead Smelter

Distance from smelter (km)	Biomass $(mg\ m^{-2})$		
	Predator	Detritivore	Fungivore
0.4	2.1[a]	2.3[a]	0.8[a]
0.8	6.8[a]	16.6[a]	2.4[a]
1.2	14.0	12.6[a]	5.8
2.0	87.6	92.6	16.1
21.0 (control)	17.3	61.1	22.1

Source: Watson et al. (1976).
[a] Significantly different from control ($P \leqslant 0.1$).

D. Trace Metal Impact on Soil Processes

Considerable evidence to support the hypothesis that trace metal contamination retards decomposition and nutrient recycling has been provided from studies that have measured soil processes in the laboratory or field.

1. Respiration

The level of aerobic respiration is the most widely used measurement for the general biological activity of soils. Changes in oxygen consumption or carbon dioxide evolution have been extensively employed to assess heavy metal impact on the soil biota. Rühling and Tyler (1973) made comparisons of the carbon dioxide evolution rates of different fractions of spruce needle litter from numerous sites around two metal processing industries in central and southeastern Sweden, emitting copper, zinc, cadmium, nickel, vanadium, and lead. Under controlled laboratory conditions highly significant negative correlations were observed between carbon dioxide release and high concentrations of lead, nickel, cadmium, and vanadium. The authors concluded that the decomposition of acid forest litter would be limited by elevated concentrations of these heavy metals during those periods of the year when soil moisture and temperature were not limiting factors. Ebregt and Boldewijm (1977) have also reported a negative correlation between

Table 8-7. Microfauna Extracted from Litter from an Abandoned Mine Contaminated with Trace Metals and a Control Pasture in Wales

Soil animal	Mine waste	Control
Acarina	30	239
Collembola	120	49
Coleoptera	3	7
Total	153	295

Source: Williams et al. (1977).

soil respiration and lead concentration in spruce forest soil. Bhuiya and Corn-field (1972) recorded carbon dioxide release from soils amended with straw and 1000 ppm of various trace metals. Zinc appeared to have no influence on soil respiration, but carbon dioxide evolution was decreased slightly by lead and to a considerable extent by copper and nickel.

The influence of cadmium on forest soil respiration has been investigated by Bond et al. (1976) using Douglas fir litter from the Oregon Coast Mountain Range in microcosms. At various moisture contents 10 ppm cadmium reduced oxygen and carbon dioxide respiration by 40%. At 0.01 ppm amendment, oxygen consumption was stimulated! Spalding (1979) has also examined carbon dioxide evolution from Douglas fir needle litter. In the laboratory mercury, cadmium, lead, nickel, zinc, and copper chlorides were added to the litter at rates of 10, 100, and 1000 $\mu g\ g^{-1}$. At 100 ppm only mercury exhibited an inhibitory effect. At 1000 ppm, however, all metals except lead inhibited respiration.

Several studies have presented contrary evidence and indicated that even relatively high concentrations of trace metals do not appear to reduce respiration in certain soils. Doelman (1978) amended two sandy soils, a clay soil, and a peat soil with lead chloride under controlled environmental conditions. While 2000 ppm lead chloride reduced oxygen consumption in the sandy soils by 50%, the clay soil decrease was only 15%, and the respiration of the peat soil was not inhibited at all. An increase of lead chloride to 10,000 ppm did not influence the respiration of the peat soil. A concentration of 5000 ppm lead did not reduce the carbon dioxide production of a clay soil enriched with glucose and ammonium nitrate in experiments reported by Mikkelsen (1974). This is consistent with the data of Fujihara et al. (1973) who could not detect any inhibition in carbon dioxide release from a silt loam soil enriched with starch and ammonium nitrate and amended with 100 ppm lead.

2. Nitrification

Nitrification is the process of ammonium oxidation to nitrate in soils. It is generally assumed that the microbes primarily responsible are chemoautotrophic bacteria belonging to the *Nitrosomonas* and *Nitrobacter* genera. Because most varieties of these bacteria have pH optima at seven or above, their importance in acid forest soils has been questioned and fungi may be important nitrifying agents in the latter situation (McLaren and Peterson, 1967; Wilde, 1958). Belser (1979) has recently reviewed the wide array of factors that may influence the population size of nitrifying bacteria. There is some evidence to suggest that trace metal contamination of soil should be added to Belser's list. Morissey et al. (1974) amended a fine sandy loam with various concentrations of lead and incubated at 30°C for 6 weeks; at 500 ppm nitrification was initially lowered but recovery occurred in 2 weeks, at 1000 ppm the inhibitory effect disappeared after 6 weeks and it took 10,000 ppm lead to permanently suppress nitrification. Bhuiya and Cornfield (1972) found the autotrophic nitrification process to be more sensitive than the heterotrophic mineralization process. If this is true, forest soils may be more resistant than higher pH soils to altered nitrification occasioned by heavy metals. Tyler et al. (1974) amended a Swedish meadow soil with various moderate concentrations of lead, cadmium, and sodium salts and observed significantly increased nitrification after laboratory incubation for 2-8 weeks. Liang and Tabatabai (1977), however, examined the laboratory response of the nitrification rate of four different soils and observed that amendment with any one of 19 trace elements did not stimulate nitrogen mineralization. Silver, mercury, and cadmium, in fact, inhibited nitrification in all four of the soils tested. In the most acid soil examined, nickel, chromium, iron, aluminum, boron, and tin also inhibited nitrification. Wilson (1977) has presented evidence that nitrification can be significantly inhibited in soils amended with sludge containing high concentrations of lead, cadmium, and zinc. At a concentration of 5 μmoles g^{-1} of soil, trace metals inhibited nitrification in three soils an average of 14 to 96% (Table 8-8) following laboratory incubation for 10 days (Liang and Tabatabai, 1978).

A few recent investigations have employed litter bags and microcosms to examine trace metal influence on soil decomposition processes. These studies supplement laboratory research and strengthen attempts to extrapolate to natural environments. Inman and Parker (1978) examined decomposition rates of black oak, quaking aspen, and starry false Solomon's seal litter in urban and rural ecosystems in northwestern Indiana. Surface soils of the urban site averaged 10, 2456, 463, and 119 ppm of cadmium, zinc, lead, and copper, respectively. These concentrations are appreciably less than those of the smelter studies discussed previously. Black oak litter at the rural site lost twice as much weight as that of the urban site after eleven months. Solomon's seal and quaking aspen litter at the rural site also lost more weight than at the urban site after 6 and 7 months, respectively. While the authors felt that trace metal contami-

Table 8-8. Average Percentage Inhibition of Nitrification in Three Soils by Various Trace Elements at 5 μmoles g^{-1} following 10 Days Incubation under Laboratory Conditions

Element	% inhibition	Element	% inhibition
Mercury	96	Cobalt	45
Silver	93	Copper	45
Selenium	91	Tin	41
Arsenic (III)[a]	83	Iron (II)	40
Chromium	81	Zinc	40
Boron	80	Vanadium	37
Cadmium	77	Arsenic (V)	37
Nickel	64	Tungsten	33
Aluminum	60	Manganese	24
Molybdenum	54	Lead	14
Iron (III)	49		

Source: Liang and Tabatabai (1978).
[a] Oxidation state.

nation, especially by copper and lead, may have been involved in the observed urban depression of decomposition rate, multiple regression analysis did not support this suggestion.

Jackson et al. (1978b) employed a microcosm to evaluate the impact of heavy metals on a forest soil. Baghouse dust from a Missouri smelter was added to a hardwood forest soil to approximate one annual deposition of metals at a distance of 0.4 km from the smelter. At this high level of contamination, export of essential elements in soil leachate was increased as a result of heavy metal impaction. Extractable nutrient pools in soil were also lowered at the end of 20 months. Using soil cores as microcosm analogs, Jackson et al. (1977) observed nutrient efflux from forest cores treated with 100 μg arsenic cm^{-2}. While they detected no disturbance to micropopulations of the soil, they did record a significant increase in the loss of calcium and nitrate nitrogen.

E. Trace Metal Influence on Soil Enzymes

At the turn of the century, the first reports of extracellular soil enzyme activity (catalase) appeared in the literature. Since that time a very large number of enzymes have been shown to participate in important extracellular soil processes. These extracellular activities include not only free extracellular enzymes and enzymes bound to inert soil components but also active enzymes within dead cells and others associated with nonliving cell fragments. While the sources of these enzymes are believed to include animal, plant, and microbial organisms, the latter are generally recognized as the most important source. Additional general information on soil enzymes may be found in the fine review edited by Burns (1978).

In recognition of the interference heavy metals may exert on enzyme activity, the relationship between selected soil enzymes and trace metals has been examined. Metal ions may inhibit enzyme reactions by complexing the substrate, by combining with the active group of the enzymes, or by reacting with the enzyme-substrate complex.

The activity of at least ten forest soil enzymes have been evaluated with respect to their susceptibility to influence by trace metal contaminants in soil (Table 8-9).

1. Dehydrogenase

Dehydrogenases are unspecific oxidative enzymes which are produced by a variety of soil organisms and catalyse the transfer of hydrogen from organic substances to molecular oxygen. Rühling and Tyler (1973) examined dehydrogenase activity in spruce needle litter they collected from various sites around metal smelters in central and southeastern Sweden. Under conditions of laboratory analysis, a strong negative correlation was measured between dehydrogenase activity and high concentrations of nickel, lead, cadmium, and vanadium. The

Table 8-9. Forest Soil Enzymes That Have Been Evaluated for Influence by Trace Metal Soil Contaminants

Enzyme	Reference
Amylase	Spalding (1979), Ebregt and Boldewijm (1977)
Arylsulfatase	Al-Khafaji and Tabatabai (1979)
β-Glucosidase	Tyler (1974), Spalding (1979)
Cellulase	Spalding (1979)
Dehydrogenase	Rühling and Tyler (1973)
Invertase	Spalding (1979)
Phosphatase	Tyler (1974), Tyler (1976), Juma and
Acid phosphatase	and Tabatabai (1977)
Alkaline Phosphatase	
Polyphenoloxidase	Spalding (1979)
Urease	Tyler (1974), Bondietti (1976), Williams et al. (1977)
Xylanase	Spalding (1979)

utility of dehydrogenase measurements as an indicator of general soil biological activity is unclear. Some investigators have found a positive correlation between dehydrogenase activity and either the rate of carbon dioxide production or oxygen uptake while others have not (Burns, 1978).

2. Phosphatase

Phosphatases are important soil enzymes that catalyse the decomposition of organic phosphorus compounds in soil. These latter compounds may constitute 30-70% of the total soil phosphate. Phosphatases apparently arise from widely diverse sources and have different pH optima (Burns, 1978).

Acid phosphatase activity of spruce litter was markedly inhibited by high concentrations of copper and zinc (Tyler, 1974,1976). A highly significant inverse linear relationship between the log of the sum of copper and zinc litter concentration and phosphatase activity was presented (Figure 8-3). Tyler (1976) judged that at equal concentration, copper was more toxic than zinc. Concentrations of vanadium added to spruce litter at 30, 50, 100, and 1000 mg kg^{-1}, inhibited phosphatase activity by 20, 40, 47, and 68%, respectively.

Juma and Tabatabai (1977) studied the influence of 20 trace elements on the activity of acid and alkaline phosphatases from three soils of various physical and chemical properties. The relative effectiveness of trace element inhibition was found to be dependent on the soil and enzyme type. At 25 μmoles g^{-1} soil, the most effective average inhibition of acid phosphatase in the three soils was by mercury, arsenic, tungsten, and molybdenum. The most effective inhibitors of alkaline phosphatase activity in soils were silver, cadmium, vanadium, and arsenic.

As in the case of dehydrogenase, attempts to relate soil phosphatase activity to microbial numbers have yielded contradictory results (Burns, 1978).

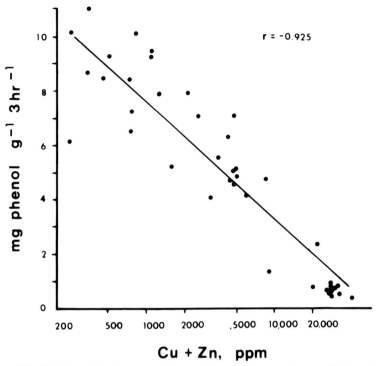

Figure 8-3. Relationship between copper + zinc concentration and phosphatase activity of the mor horizon of a conifer forest surrounding a brass mill in Sweden. (From Tyler, 1976.)

3. β-Glucosidase

β-Glucosidase is an important member of the soil carbohydrases. In his investigation of spruce litter enzymes, Tyler (1974) recorded that β-glucosidase activity was not measurably reduced at concentrations of copper and zinc up to 40 mg g^{-1}. Spalding (1979) measured extractable enzyme activity of Douglas fir needle litter subject to mercury, cadmium, lead, nickel, zinc, and copper chlorides at rates of 10, 100, and 1000 $\mu g\, g^{-1}$ at 1-day, 2-week, and 4-week intervals following treatment. β-Glucosidase activity was elevated at each time interval after treatment with 1000 $\mu g\, g^{-1}$ mercury and after 4 weeks by the 100 $\mu g\, g^{-1}$ mercury amendment. Initially 1000 $\mu g\, g^{-1}$ cadmium stimulated β-glucosidase activity, but this effect was not maintained at 2 and 4 weeks.

4. Urease

Because of the agricultural importance of urease as a decomposing agent for urea which is widely employed as a fertilizer, urease has been the most widely studied soil enzyme. It occurs widely in higher plants and microbes and catalyses the hydrolysis of urea to carbon dioxide and ammonia. The primary source of soil urea is presumed to be microbial (Burns, 1978).

As in the case of acid phosphatase, Tyler (1974) found a highly significant negative correlation between urease activity in spruce litter and high concentrations of copper and zinc. In their comparison of mine waste soils, contaminated with excessive zinc and lead, and uncontaminated pasture soil, Williams et al. (1977) evaluated urease activity. They observed no significant difference in enzyme activity of the two soils. If urea was added to the two soils, however, the ammonium nitrogen released was significantly greater from the pasture relative to the mine waste soil. Even though the Williams study did not include a forest soil, it is pertinent because of the increasing use of urea as a forest fertilizer.

Several studies employing water-soluble metallic salts to determine influence on urease activity generally rank, in decreasing order of impact on the enzyme, as follows (50 ppm soil basis): silver > mercury > gold > copper > cobalt > lead > arsenic > chromium > nickel (Bremner and Mulvaney, 1978).

5. Amylase

Alpha- and β-amylases accumulate in soil and hydrolyse starch. Spalding (1977) has observed significant correlations between carbon dioxide evolution rates from coniferous litter samples and amylase activity. Ebregt and Boldewijm (1977) assessed heavy metal contamination from a brass foundry at Gusum, Sweden, on starch decomposition in spruce litter. A linear, negative correlation was determined between amylase activity and the sum of copper, zinc, lead, and cadmium concentrations of the soil. Although amylase activity exhibited the best correlation with respiration of coniferous leaf litter (Spalding, 1977) it was influenced only by cadmium or lead at $1000 \mu g \, g^{-1}$ after 4 weeks in subsequent studies by this author (Spalding, 1979).

6. Cellulase

Cellulolytic activity appears to be inducible in soils and soils may not have measurable amounts of extracellular cellulases. In Spalding's (1979) study, cellulase activity in Douglas fir litter decreased in samples treated with $1000 \mu g \, g^{-1}$ mercury. Reduced activity was again observed after 4 weeks at $1000 \mu g \, g^{-1}$ of both mercury and cadmium.

7. Xylanase

Xylanase activity, which is strongly correlated with cellulase activity (Spalding, 1977), was depressed in Douglas fir litter after 4 weeks, by mercury at a concentration of $1000 \mu g \, g^{-1}$ (Spalding, 1979).

8. Invertase

Invertase mediates the hydrolysis of glucose to fructose. Invertase activities in soil are not consistently related to microorganisms present in soils (Burns, 1978). Spalding (1977) found that extractable invertase activity was unrelated to respiration rates of coniferous litter. In his trace metal tests with Douglas fir litter, Spalding (1979) found invertase activity severely decreased by mercury at both 100 and 1000 $\mu g\,g^{-1}$ and was sustained for 2 and 4 weeks in the latter treatment. Invertase activity was initially stimulated by 10 or 1000 $\mu g\,g^{-1}$ zinc and cadmium, but this effect was not sustained at 2 and 4 weeks.

9. Arylsulfatase

Arylsulfatase catalyzes the hydrolysis of the arylsulfate anion. Its presence in soils is believed reponsible for sulfur cycling. Al-Khafaji and Tabatabai (1979) studied the influence of 21 trace elements on arylsulfatase activity in four soils. When the trace elements were compared by employing 25 μmoles g^{-1} of soil, the average inhibition was highest with silver, mercury, boron, vanadium, and molybdenum.

It should be clearly kept in mind that the specific abundance and activity of extracellular enzymes in soil are extremely variable and dependent on vegetation, season, depth of sampling, fertilization, and pesticide use (Burns, 1978). Doelman (1978) has further observed that enzymes in soil are known to be more resistant to inactivation by various inhibitory agents than enzymes tested in vitro.

F. Summary

The high productivity of forest ecosystems relative to other terrestrial ecosystems is largely the result of efficient recycling of the elements essential for growth. Forest ecosystems are nutrient element accumulating systems in which elements are continuously recombined in a variety of organic and inorganic compounds. The release of nutrients from organic compounds is accomplished by a large number of soil microbes and animals via a complex series of decomposition and mineralization processes. The rate of litter decomposition appears to control the rate of nutrient release. As nutrients are probably always limiting, the rate of decomposition also controls the rate of primary production in forests (Witkamp and Ausmus, 1976). Any significant reduction in the rate of decomposition, therefore, has the potential to importantly impact forest growth.

Over the last decade a large amount of evidence has been provided to address the hypothesis that trace metal contaminants of the soil reduce forest litter decomposition and mineralization rates. We judge that this hypothesis has been supported by the evidence provided, but only for *excessively* contaminated environments in the immediate vicinity of metal processing industries or other extreme sources of metal contaminants such as some urban and roadside environments. Interference with decomposition processes is certainly a Class II relationship in that its influence would be very subtle at moderate dose levels. It turns out, however, that current evidence can only support its importance in Class III, or high air pollution load, situations.

The thresholds of toxicity for numerous bacterial, fungal, insect, and other components of the soil biota are in the range of 1000 to 10,000 ppm metal cation on a soil dry weight basis. These concentrations are 2 and 3 orders of magnitude greater than the concentrations of heavy metals throughout most temperate forest ecosystems. In addition, fungi probably play a larger role in nutrient cycling in forests than other microbial groups and they appear more resistant to trace metal influence than other microbes. In addition, there is considerable evidence for microbial adaptation to high trace metal exposure. While components of the forest floor soil biota may be impacted by trace metals, decomposition rates may remain unchanged due to adaptation or shifts in species composition unreflected in gross soil process activities.

Many of the investigations reviewed report data from studies that have amended soils or laboratory media with soluble salts containing the trace cations of interest. In natural soils, trace metals may not be contained in soluble compounds and natural availability and hence dissolved concentration may be appreciably less than in experimental designs. The experimental protocols were also seen to frequently employ optimal temperature and moisture conditions for decomposition. In nature, soil climate may impose more severe restrictions on decomposition than metal pollutants.

Future research efforts should concentrate on field observations that introduce as little artificiality as possible. Investigators should examine several parameters, for example, soil biomass, soil process, and soil enzyme concentration or activity, concurrently; they should also consider the influence of gaseous contaminants on decomposition and mineralization processes. Grant et al. (1979) have recently observed that exposure of a forest soil to 1.0 ppm ($2620 \, \mu g \, m^{-3}$) sulfur dioxide reduced the rate of glucose decomposition and that nitrite (at 5 mg N g^{-1} soil) inhibited oxygen consumption and carbon dioxide evolution. Continuous fumigation of an acid soil with excessive sulfur dioxide ($26,200 \, \mu g \, m^{-3}$, 10 ppm) reduced nitrification in the experiments of Labeda and Alexander (1978). These investigators also found that sustained fumigation with $9400 \, \mu g \, m^{-3}$ (5 ppm) nitrogen dioxide inhibited the rate of ammonium disappearance, led to greater rates of nitrate formation, and resulted in nitrite accumulation. Can ambient concentrations of sulfur and nitrogen dioxides adversely impact forest nutrient cycling?

References

Al-Khafaji, A. A., and M. A. Tabatabai. 1979. Effects of trace elements on arylsulfatase activity in soils. Soil Sci. 127:129-133.

Atlas, R. M., D. Pramer, and R. Bartha. 1978. Assessment of pesticide effects on non-target soil microorganisms. Soil Biol. Biochem. 10:231-239.

Ausmus, B. S. 1973. The use of ATP assay in terrestrial decomposition studies. Bull. Ecol. Res. Commun. (Stockholm) 17:223-234.

Ausmus, B. S. 1977. Regulation of wood decomposition rates by arthropod and annelid populations. In: U. Lohm and T. Persson (Eds.), Soil Organisms as Components of Ecosystems. Ecolog. Bull. (Stockholm) 25:180-192.

Babich, H., and G. Stotzky. 1978. Effects of cadmium on the biota: Influence of environmental factors. Adv. Appl. Microbiol. 23:55-117.

Belser, L. W. 1979. Population ecology of nitrifying bacteria. Annu. Rev. Microbiol. 33:309-333.

Bhuiya, M. R. H., and A. H. Cornfield. 1972. Effects of addition of 1000 ppm Cu, Ni, Pb and Zn on carbon dioxide release during incubation of soil alone and after treatment with straw. Environ. Pollut. 3:173-177.

Bond, H., B. Lighthart, R. Shimabuku, and L. Russell. 1976. Some effects of cadmium on coniferous forest soil and litter microcosms. Soil Sci. 121:278-287.

Bondietti, E. A. 1976. Percent amino sugars and urease enzyme activity in litter as a function of distance from the smelter stack on Crooked Creek Watershed. In: R. I. Van Hook and W. D. Shults (Eds.), Ecology and Analysis of Trace Contaminants. Progress Report. Oct. 1974-Dec. 1975. ORNL/NSF/EATC-22 Oak Ridge Nat. Lab., Oak Ridge, Tennessee, p. 102.

Bormann, F. H., and G. E. Likens. 1979. Pattern and Process in a Forested Ecosystem. Springer-Verlag, New York, 253 pp.

Bremner, J. M., and R. L. Mulvaney. 1978. Urease activity in soils. In: R. G. Burns (Ed.), Soil Enzymes. Academic Press, New York, pp. 149-196.

Burns, R. G. 1978. Soil Enzymes. Academic Press, New York, 380 pp.

Cromack, K., Jr., and C. D. Monk. 1975. Litter production, decomposition, and nutrient cycling in a mixed hardwood watershed and a white pine watershed. In: F. G. Howell, J. B. Gentry, and M. H. Smith (Eds.), Mineral Cycling in Southeastern Ecosystems. ERDA Symposium Series No. CONF-740513, pp. 609-624.

Devi, L. S. 1962. Nutritional Requirements of Fungi. University of Madras, Madras, India, 29 pp.

Doelman, P. 1978. Lead and terrestrial microbiota. In: J. O. Nriagu (Ed.), The Biogeochemistry of Lead in the Environment. Part B. Biological Effects. Elsevier-North-Holland Biomedical Press, New York, pp. 343-353.

Ebregt, A., and J. M. A. M. Boldewijm. 1977. Influence of heavy metals in spruce forest soil on amylase activity, CO_2 evolution from starch and soil respiration. Plant Soil 47:137-148.

Etherington, J. R. 1975. Environment and Plant Ecology. Wiley, New York, 347 pp.

Flanagan, P. W., and K. Van Cleve. 1977. Microbial biomass, respiration and nutrient cycling in a black spruce taiga ecosystem. In: V. Lohnn and T. Persson (Eds.), Soil Organisms as Components of Ecosystems. Ecolog. Bull. (Stockholm) 25:261-273.

Frankland, J. C., and D. K Lindley. 1978. A comparison of two methods for the estimation of mycelial biomass in leaf litter. Soil Biol. Biochem. 10:323-333.

Fujihara, M. P., T. R. Gorland, R. E. Wildung, and H. Drucker. 1973. Response of microbiota to the presence of heavy metals in soils. Proc. 1973 Annu. Meet. Amer. Soc. Microbiol.

Garrett, S. D. 1963. Soil Fungi and Soil Fertility. Pergamon Press, New York, 165 pp.

Goodfellow, M., and D. Dawson. 1978. Qualitative and quantitative studies of bacteria colonizing Picea sitchensis litter. Soil Biol. Biochem. 10:303-307.

Grant, I. F., K. Bancroft, and M. Alexander. 1979. SO_2 and NO_2 effects on microbial activity in an acid forest soil. Microbiol. Ecol. 5:85-89.

Griffin, D. M. 1972. Ecology of Soil Fungi. Syracuse Univ. Press, Syracuse, New York, 193 pp.

Horsfall, J. G. 1956. Principles of Fungicidal Action. Chronicá Botanica Co., Waltham, Massachusetts, 279 pp.

Inman, J. C., and G. R. Parker. 1978. Decomposition and heavy metal dynamics of forest litter in northwestern Indiana. Environ. Pollut. 17:39-51.

Jackson, D. R., and A. P. Watson. 1977. Disruption of nutrient pools and transport of heavy metals in a forested watershed near a lead smelter. J. Environ. Qual. 6:331-338.

Jackson, D. R., C. D. Washburne, and B. S. Ausmus. 1977. Loss of Ca and NO_3-N from terrestial microcosms as an indicator of soil pollution. Water, Air, Soil Pollut. 8:279-284.

Jackson, D. R., W. J. Selvidge, and B. S. Ausmus. 1978a. Behavior of heavy metals in forest microcosms: I. Transport and distribution among components. Water, Air, Soil Pollut. 10:3-11.

Jackson, D. R., W. J. Selvidge, and B. S. Ausmus. 1978b. Behavior of heavy metals in forest microcosms: II. Effects on nutrient cycling processes. Water, Air, Soil Pollut. 10:13-18.

Jensen, V. 1976. Effects of lead on biodegradation of hydrocarbons in soil. Oikos 28:220-224.

Jernelöv, A., and A. L. Martin. 1975. Ecological implications of metal metabolism by microorganisms. Annu. Rev. Microbiol. 29:61-77.

Joosse, E. N. G., and J. B. Buker. 1979. Uptake and excretion of lead by litter-dwelling collembola. Environ. Pollut. 18:235-240.

Jordan, M. J., and M. P. Lechevalier. 1975. Effects of zinc-smelter emissions on forest soil microflora. Can. J. Microbiol. 21:1855-1865.

Juma, N. G., and M. A. Tabatabai. 1977. Effects of trace elements on phosphatase activity in soils. Soil Sci. Soc. Am. J. 41:343-346.

Kühnelt, W., and N. Walker. 1976. Soil Biology. Michigan State Univ. Press, East Lansing, Michigan, 483 pp.

Labeda, D. P., and M. Alexander. 1978. Effects of SO_2 and NO_2 on nitrification in soil. J. Environ. Qual. 7:523-526.

Lang, G. E., and R. T. T. Forman. 1978. Detrital dynamics in a mature oak forest: Hutcheson Memorial Forest, New Jersey. Ecology 59:580-595.

Lewis, J. A., G. C. Papavizas, and T. S. Hora. 1978. Effect of some herbicides on microbial activity in soil. Soil Biol. Biochem. 10:137-141.

Liang, C. N., and M. A. Tabatabai. 1977. Effects of trace elements on nitrogen mineralization in soils. Environ. Pollut. 12:141-147.

Liang, C. N., and M. A. Tabatabai. 1978. Effects of trace elements on nitrification in soils. J. Environ. Qual. 7:291-293.

Likens, G. E., F. H. Bormann, R. S. Pierce, J. S. Eaton, and N. M. Johnson. 1977. Biogeochemistry of a Forested Ecosystem. Springer-Verlag, New York, 146 pp.

Lutz, J. H., and R. F. Chandler, Jr. 1946. Forest Soils. Wiley, New York, 514 pp.

Lynch, J. M. 1979. Micro-organisms in their natural environments. The terrestrial environment. In: J. M. Lynch and N. J. Poole (Eds.), Microbial Ecology: A Conceptual Approach. Wiley, New York, pp. 67-91.

Mader, D. L., H. W. Lull, and E. I. Swenson. 1977. Humus Accumulation in Hardwood Stands in the Northeast. Mass. Exp. Sta. Res. Bull. No. 648, Univ. of Massachusetts, Amherst, Massachusetts, 37 pp.

Mason, C. F. 1977. Decomposition. Institute of Biology's, Studies in Biology No. 74, Edward Arnold, London, 58 pp.

McLaren, A. D., and G. H. Peterson. 1967. Soil Biochemistry. Dekker, New York, 509 pp.

Meentemeyer, V. 1978. Macroclimate and lignin control of litter decomposition rates. Ecology 59:465-472.

Mikkelsen, J. P. 1974. Effects of lead on the microbiological activity in soil. Tidster Plant 78:509-516.

Morissey, R. F., E. P. Dugan, and J. S. Koths. 1974. Inhibition of nitrification by incorporation of select heavy metals in soil. Proc. Annu. Meet. Am. Soc. Microbiol. 74:2.

Nannipieri, P., R. L. Johnson, and E. A. Paul. 1978. Criteria for measurement of microbial growth and activity in soil. Soil Biol. Biochem. 10:223-229.

Odum, E. P. 1971. Fundamentals of Ecology. Saunders, Philadelphia, 574 pp.

Perlman, D. 1949. Effects of minor elements on the physiology of fungi. Bot. Rev. 15:195-220.

Riechle, D. E. 1977. The role of soil invertebrates in nutrient cycling. In: U. Lohm and T. Persson (Eds.), Soil Organisms as Components of Ecosystems. Ecolog. Bull. (Stockholm) 25:145-156.

Rühling, Å., and G. Tyler. 1973. Heavy metal pollution and decomposition of spruce needle litter. Oikos 24:402-416.

Saxena, J., and P. H. Howard. 1977. Environmental transformation of alkylated and inorganic forms of certain metals. Adv. Appl. Microbiol. 21:185-226.

Singh, U. R., and B. D. Tripathi. 1978. Effects of industrial effluents on the population density of soil microarthropods. Environ. Conserva. 5:229-231.

Somers, E. 1961. The fungitoxicity of metal ions. Annu. Appl. Biol. 49:246-253.

Spalding, B. P. 1977. Enzymatic activities related to the decomposition of coniferous leaf litter. Soil Sci. Soc. Am. J. 41:622-627.

Spalding, B. P. 1979. Effects of divalent metal chlorides on respiration and extractable enzymatic activities of Douglas-fir needle litter. J. Environ. Qual. 8:105-109.

Summers, A. O. 1978. Microbial transformations of metals. Annu. Rev. Microbiol. 32:637-672.

Todd, R. L., K. Cromack, and J. C. Stormer. 1973. Chemical exploration of the microhabitat by electron probe microanalysis of decomposer organisms. Nature 243:544-546.

Tyler, G. 1972. Heavy metals pollute nature, may reduce productivity. Ambio 1:52-59.

Tyler, G. 1974. Heavy metal pollution and soil enzymatic activity. Plant Soil 41:303-311.

Tyler, G. 1976. Heavy metal pollution, phosphatase activity, and mineralization or organic phosphorus in forest soils. Soil Biol. Biochem. 8:327-332.

Tyler, G., B. Mörnsjö, and B. Nilsson. 1974. Effects of cadmium, lead, and sodium salts on nitrification in a mull soil. Plant Soil 40:237-242.

Watson, A. P., R. I. Van Hook, and D. E. Reichle. 1976. Impact of lead mining smelting complex on the forest-floor litter arthropod fauna in the new lead belt region of southeast, Missouri. Environ. Sci. Div. Publica. No. 881, Oak Ridge National Laboratory, Oak Ridge, Tennessee, 163 pp.

Weinberg, E. D. 1970. Biosynthesis of secondary metabolites: Roles of trace metals. Adv. Microbiol. Physiol. 4:1-44.

Whittaker, R. H. 1975. Communities and Ecosystems. Macmillan, London, 385 pp.

Wilde, S. A. 1958. Forest Soils. Ronald Press, New York, 537 pp.

Williams, S. T., T. McNeilly, and E. M. H. Wellington. 1977. The decomposition of vegetation growing on metal mine waste. Soil Biol. Biochem. 9:271-275.

Wilson, D. O. 1977. Nitrification in soil treated with domestic and industrial sewage sludge. Environ. Pollut. 12:73-82.

Witkamp, M. 1971. Soils as components of ecosystems. Annu. Rev. Ecol. Syst. 2:85-110.

Witkamp, M., and B. S. Ausmus. 1976. Processes in decomposition and nutrient transfer in forest systems. In: J. M. Anderson and A. Macfadyen (Eds.), The Role of Terrestrial and Aquatic Organisms in Decomposition Processes. 17th Symp. Brit. Ecol. Soc. Blackwell, London, pp. 375-376.

Zajic, J. E. 1969. Microbial Biogeochemistry. Academic Press, New York, 345 pp.

9
Forest Nutrient Cycling: Influence of Acid Precipitation

Acid precipitation is defined as rain or snow having a pH of less than 5.6. The pH parameter is a measurement of the difference in hydrogen ion activity between an unknown solution and a standard buffer of assigned pH value. Upon ionization water yields hydrogen and hydroxyl ions. When the activity of these ions is equal, water is neutral and the pH recorded will be 7. At pH values below 7 water becomes increasingly acid, and above 7 increasingly alkaline. In the absence of air pollutants, the pH of precipitation is presumed to be dominated by carbonic acid formed from ambient carbon dioxide, which produces a pH of approximately 5.6-6.0. The pH of precipitation presently falling in North and Central Europe and in the northeastern United States and adjacent portions of Canada is commonly in the range of 3-5.5. Individual storm events have been recorded with pH values between 2.0 and 3.0.

European areas receiving precipitation between pH 4.0 and 4.5 have increased considerably in the past few decades and now comprise substantial portions of northern and central regions (Odén, 1976). Precipitation falling in central New Hampshire in the early 1960s was found to have pH values between 4.0 and 4.5 (Likens et al., 1972) and led to numerous determinations substantiating the widespread occurrence of pH 5.0-5.5 precipitation throughout the United States east of the Mississippi River, and pH 4.0-5.0 in certain northeastern regions, particularly the Adirondacks (Figure 9-1). In western sections of the United States scattered pockets of pH 4.0-5.0 precipitation have been recorded in Tucson, the Los Angeles basin, the San Francisco Bay area, Spokane, and the Willamette Valley-Portland area (Pack, 1980; Rambo, 1978).

The precursors of acid precipitation are presumed to be gaseous sulfur and nitrogen compounds of the atmosphere. The oxidation of sulfur dioxide and

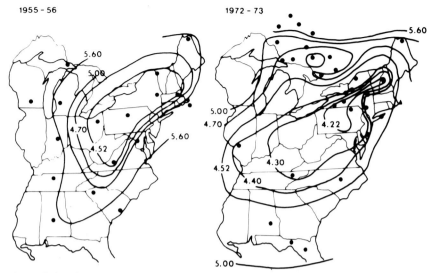

Figure 9-1. The weighted annual average of pH of precipitation in the eastern United States in 1955-1956 and 1972-1973. The solid dots indicate locations of pH sample stations. (From Likens, 1976.)

nitrogen oxides leads to the formation of sulfuric and nitric acids. Analyses of more than fifteen hundred precipitation samples, with a median pH of 4.0, from New York and New Hampshire, revealed that in 80-100% of the cases low pH was attributable to sulfuric and nitric acid (Galloway et al., 1976). In central New Hampshire, Likens (1975) observed that precipitation hydrogen-ion content was 60% due to sulfuric acid, 34% due to nitric acid, and 6% the result of various organic acids. In theory it is possible that acid precipitation precursors may have natural or anthropogenic sources. In the European (Odén, 1976) and United States (Likens, 1976; Hitchcock, 1976) locations of greatest precipitation acidity, however, anthropogenic sources are judged to play a dominant role as air mass trajectories pass over high human emission sources prior to reaching these most seriously impacted regions.

Additional consideration of the general nature of acid precipitation is beyond the scope of this book. Excellent reviews on history (Cogbill, 1975; Rambo, 1978), chemistry and distribution (Husar et al., 1978; Liljestrand and Morgan, 1978; MacCracken, 1978; Perhac, 1978; McColl and Bush, 1978; Wilson, 1978), and general terrestrial and aquatic impacts of acid precipitation (Dochinger and Seliga, 1976; Galloway and Cowling, 1978; Wright and Gjessing, 1976) are available.

For our purposes, the most important aspect of acid precipitation is the potential low pH precipitation has to alter the processes of nutrient cycling (Figure 8-1) in forest ecosystems. It has been hypothesized that acid precipitation may accelerate leaching of nutrients from forest foliage and forest soils and alter weathering rates of forest soil minerals (Likens et al., 1972; U.S.D.A. Forest Service, 1976).

A. Acid Precipitation and Vegetative Leaching

Vegetative leaching refers to the removal of substances from plants by the action of aqueous solutions, such as rain, dew, mist, and fog. Precipitation washout of chemicals from trees has been appreciated for some time (for example, Ovington, 1962). The review of Tukey (1970) has presented numerous, pertinent generalizations. Inorganic chemicals leached from plants include all the essential macro- and microelements. Potassium, calcium, magnesium, and manganese are typically leached in greatest quantities. A variety of organic compounds, including sugars, amino acids, organic acids, hormones, vitamins, pectic and phenolic substances, and others, is also leached from vegetation. As the maturity of leaves increases, susceptibility to nutrient loss via leaching also increases and peaks at senescence. Leaves from healthy plants are more resistant to leaching than leaves that are injured, infected with microbes, infested with insects, or otherwise under stress.

Deciduous trees lose more nutrients from foliage than do coniferous species during the growing season. Conifers, however, continue to lose nutrients throughout the dormant season. The stems and branches of all woody plants lose nutrients during both the growing and dormant seasons. Tamm (1951) has estimated that 2-3 kg ha^{-1} each of potassium, sodium, and calcium were transferred by rain from spruce and pine foliage to the forest floor during 30 days in the autumn. In his studies of the coniferous forests of southern, coastal British Columbia, Feller (1977) has judged that the extent to which precipitation leached nutrients from standing trees decreased in the order: potassium > calcium > sodium > magnesium.

The mechanism of leaching is presumed to be primarily a passive process. Cations are lost from "free space" areas within the plant. Under uncontaminated natural environmental conditions little if any cations are thought to be lost from within cells or cell walls. On the leaf surface it has been demonstrated that leaching of cations involves exchange reactions in which cations on exchange sites of the cuticle are exchanged by hydrogen from leaching solutions. Cations may move directly from the translocation stream within the leaf into the leaching solution by diffusion and mass flow through areas devoid of cuticle (Tukey, 1970). Since acid precipitation may increase hydrogen ion activity by 1-2 orders of magnitude (pH 5.6 to pH 3.6) due to increasing concentrations of sulfuric and nitric acids in precipitation (Galloway and Cowling, 1978) and since damage to cuticles and epidermal cells may also result from exposure to acid precipitation (Tamm and Cowling, 1976) the potential for accelerated leaching under this stress is obvious.

Grennfelt et al. (1978) have very appropriately pointed out that the phrase "acid precipitation" infers that acid in rain and snow is the sole acidifying agent input to ecosystems. They correctly point out that any substance capable of increasing the hydrogen ion activity in the ecosystem is important and that all compounds capable of this should be termed "acidifying substances." For coniferous forests of southern Sweden a comprehensive inventory of all acidifying substances would include all materials listed in Table 9-1. What is the evidence that these acidifying substances enhance forest tree leaching?

Table 9-1. Estimation of Acidifying Substances Deposited to a Coniferous Forest Ecosystem in a Rural Area of Southern Sweden

Substance	Deposition (mmole m^{-2} yr^{-1})	S deposition (kg ha^{-1} yr^{-1})	N deposition (kg ha^{-1} yr^{-1})
Gases			
SO_2	32	10.2	
H_2S	<0.5	<0.2	
NO	<3		<0.4
NO_2	24		3.4
HNO_2	<0.5		<0.1
HNO_3	8		1.1
NH_3	<2		<0.3
Particles			
SO_4^{2-}	9.5	3.0	
NH_4^+	16		2.2
NO_3^-	3.2		0.4
Mist and fog			
SO_4^{2-}	2.5	0.8	
NH_4^+	1.7		0.2
NO_3^-	1.7		0.2
Precipitation			
SO_4^{2-}	30	9.6	
NH_4^+	20		2.8
NO_3^-	20		2.8
Total		23.6-23.8	13.1-13.9

Source: Grennfelt et al. (1978).

One of the few efforts that attempted to quantify increases in tree leaching by acidifying substances was conducted in Norway by Abrahamsen et al. (1976a). These investigators compared throughfall collected in forest ecosystems in southern Norway (higher pollution loads) with systems in northern Norway (lower pollution loads). Throughfall enrichment of sulfate, calcium, and potassium was greater in southern Norway than in northern Norway. The authors judged, however, that conclusive interpretation of their data was impossible, and that it was probable that a larger part of the throughfall enrichment in chloride, sulfate, calcium, and sodium was derived from dry deposition than from leached metabolites.

There is evidence of acidification of throughfall, stemflow, and tree surfaces in natural forests. Baker et al. (1976) have examined forest ecosystems in Alberta, Canada, adjacent to large industrial sources of sulfur dioxide. The latter gas appeared to have an acidifying effect on open rainfall, throughfall, and stemflow near the sources relative to sampling sites more distant from the sources. Grodzinska (1976) was able to find a positive correlation between tree bark pH and regional sulfur dioxide concentration in Poland, and Staxäng (1969) between tree bark pH and regional pollution in Sweden. Grether (1976), on the other hand, was unable to find significant increases in bark acidity of trees in the vicinity of an electric generating facility in Minnesota.

Colleagues at the School of Forestry and Environmental Studies, Yale University, have employed simulated acid rain applied to seedling forest species in order to evaluate the potential for foliar leaching. Field collected first-year sugar maple seedlings were exposed to simulated rain ranging in acidity from pH 5.0 to pH 2.3. Foliar leachate was collected and analyzed for sodium, potassium, magnesium, and calcium cations. As the acidity of the artificial mist increased the loss of cations increased. Significant increases in leaching were measured at pH 3.3 and pH 4.0 where no foliar injury symptoms appeared. At pH 3.0 and below the seedlings were visibly damaged and tissue destruction may have contributed to foliar leaching (Figure 9-2) (Wood and Bormann, 1975). Additional studies have established threshold doses of acid precipitation necessary to cause foliar tissue damage for forest tree seedlings. It is presumed that this damage would increase the loss of cations via leaching (Table 9-2).

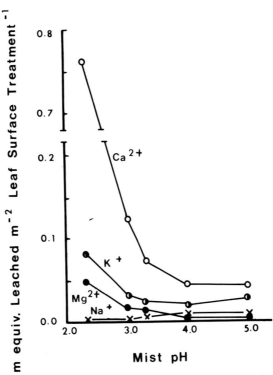

Figure 9-2. Effect of artificial acid mist on the leaching of Na^+, K^+, Mg^{2+}, and Ca^{2+} from first-year sugar maple seedlings. (From Wood and Bormann, 1975.)

Table 9-2. Threshold of Acute Damage of Acid Rain to Forest Tree Seedling Foliage

Species	Simulant pH	Exposure	Period	Reference
Yellow birch	3.0	Intermittent	11 weeks	Wood and Bormann (1974)
White pine	2.3	Intermittent	20 weeks	Wood and Bormann (1976, 1977)
Hybrid poplar	3.4	–	5 days	Evans et al. (1978)

B. Acid Precipitation and Soil Leaching

The chemistry of soil solutions is very variable and depends on a complex series of equilibrium adsorption, displacement, immobilization, weathering, and decomposition reactions well beyond the scope of this book. It is critically important, however, to consider soil leaching as changes in soil solution chemistry have been judged to depend *mainly* on transfers by leaching (Feller, 1977) and leaching rates have been hypothesized to be influenced by acid precipitation.

Most metallic nutrient elements taken up by trees are absorbed as cations and exist in three forms in the soil: (1) as slightly soluble components of mineral or organic material; (2) adsorbed into the cation exchange complex of clay and organic matter; and (3) in small quantities in the soil solution (Etherington, 1975). As water migrates through the soil profile the movement of cations from any of the above three compartments is known as cation desaturation or leaching. An excellent overview of the leaching process and its role in nutrient cycling has been provided by Trudgill (1977).

The relative mobilities of cations leached from decomposing litter are typically sodium > potassium > calcium > magnesium. The leaching rate is largely dependent on the generation of a supply of mobile anions. In forest soils the anions are produced along with hydrogen cations and occur as acids. The hydrogen cations have a very powerful substituting capability and can readily replace other cations adsorbed to the soil. The mobile anions function to move released cations through the soil (Feller, 1977). In forest soils not subject to air pollution, carbonic acid and organic acids produced by biological activity are the primary agents presumed responsible for the replacement and transport of cations. In regions with sufficient moisture cations will be transported from the forest floor in proportion to the availability of HCO_3^- or organic acid anions, or both (Cronan et al., 1978; Feller, 1977; Graustein et al., 1977). In forest soils that are subject to air pollution, it has been proposed that sulfuric and nitric acids may provide the primary or a significant source of H^+ for cation displacement and mobile anions for cation transport. What is the evidence?

1. Lysimeter Studies

Overrein (1972) subjected 40-cm deep forest soil profiles, with differing physical and chemical properties collected from southern Norway, to precipitation with pH levels adjusted from 2.0 to 5.0. When the pH of simulated precipitation was less than 4 he observed a sharp increase in the rate of calcium leaching relative to distilled water controls. At pH 2, soil analyses made over the 40-day experimental period revealed a gradual acidification of the soil starting at the profile top and gradually progressing through the entire column. Mayer and Ulrich (1976) have monitored for several years lysimeters subject to natural precipitation in spruce and beech forests in Central Germany. The hydrogen ion input to the mineral soil of the beech forest was determined to be 2 kiloequivalents ha^{-1} $year^{-1}$ of which 60% was from precipitation input and 40% from mineralization of nitrogen and sulfur compounds via litter decomposition. Aluminum, manganese, sodium, potassium, calcium, and magnesium were lost from the forest. Johnson and Cole (1976) have leached lysimeters established in second-growth Douglas fir in Washington with aliquots of sulfuric acid ranging in concentration from 10 to 1000 times higher (10^{-3}, 10^{-2}, and 10^{-1} N H_2SO_4) than ambient precipitation at present. Cation and sulfate concentrations were monitored for three months after treatment. Approximately two-thirds of the estimated supply of sodium, calcium, and potassium were removed from the forest floor and A horizon but almost none of the cations were moved beyond 50 cm in the B horizon. This restriction may have been due to the limited mobility of sulfate in this horizon and the authors cautioned that failure to consider sulfate mobility could lead to overestimations of cation transfer due to acid precipitation. Abrahamsen et al. (1976b) have recorded decreased cation saturation, mainly due to leaching of calcium and magnesium, in lysimeters containing Norwegian forest soil and subject to 50 mm $month^{-1}$ of simulated rain water of pH 3. Tamm et al. (1976) applied sulfuric acid to the soil of a young pine forest north of Stockholm at a rate equivalent to 100 kg of acid ha^{-1}. The soil appeared relatively resistant to leaching and much of the sulfate ions and cations were retained in the lysimeter soil. The addition of ammonium nitrate to fertilized lysimeters led to an increased leaching of hydrogen ions and cations.

Cronan et al. (1978) have analyzed soil water and groundwater samples in subalpine balsam fir forests in New Hampshire for HCO_3^-, SO_4^{2-}, and organic acid anions in an effort to determine the relative importance of these compounds as sources of H^+ and mobile anions. Their results indicated that sulfate anions supplied 76% of the electrical charge balance in the leaching solutions suggesting that atmospheric input of sulfuric acid provided the primary mechanism of cation replacement and transport in this region. Preliminary analyses of additional field and laboratory data led these authors to suggest that cation leaching in the subalpine New England fir forest increases as the rainfall pH drops below 4-4.5.

Pots containing eastern white pine seedlings were subjected to one weekly six-hour application of artificial acid rain adjusted to pH 5.6, 4.0, 3.3, 3.0, and 2.3 for a 20-week period (Wood and Bormann, 1977). Leaching of magnesium and calcium steadily increased with "rain" acidity through the pH range 5.6 to

2.3. Potassium losses did not increase until the pH was lowered to 3.0 or 2.3 (Figure 9-3). Declines in exchangeable potassium, magnesium, and calcium were observed at pH levels 3.0 and below.

2. Forest Studies

Baker et al. (1976) have studied the impact of sulfur dioxide release from natural gas treatment plants in Alberta, Canada, on the soils of the surrounding forest ecosystem. Soils collected near (within 18 km or less) and far (32-78 km) from the point source were compared for exchangeable cations (Table 9-3). While the differences in cation characteristics between the sites were not extreme, the decreased calcium and increased aluminum close to the source are significant. Aluminum ions rapidly react with water to form insoluble aluminum hydroxide and hydrogen ions (Etherington, 1975). This highlights the difficulty of assigning relative importance to the various sources of hydrogen ions in acid forest soils subject to air pollution stress. Exchangeable calcium and magnesium concentrations were relatively low in soils of lowland forests in southeastern Ohio most affected by acid mine drainage, but were significantly higher in soils less affected by this drainage (Cribbin and Scacchetti, 1976). The authors judged that altered nutrient status along with high acidity and aluminum concentrations of the soil exerted strong selection pressures in certain Ohio river bottom forest communities.

Cronan and Schofield (1979) have recently pointed out an additional aspect of soil aluminum chemistry in fir zone soils in New England that may be important in other noncalcareous forest ecosystems subject to acid precipitation. In most podzolized soils, acidification of the soil solution and organometallic com-

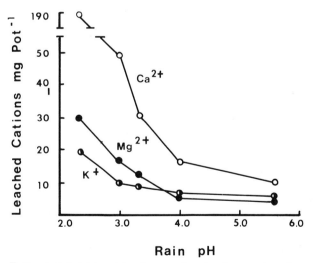

Figure 9-3. Estimated total leaching losses of potassium, magnesium, and calcium from pots containing white pine seedlings following 6-hour weekly treatments of artificial rain of various pH's for 20 weeks. (From Wood and Bormann, 1977.)

Table 9-3. pH and Exchangeable Cations in 1 N KCl Extracts of Forest Soils Sampled Close to and Far from a Sulfur Dioxide Point Source in Alberta, Canada

Location	Horizon	pH	Al	Ca	Mg	Fe	Mn	Total bases
				Cations (meq 100 g^{-1})				
Distant sites	L-F-H	6.0	1.1	47.0	7.5	0.3	0.1	54.9
(mean)	0-5 cm	5.2	0.5	10.6	1.9	nil	nil	12.5
	5-15 cm	5.1	0.7	7.7	1.4	nil	nil	9.1
	15-30 cm	5.0	0.5	9.4	2.0	nil	nil	11.4
Near sites	L-F-H	4.5	1.5	29.4	5.2	0.6	1.6	36.8
(mean)	0-5 cm	4.2	3.2	8.5	1.8	0.1	0.1	10.5
	5-15 cm	4.4	2.2	6.4	1.5	0.1	nil	8.0
	15-30 cm	4.4	1.7	8.6	2.2	nil	nil	10.8

Source: Baker et al. (1976).

plexation mediated by organic acids results in movement of aluminum and iron into lower horizons. In New Hampshire balsam fir forests, however, the soils exhibited limited increases in soil solution pH with depth and amorphous and exchangeable aluminum were mobilized and tended to be transported in solution out of the profile and ultimately into streams.

C. Acid Precipitation and Soil Weathering

Chemical weathering involves hydrolysis, hydration, oxidation-reduction reactions, carbonation, and solution of compounds and elements from parent soil material (Frink and Voigt, 1976). Numerous reactants participate in weathering processes. Commonly the most significant reactant is the hydrogen ion. The latter in solution reacts with a cation in a mineral and replaces it in the crystal lattice of the mineral. Soil acids are the primary source of hydrogen ions for weathering (Trudgill, 1977). The hypothesis that mineral acids resulting from air pollution can importantly supplement weathering reactions principally due to carbonic acid and organic acids is not supported by substantial evidence. In fact in the forest ecosystems in the United States known to have soil leaching processes dominated by mineral acids, the New England region, the role of sulfuric and nitric acids in chemical weathering has been judged to be relatively small (Johnson et al., 1972; Johnson, 1979). By including biomass accumulation in the estimation of cationic denudation rates for New Hampshire forests, however, the values of Johnson et al. (1972) are doubled and approximate 2.0 $\times$ 10^3 equivalents ha^{-1} yr^{-1} (Likens et al., 1977). The latter authors have estimated that this cationic denudation is balanced by H$^+$ ions produced in equal amounts by internal generation (chemical reactions of soil nitrogen, carbon, and sulfur) and external generation (meteorologic input of mineral acids). This infers considerable importance of weathering associated with acid precipitation. Unique characteristics of biogeochemical cycling associated with upland New England forest systems and the unusually high input of acid precipitation in this region preclude extrapolation of the circumstance of this location to the more general temperate forest situation.

In their pot trials of seedling white pine subjected to artificial precipitation acidified to various pH levels, Wood and Bormann (1977) observed that the treatment appeared to accelerate the weathering rate of their sandy loam greenhouse soil. Estimated weathering inputs of potassium and magnesium increased linearly with logarithmic increases in the acidity of the simulated rain. Three- to fourfold increases coincided with 1500-fold increases in acidity from pH 5.6 to 2.3. The weathering of calcium responded more logarithmically with an approximate 22-fold increase over the same pH range.

D. Summary

Anthropogenic release of sulfur and nitrogen oxides has increased the acidity of precipitation 10 to 100 times preindustrial levels. In certain sections of the world, most notably the northeastern United States, northern Europe, and Scandinavia, large areas of forest are currently subjected to precipitation of dramatically reduced pH.

The data reviewed to support the hypothesis that acid precipitation accelerates leaching loss of nutrients from forest foliage are not convincing. The limited analyses of throughfall in natural environments have not adequately differentiated material actually leached from the plants from material deposited on and subsequently washed from the foliage. Studies that have employed seedling trees subjected to artificially acidified rain have revealed a potential for foliar nutrient loss but only at pH levels of 4.0 or less. In the presence of foliar damage to the cuticle occasioned by acid rain, nutrient loss from leaves could be substantial. The threshold for this damage, however, appears to be approximately pH 3 for numerous forest trees and this intensity of precipitation acidification is not widespread in natural environments at the present time.

The movement of nutrient cations via leaching in forest soil profiles is an extremely important component of forest nutrient cycling. The evidence that has been provided by numerous experiments subjecting soil lysimeters to natural or artificially acidified precipitation indicates a potential for meaningful acid precipitation influence on the soil leaching process. The threshold for significant increases in the rate of movement of calcium, potassium, and magnesium appears to require precipitation in the pH range of 3-4 for most systems examined. For certain forests, for example, subalpine balsam fir in New England, the threshold of increased leaching may be higher and in the range of pH 4.0-4.5. Lysimeter data reflects an integration of many complex soil processes. As a result, this evidence does not provide answers to numerous important questions concerning acid rain impact on leaching. What are the relative efficiencies of cation ion transfer between the forest floor and the A soil horizon and between the A and B soil horizons? What are the specific sources of H^+ ions in the soil profile?

In summarizing their field evidence obtained from experimentation with coniferous forests in southern Norway, Abrahamsen et al. (1976b) have concluded that increasing rain acidity does cause increased foliar leaching of calcium, magnesium, and potassium. They further concluded that increasing acidity (pH 6 to 2) of simulated acid rain reduced the pH, the content of exchangeable cations, and the base saturation in the top layers of podzol soils.

An important aspect of soil leaching in areas subject to acid precipitation is the possibility of increased concentrations of soluble aluminum in the soil profile. What is the influence of this aluminum on components of the soil biota? Can aluminum concentrations directly or indirectly toxic to tree roots be reached?

The possibility of enhanced cation availability in, or loss from, soils due to increased weathering of parent soil materials by hydrogen ions from acid precipitation is not supported by evidence presently available.

Forest soils have been judged more vulnerable to influence by acid precipitation than agricultural soils. Forest soils supporting early regeneration following harvest or severe natural events may be especially vulnerable to adverse impact on nutrient cycling by acid rain as the system "controls" on nutrient conservation are weakest at this time (Tamm, 1976; Likens et al., 1977). The bulk of the current evidence, however, is consistent with the conclusion of Frink and Voigt (1976) that unless the acidity of precipitation increases substantially or the buffering capacity of forest soils declines significantly, acid rain influence will not quickly nor dramatically alter the productivity of most temperate forest soils.

References

Abrahamsen, G., R. Horntveldt, and B. Tveite. 1976a. Impacts of acid precipitation on coniferous forest ecosystems. In: L. S. Dochinger and T. A. Seliga (Eds.), Proc. 1st Internat. Symp. Acid Precipitation and the Forest Ecosystem. U.S.D.A. Forest Service, Gen. Tech. Rep. No. NE-23, Upper Darby, Pennsylvania, pp. 991-1009.

Abrahamsen, G., K. Bjor, R. Horntveldt, and B. Tveite. 1976b. Effects of acid precipitation on coniferous forests. In: F. H. Braeke (Ed.), Impact of Acid Precipitation on Forest and Freshwater Ecosystems in Norway. Research Report No. 6. SNF Project, Olso, Norway, pp. 37-63.

Baker, J., D. Hocking, and M. Nyborg. 1976. Acidity of open and intercepted precipitation in forests and effects on forest soils in Alberta, Canada. In: L. S. Dochinger and T. A. Seliga (Eds.), Proc. 1st Internat. Symp. Acid Precipitation and the Forest Ecosystem. U.S.D.A. Forest Service, Gen. Tech. Rep. No. NE-23, Upper Darby, Pennsylvania, pp. 779-790.

Cogbill, C. V. 1975. The history and character of acid precipitation in eastern North America. In: L. S. Dochinger and T. A. Seliga (Eds.), Proc. 1st Internat. Symp. Acid Precipitation and the Forest Ecosystem. U.S.D.A. Forest Service, Gen. Tech. Rep. No. NE-23, Upper Darby, Pennsylvania, pp. 363-370.

Cribbin, L. D., and D. D. Scacchetti. 1976. Diversity in tree species in southeastern Ohio Betula nigra L. communities. In: L. S. Dochinger and T. A. Seliga (Eds.), 1st Internat. Symp. Acid Precipitation and the Forest Ecosystem, U.S.D.A. Forest Service, Gen. Tech. Rep. No. NE-23, Upper Darby, Pennsylvania, pp. 779-790.

Cronan, C. S., W. A. Reiners, R. C. Reynolds, Jr., and G. E. Lang. 1978. Forest floor leaching: Contributions from mineral, organic and carbonic acids in New Hampshire subalpine forests. Science 200:309-311.

Cronan, C. S., and C. L. Schofield. 1979. Aluminum leaching response to acid precipitation: Effects on high-elevation watersheds in the Northeast. Science 204:304-306.

Dochinger, L. S., and T. A. Seliga (Eds.). 1976. Proc. 1st International Symposium on Acid Precipitation and the Forest Ecosystem. U.S.D.A. Forest Service, Gen. Tech. Rep. No. NE-23, Upper Darby, Pennsylvania, 1074 pp.

Etherington, J. R. 1975. Environment and Plant Ecology. Wiley, New York, 347 pp.

Evans, L. S., N. F. Gmur, and F. DaCosta. 1978. Foliar response of six clones of hybrid poplar. Phytopathology 68:847-856.

Feller, M. C. 1977. Nutrient movement through western hemlock-western red cedar ecosystems in southwestern British Columbia. Ecology 58:1269-1283.

Frink, C. R., and G. K. Voigt. 1976. Potential effects of acid precipitation on soils in the humid temperate zone. *In*: L. S. Dochinger and T. A. Seliga (Eds.), 1st Internat. Symp. Acid Precipitation and the Forest Ecosystem, U.S.D.A. Forest Service, Gen. Tech. Rep. No. NE-23, Upper Darby, Pennsylvania, pp. 685-709.

Galloway, J. N., and E. B. Cowling. 1978. The effects of precipitation on aquatic and terrestrial ecosystems. A proposed precipitation chemistry network. J. Air Pollut. Control Assoc. 28:229-235.

Galloway, J. N., G. E. Likens, and E. S. Edgerton. 1976. Acid precipitation in the Northeastern United States: pH and acidity. Science 194:722-724.

Graustein, W. C., K. Cromack, Jr., and P. Sollins. 1977. Calcium oxalate: Occurrence in soils and effect on nutrient and geochemical cycles. Science 198:1252-1254.

Grennfelt, P., C. Bengtson, and L. Skärby. 1978. An estimation of the atmospheric input of acidifying substances to a forest ecosystem. Swedish Water and Air Pollution Res. Instit. No. B438, Gothenburg, Sweden, 12 pp.

Grether, D. F. 1976. The effects of a high-stack coal-burning power plant on the relative pH of the superficial bark of hardwood trees. *In*: L. S. Dochinger and T. A. Seliga (Eds.), Proc. 1st Internat. Symp. Acid Precipitation and the Forest Ecosystem. U.S.D.A. Forest Service, Gen. Tech. Rep. No. NE-23, Upper Darby, Pennsylvania, pp. 913-918.

Grodzinska, K. 1976. Acidity of tree bark as a bioindicator of forest pollution in southern Poland. *In*: L. S. Dochinger and T. A. Seliga (Eds.), Proc. 1st Internat. Symp. Acid Precipitation and the Forest Ecosystem. U.S.D.A. Forest Service, Gen. Tech. Rep. No. NE-23, Upper Darby, Pennsylvania, pp. 905-911.

Hitchcock, D. R. 1976. Atmospheric sulfates from biological sources. J. Air Pollut. Control Assoc. 26:210-215.

Husar, R. B., J. P. Lodge, Jr., and D. J. Moore. 1978. Sulfur in the Atmosphere. Proc. Internat. Symp., Dubrovnik, Yugoslavia, 7-14 Sept. 1977. Atmos. Environ. 12:1-796.

Johnson, D. W., and D. W. Cole. 1976. Sulfate mobility in an outwash soil in western Washington. *In*: L. S. Dochinger and T. A. Seliga (Eds.), 1st Internat.

Symp. Acid Precipitation and the Forest Ecosystem. U.S.D.A. Forest Service, Gen. Tech. Rep. No. NE-23, Upper Darby, Pennsylvania, pp. 827-835.

Johnson, N. M. 1979. Acid rain: Neutralization within the Hubbard Brook ecosystem and regional implications. Science 204:497-499.

Johnson, N. M., R. C. Reynolds, and G. E. Likens. 1972. Atmospheric sulfur: Its effect on the chemical weathering of New England. Science 177:514-515.

Likens, G. E. 1975. Acid precipitation: Our understanding of the phenomenon. Proc. Conf. Emerging Environmental Problems: Acid Precipitation, May 1975, Renssalaerville, N. Y. EPA-902/9-75-001. U.S. Environmental Protection Agency, New York, 115 pp.

Likens, G. E. 1976. Acid Precipitation. Chem. Eng. News 54:29-44.

Likens, G. E., F. H. Bormann, and N. M. Johnson. 1972. Acid rain. Environment 14:33-40.

Likens, G. E., F. H. Bormann, R. S. Pierce, J. S. Eaton, and N. M. Johnson. 1977. Biogeochemistry of a Forested Ecosystem. Springer-Verlag, New York, 146 pp.

Liljestrand, H. M., and J. J. Morgan. 1978. Chemical composition of acid precipitation in Pasadena, California. Environ. Sci. Technol. 12:1271-1273.

MacCracken, M. C. 1978. MAP3S: An investigation of atmospheric energy related pollutants in the northeastern United States. Atmos. Environ. 12:649-660.

Mayer, R., and B. Ulrich. 1976. Acidity of precipitation as influenced by the filtering of atmospheric sulfur and nitrogen compounds—its role in the element balance and effect on soil. In: L. S. Dochinger and T. A. Seliga (Eds.), 1st Internat. Symp. Acid Precipitation and the Forest Ecosystem. U.S.D.A. Forest Service, Gen. Tech. Rep. No. NE-23, Upper Darby, Pennsylvania, pp. 737-743.

McColl, J. G., and D. S. Bush. 1978. Precipitation and throughfall chemistry in the San Francisco Bay area. J. Environ. Qual. 7:352-357.

Odén, S. 1976. The acidity problem—An outline of concepts. In: L. S. Dochinger and T. A. Seliga (Eds.), Proc. 1st Internat. Symp. Acid Precipitation and the Forest Ecosystem. U.S.D.A. Forest Service, Gen. Tech. Rep. No. NE-23, Upper Darby, Pennsylvania, pp. 1-36.

Overrein, L. N. 1972. Sulfur pollution patterns observed; leaching of calcium in forest soil determined. Ambio 1:145-147.

Ovington, J. D. 1962. Quantitative ecology and the woodland ecosystem concept. Adv. Ecol. Res. 1:103-192.

Pack, D. H. 1980. Precipitation chemistry patterns: A two-network data set. Science 208:1143-1145.

Perhac, R. M. 1978. Sulfate regional experiment in the northeastern United States: The SURE program. Atmos. Environ. 12:641-648.

Rambo, D. L. 1978. Interim Report: Acid precipitation in the United States, history, extent, sources, prognoses. U.S. Environmental Protection Agency, Contract No. 68-03-2650. Corvallis, Oregon, 24 pp.

Staxäng, B. 1969 Acidification of bark of some deciduous trees. Oikos 20:224-230.

Tamm, C. O. 1951. Removal of plant nutrients from tree crowns by rain. Physiol. Plant 4:184-188.

Tamm, C. O. 1976. Acid precipitation and forest soils. In: L. S. Dochinger and T. A. Seliga (Eds.), Proc. 1st Internat. Symp. Acid Precipitation and the Forest Ecosystem. U.S.D.A. Forest Service, Gen. Tech. Rep. No. NE-23, Upper Darby, Pennsylvania, pp. 681-683.

Tamm, C. O., and E. B. Cowling. 1976. Acidic precipitation and forest vege-
tation. *In*: L. S. Dochinger and T. A. Seliga (Eds.), Proc. 1st Internat. Symp.
Acid Precipitation and the Forest Ecosystem. U.S.D.A. Forest Service, Gen.
Tech. Rep. No. NE-23, Upper Darby, Pennsylvania, pp. 845-855.

Tamm, C. O., G. Wiklander, and B. Popovic. 1976. Effects of application of sul-
phuric acid to poor pine forests. *In*: L. S. Dochinger and T. A. Seliga (Eds.),
1st Internat. Symp. Acid Precipitation and the Forest Ecosystem. U.S.D.A.
Forest Service, Gen. Tech. Rep. No. NE-23, Upper Darby, Pennsylvania, pp.
1011-1024.

Trudgill, S. T. 1977. Soil and Vegetation Systems. Clarendon Press, Oxford, 180
pp.

Tukey, H. B. Jr. 1970. The leaching of substances from plants. Annu. Rev. Pl.
Physiol. 21:305-324.

U.S.D.A. Forest Service. 1976. Workshop report on acid precipitation and the
forest ecosystem. U.S.D.A. Forest Service, Gen. Tech. Rep. No. NE-26, U.S.
D.A. Forest Service, Upper Darby, Pennsylvania, 18 pp.

Wilson, W. E. 1978. Sulfates in the atmosphere: A progress report on project
MISTT. Atmos. Environ. 12:537-548.

Wood, T., and F. H. Bormann. 1974. The effects of an artificial acid mist upon
the growth of *Betula alleghaniensis* Britt. Environ. Pollut. 7:259-268.

Wood, T., and F. H. Bormann. 1975. Increases of foliar leaching caused by acidification of an artificial mist. Ambio 4:169-171.

Wood, T., and F. H. Bormann. 1976. Short-term effects of a simulated acid rain
upon the growth and nutrient relations of *Pinus strobus* L. *In*: L. S. Dochinger
and T. A. Seliga (Eds.), Proc. 1st Internat. Symp. Acid Precipitation and the
Forest Ecosystem. U.S.D.A. Forest Service, Gen. Tech. Rep. No. NE-23,
Upper Darby, Pennsylvania, pp. 815-825.

Wood, T., and F. H. Bormann. 1977. Short-term effects of a simulated acid rain
upon the growth and nutrient relations of *Pinus strobus* L. Water, Air, Soil
Pollut. 7:479-488.

Wright, R. F., and E. T. Gjessing. 1976. Acid precipitation: Changes in the chemical composition of lakes. Ambio 5:219-223.

10

Forest Nutrient Cycling: Influence of Air Pollutants on Symbiotic Microorganisms

Symbiotic microorganisms have roles of very great importance in nutrient relations in forest ecosystems. Forests frequently flourish in regions of low, marginal, or poor soil nutrient status. In addition to nutrient conservation and tight control over nutrient cycling, trees have evolved critically significant symbiotic relationships with soil fungi and bacteria that enhance nutrient supply and uptake. The interaction between air contaminants, symbiotic microbes, and their relationship with host trees is of critical importance. Adverse impact on mycorrhizae by air pollution has been hypothesized (Sobotka, 1968).

A. Fungal Symbionts

An enormous number of soil fungi infect fine roots of forest trees and form mycorrhizal or "fungus roots." It is widely held that these roots colonized by beneficial fungi are essential for the growth of essentially all woody plants in natural forest environments. Morphological differences conveniently place mycorrhizal associations into one of two groups; ectomycorrhizae and endomycorrhizae. The latter, formed by fungal species of the Endogonaceae (Phycomycete class), are the most common mycorrhizal associates of forest trees. Infection by these fungi does not alter the gross morphology of the root but mycelia do emanate from the roots to form a loose network in the rhizosphere. The formation of vesicles and arbuscules on hyphae inside the root has led to the designation VA mycorrhizae. Ectomycorrhizae, while less abundant than endomycorrhizae, form on a variety of extremely important forest tree species including pine, hemlock, spruce, fir, oak, birch, beech, eucalyptus, willow, and poplar among others.

Ectomycorrhizal fungi are taxonomically diverse and may be Basidiomycetes, Ascomycetes, or family Endogonaceae of Phycomycetes. Infection in this case does morphologically alter the external appearance of fine roots by the development of a distinctive mass of external hyphae (Ruehle and Marx, 1979; Trappe, 1977) (Figure 10-1).

Mycorrhizal roots are presumed to confer multiple advantages to host trees. More efficient water uptake and increased resistance to infection by soil pathogens are two of the most important. Traditionally, however, the advantages associated with nutrient availability and uptake have been judged to be paramount. The uptake of nutrients from the forest floor (litter) and soil throughout the relatively large interroot distances of forest vegetation is achieved by the longevity of mycorrhizal roots and especially by the growth of hyphal strands into rhizosphere soil and beyond. Radio-tracer investigations with essentially all the macronutrients and numerous micronutrients have invariably confirmed uptake by the fungus and translocation to tree hosts. The significance of these specialized roots is particularly great for relatively immobile ions such as phosphate, zinc, copper, molybdenum, and occasionally ammonium (Bowen, 1973; Trappe and Fogel, 1977). Excellent reviews of mycorrhizal physiology and ecology are available (Hacskaylo, 1971; Harley, 1969; Lobanow, 1960; Marks and Kozlowski, 1973; Sanders et al., 1975).

1. Mycorrhizae and Heavy Metals

Ninety-five percent of the active ectomycorrhizae of mature Douglas fir and larch forests were found to be associated with the humus, decayed wood, and charcoal of the forest floor in Montana (Harvey et al., 1976). Concentration of mycorrhizal roots in the upper organic horizon of the soil profile places them in the zone of excessive heavy metal accumulation in those forest regions subject to trace metal input from the atmosphere (Chapter 4). Limited evidence for increased heavy metal uptake by mycorrhizal roots has been presented for zinc, copper, and manganese (Bowen, 1973). Unfortunately little is known concerning the relative tolerance of symbiotic fungi to heavy metal contamination. Fungi are known to accumulate metal ions and it has been hypothesized that they may combine these with oxalic acid and then dispose of them as insoluble oxalate salts (Cromack et al., 1975).

Under laboratory conditions, but with field collected root material, Bowen et al. (1974) have presented data indicating enhancement of zinc uptake from solution with both ectomycorrhizas of Monterey pine and with VA mycorrhizas of hoop pine compared with uninfected short roots. Zinc amendments of 45 and 135 μg g^{-1} of soil decreased both nodulation and mycorrhizae of Pinto bean when compared to amounts in nonamended soil (McIlveen et al., 1975). Smelter emissions and litter contaminated with heavy metals were added to intact forest microcosms and the influence on litter-soil carbon metabolism was monitored by recording daily efflux of carbon dioxide by Ausmus et al. (1978). Heavy metal contamination increased the rate of loss and daily pattern of carbon dioxide ef-

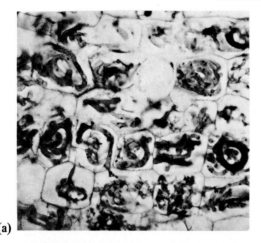

(a)

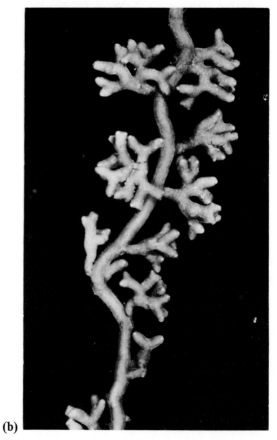

(b)

Figure 10-1. Mycorrhizal associations of yellow poplar and eastern white pine. (a) Fungal mycelium inside the endomycorrhizal roots of yellow poplar. (b) Mantle of fungal mycelium on the morphologically distinctive ectomycorrhizal roots of eastern white pine.

flux from the microcosms. Bacterial populations at the end of the experiment were greater in treated microcosms than in controls. The authors hypothesized, but did not provide specific evidence, that the disruption of mycorrhizal associations may have increased available substrate from fungal cells and stimulated carbon dioxide efflux.

With current information we cannot even approximate the threshold levels of any heavy metal that might exert an adverse influence on a mycorrhizal fungus in soil. In view of the extraordinarily large number of fungi capable of forming mycorrhizal associations, estimated to be 2000 species for Douglas fir alone (Trappe, 1977), and general variation of fungi in response to heavy metal exposure, for example the decrease in formation of ectomycorrhizae by some and the increase by others following application of a copper fungicide to seedlings (Göbl and Pümpel, 1973), it is assumed that the threshold range would be very broad, variable, and species specific.

2. Mycorrhizae and Acid Precipitation

A soil pH of approximately 5 appears optimal for many mycorrhizal fungi (Meyer, 1973). Alkaline soil pH is known to be associated with poor mycorrhizal formation (Bowen and Theodorou, 1973). Could soil acidification and increased H^+ ion concentration resulting from acid precipitation have an adverse influence on mycorrhizal associations?

The influence of simulated acid rain on nitrogen uptake by the fungus *Glomus mosseae* endomycorrhizal with the roots of sweetgum seedlings has been studied by Haines and Best (1976). Natural forest soil profiles collected in North Carolina and supporting the growth of sweetgum seedlings (originating from planted seeds) were treated with solutions of pH 5.9 and 3.0 and artificial eastern United States rain. Application of the pH 5.9 and 3.0 solutions produced greater nitrate- and ammonium-nitrogen concentrations in the soil solution in columns with mycorrhizae relative to columns containing either soil alone or soil plus tree seedlings. When artificial acid rain acidified the top 5 cm of soil to a soil solution pH of 2.0, nitrate-nitrogen concentrations were unchanged by mycorrhizal columns while ammonia appeared excluded from soil exchange sites, presumably by H^+ ions. The authors judged that if mycorrhizal roots are more subject to competitive inhibition of cation uptake by H^+ ions than nonmycorrhizal roots, then acidification of soil solutions may result in decreased cation uptake by mycorrhizal roots.

B. Bacterial Symbionts

While generally much less abundant than fungal symbioses, mutually beneficial relationships with bacteria are important in nutrient relationships in certain forest ecosystems. The influence air contaminants may have on these microbes has been the topic of a very limited number of studies.

1. Rhizobia and Air Pollution Stress

Members of the *Rhizobium* genus can infect the root hairs of leguminous plants and initiate the formation of root nodules, within which they develop as intracellular symbionts and fix atmospheric nitrogen. The 12,000 leguminous species include trees, shrubs, and woody vines of which less than 10% have been studied for nodulation (Postgate and Hill, 1979). Globally important tree genera include *Dalbergia, Pterocarpus, Albizzia, Peltogyne, Acacia, Cassia, Castanospermum, Haemotoxylon,* and *Caragana. Gleditsia* and *Robinia* are the two most important native American genera and along with sixteen other arborescent genera contain 40 species.

The environmental factors that influence nitrogen fixation by rhizobia have been reviewed by Postgate and Hill (1979). It has been suggested by Tamm (1976) that microbial nitrogen fixation may be impaired by air pollution.

Ladino clover was treated with filtered air, 30 pphm (588 μg m^{-3}) ozone or 60 pphm (1176 μg m^{-3}) ozone for two 2-hr exposures, 1 week apart in controlled environment chambers (Letchworth and Blum, 1977). Plants of various ages were tested. The influence of ozone varied with gas concentration and plant age. Ozone reduced the growth and nodulation of test plants. Nitrogenase activity per nodule and per plant was not significantly different from the control. Total nitrogen content per plant was correlated with plant biomass, but not with nodule number per plant (Table 10-1).

Heavy metals have been implicated as capable of altering nitrogen fixation in legumes (Döbereiner, 1966; Huang et al., 1974). Vesper and Weidensaul (1978) subjected soybeans to cadmium and nickel at 1, 2.5, and 5 ppm and copper and zinc at 1, 5, and 10 ppm in sand culture. The degree of toxicity was cadmium > nickel > copper > zinc. Cadmium dramatically reduced nodule number, dry weight, and nitrogen fixation. Fixation by nickel treated plants was very low. Copper suppressed nodulation, but inhibited nitrogen fixation directly only at 5 and 10 ppm. Zinc reduced nodulation, but only slightly inhibited fixation (Table 10-2).

The clover exposures to ozone conducted by Letchworth and Blume (1977) employed ozone doses that were high relative to typical ambient levels. The heavy metal experiments of Vesper and Weidensaul (1978), on the other hand, used quite realistic metal concentrations relative to natural areas subject to modest trace metal pollution. The results of the latter investigation more than justify additional efforts to gather data from more legumes grown in natural environments. Unfortunately we were unable to find any studies concerned with nitrogen fixation and trees as affected by air pollution.

2. Alder-Type Endophytes (*Frankia*)

Although most species of *Alnus* do not attain a size nor form that make them valuable for wood production, their nitrogen-fixing capabilities make them extremely valuable components of the extensive temperate and boreal forests where they naturally occur. The endophyte of *Alnus* nodules has been identified

Table 10-1. Effects of Ozone on Dry Weight, Nodule Number, and Plant Nitrogen of Ladino Clover

	Top dry weight (g)	Root dry weight (g)	Nodule number	Nitrogen			
				Top (%)	Root (%)	Top (mg)	Root (mg)
Control	10.37(a)[a]	1.75(a)	483(a)	2.85(b)	2.66(b)	265.5(a)	44.5(a)
30 pphm ozone	9.65(b)	1.48(b)	317(b)	3.02(a)	2.79(ab)	247.5(ab)	38.0(b)
60 pphm ozone	8.15(c)	1.16(c)	309(b)	3.08(a)	2.87(a)	214.8(b)	30.8(c)
LSD (0.05)	0.650	0.17	41	0.14	0.14	18.5	4.3

Source: Letchworth and Blum (1977).
[a]Means followed by the same letter are not significantly different at the 5% level.

Table 10-2. Influence of Cadmium, Nickel, Copper, and Zinc on Soybean Nodule Size, Number, and Nitrogen-Fixing Ability

Treatment (ppm)	Number of nodules per plant	Nodule size (mg)	Ethylene produced (nmoles plant^{-1} 0.5 h^{-1})
Cd			
1.0	12(g)[a]	25.31(d)	144(b)
2.5	12(g)	24.58(d)	87(c)
5.0	3(h)	13.47(e)	38(d)
Ni			
1.0	44(ab)	58.86(a)	824(e)
2.5	38(cd)	48.33(c)	611(f)
5.0	28(e)	47.41(c)	527(g)
Cu			
1.0	28(e)	54.09(b)	1509(h)
5.0	27(ef)	54.06(b)	991(i)
10.0	24(f)	47.85(c)	879(j)
Zn			
1.0	40(bc)	58.40(a)	1659(a)
5.0	38(c)	53.69(a)	1585(k)
10.0	36(d)	52.90(b)	1252(l)
Control	46(a)	62.30(f)	1640(a)

Source: Vesper and Weidensaul (1978).
[a] Values in a given column followed by the same letter are not significantly different (P = 0.05).

as *Frankia*, an actinomycete (Becking, 1974). The quantity of nitrogen fixed in alder nodules exceeds that of legume nodules and may range from 60 to 209 kg ha^{-1} yr^{-1} under favorable conditions. Growth of trees of the following genera have been shown to increase under the influence of associated alder: *Fraxinus, Liquidambar, Liriodendron, Picea, Pinus, Platanus, Populus,* and *Pseudotsuga* (Tarrant, 1968). Seven additional genera of woody, nitrogen-fixing angiospermous plants, including *Cosuarina, Hippophal, Purshia, Ceanothus, Myrica, Coriaria* and *Dryas,* have "alder-type" nodules and are of varied importance and distribution throughout temperate wooded ecosystems.

Our ignorance of the ecology and physiology of the endophytes associated with woody plants with alder type nodules is very great. The limited number of investigators working in this area are hampered by serious cultural limitations. *Frankia*, for example, has not been isolated from soil and Koch's postulates have not been fulfilled. No nodule isolate, grown in pure culture, has been successfully employed to yield nodules on artificially infected plants (Postgate and Hill, 1979). It is important to continue studies directed to improvement of our understanding of nutrient and cultural requirements of alder-type endophytes, since in vitro cultivation experiments have shown actinomycete growth to be influenced by heavy metals in growth media. Cadmium has been shown to be toxic over a wide range of media concentrations. Aluminum and nickel are generally toxic above 10 ppm, while lead and vanadium appear to be relatively nontoxic (Waksman, 1967).

C. Other Nitrogen-Fixing Organisms

The relative importance of other nitrogen-fixing organisms, for example, free-living bacteria, algae, lichens and liverworts, mosses, and ferns with blue-green algal symbionts, in forest ecosystems is very poorly appreciated. The influence of air contaminants on these potentially significant organisms is, of course, equally poorly understood. Demison et al. (1976) have claimed that acid rain has had an adverse impact on nitrogen-fixing lichens of western Washington coniferous forests, but the laboratory data they provided to support their contention were modest and quite variable. Hällgren and Huss (1975) have evaluated the influence of sodium hyposulfite on photosynthesis and nitrogen fixation of a lichen, collected from a Swedish pine forest, and the blue-green alga *Anabaena cylindrica.* Treatment with 5×10^{-4} M NaHSO$_3$ at pH 5.8 caused no reduction of photosynthesis in the lichen, while inhibition of nitrogen fixation was 97%. For the alga, the corresponding values were 40 and 75%, respectively. The authors speculated that sulfite may have some specific inhibitory action on nitrogenase enzyme activity.

D. Summary

The research attention given the interaction between mycorrhizal fungi, the specialized roots they form, and air pollution is grossly out of scale with the significance of this symbiotic relationship in forest ecosystems. The potential for heavy metal adverse impact on mycorrhizal associations appears particularly great due to the physical juxtaposition of the two entities in the forest floor. Substrate characteristics are also important to nitrogen fixation dynamics. Sheridan (1979) has proposed that anthropogenic input of nitrogen to ecosystems may result in a decline in the abundance of nitrogen fixing microbes. This decline might not recover to precontamination levels following ultimate removal of the air pollution stress.

Unfortunately research efforts to describe the influence of air contaminants on alder-type endophytes will be restricted until our appreciation of isolation, inoculation, and in vitro cultivation techniques is improved. In any case, efforts directed toward clarification of air pollution influences on all forest organisms capable of nitrogen fixation are justified due to the potentially important role these organisms play in forest nutrient dynamics.

References

Ausmus, B. S., G. J. Dodson, and D. R. Jackson. 1978. Behavior of heavy metals in forest microcosms. III. Effects of litter-soil carbon metabolism. Water, Air, Soil Pollut. 10:19-26.

Becking, J. H. 1974. Frankiaceae Becking. *In*: R. E. Buchanan and N. E. Gibbons (Eds.), Bergey's Manual of Determinative Bacteriology. Williams and Wilkens, Baltimore, Maryland, p. 701.

Bowen, G. D. 1973. Mineral nutrition in ectomycorrhizae. *In*: G. C. Marks and T. T. Kozlowski (Eds.), Ectomycorrhizae. Their Ecology and Physiology. Academic Press, New York, pp. 151-205.

Bowen, G. D., and C. Theodorou. 1973. Growth of ectomycorrhizal fungi around seeds and roots. *In*: G. C. Marks and T. T. Kozlowski (Eds.), Ectomycorrhizae. Their Ecology and Physiology. Academic Press, New York, pp. 107-150.

Bowen, G. D., M. F. Skinner, and D. I. Bevege. 1974. Zinc uptake by mycorrhizal and uninfected roots of *Pinus radiata* and *Araucaria cunninghamii.* Soil Biol. Biochem. 6:141-144.

Cromack, K., R. L. Todd, and C. D. Monk. 1975. Patterns of Basidiomycete nutrient accumulation in conifer and deciduous forest litter. Soil Biol. Biochem. 7:265-268.

Demison, R., B. Caldwell, B. Bormann, L. Eldred, C. Swanberg, and S. Anderson. 1976. The effects of acid rain on nitrogen fixation in western Washington coniferous forests. *In*: L. S. Dochinger and T. A. Seliga (Eds.), Proc. 1st Internat. Symp. Acid Precipitation and the Forest Ecosystem. U.S.D.A. Forest Service, Gen. Tech. Rep. No. NE-23, Upper Darby, Pennsylvania, pp. 933-949.

Döbereiner, J. 1966. Manganese toxicity effects on nodulation and nitrogen fixation of beans (*Phaseolus vulgaris* L.) in acid soils. Plant Soil 24:153-166.

Göbl, F., and B. Pümpel. 1973. Einfluss von "Grünkupfer Linz" auf Pflanzenausbildung, Mykorrhizabesatz sowie Frosthärte von Zirbenjungpflanzen. Eur. J. For. Pathol. 3:242-245.

Hacskaylo, E. (ed.). 1971. Mycorrhizae. U.S.D.A. Forest Service, Misc. Publ. No. 1189, Washington, D.C., 255 pp.

Haines, B., and G. R. Best. 1976. The influence of an endomycorrhizal symbiosis on nitrogen movement through soil columns under regimes of artificial throughfall and artificial acid rain. *In*: L. S. Dochinger and T. A. Seliga (Eds.), Proc. 1st Internat. Symp. Acid Precipitation and the Forest Ecosystem. U.S. D.A. Forest Service, Gen. Tech. Rep. No. NE-23, Upper Darby, Pennsylvania, pp. 951-961.

Hällgren, J. E., and K. Huss. 1975. Effects of SO_2 on photosynthesis and nitrogen fixation. Physiol. Plant 34:171-176.

Harley, J. L. 1969. The Biology of Mycorrhiza. Leonard Hill, London, 334 pp.

Harvey, A. E., M. J. Larsen, and M. F. Jurgensen. 1976. Distribution of ectomycorrhizae in a mature Douglas-fir/larch forest soil in western Montana. For. Sci. 22:393-398.

Huang, C., F. A. Bazzaz, and L. N. Vanderhoef. 1974. The inhibition of soybean metabolism by cadmium and lead. Plant Physiol. 54:122-124.

Letchworth, M. B., and V. Blum. 1977. Effects of acute ozone exposure on growth, nodulation and nitrogen content of ladino clover. Environ. Pollut. 14:303-312.

Lobanow, N. W. 1960. Mykotrophie der Holzpflanzen. Springer Verlag, Berlin, 352 pp.

Marks, G. C., and T. T. Kozlowski (eds.). 1973. Ectomycorrhizae. Their Ecology and Physiology. Academic Press, New York, 444 pp.

McIlveen, W. D., R. A. Spotts, and D. D. Davis. 1975. The influence of soil zinc on nodulation, mycorrhizae, and ozone-sensitivity of Pinto bean. Phytopathology 65:645-647.

Meyer, F. H. 1973. Distribution of ectomycorrhizae in native and man-made forests. *In*: G. C. Marks and T. T. Kozlowski (Eds.), Ectomycorrhizae. Their Ecology and Physiology. Academic Press, New York, pp. 79-105.

Postgate, J. R., and S. Hill. 1979. Nitrogen fixation. *In*: J. M. Lynch and N. J. Poole (Eds.), Microbial Ecology: A Conceptual Approach. Wiley, New York, pp. 191-123.

Ruehle, J. L., and D. H. Marx. 1979. Fiber, food, fuel, and fungal symbionts. Science 206:419-422.

Sanders, F. E., B. Mosse, and P. B. Tinker. 1975. Endomycorrhizas. Academic Press, New York, 626 pp.

Sheridan, R. P. 1979. Effects of airborne particulates on nitrogen fixation in legumes and algae. Phytopathology 69:1011-1018.

Sobotka, A. 1968. Wurzeln von *Picea excelsa* L. unter dem Einfluss der Industrie-exhalate im Gebiet des Erzgebirges in der CSSR. Immissionen und Waldzönosen. Cesk. Akad. Ved. Ustav pro Tvorlu a Ochr. Kran., Praha, p. 45.

Tamm, C. F. 1976. Acid precipitation—biological effects on soil and on forest vegetation. Ambio 5:235-238.

Tarrant, R. F. 1968. Some effects of alder on the forest environment. *In*: J. M. Trappe, J. F. Franklin, R. F. Tarrant, and G. M. Hansen (Eds.), Biology of Alder. U.S.D.A. Forest Service, Pac. Northwest Forest Range Exp. Sta., Portland, Oregon, pp. 193.

Trappe, J. M. 1977. Selection of fungi for ectomycorrhizal inoculation in nurseries. Annu. Rev. Phytopathol. 15:203-222.

Trappe, J. M., and R. D. Fogel. 1977. Ecosystematic functions of mycorrhizae. *In*: The Belowground Ecosystem: A Synthesis of Plant-Associated Processes. Range Sci. Dept., Science Ser. No. 26, Colorado State Univ., Fort Collins, Colorado, pp. 205-214.

Vesper, S. J., and T. C. Weidensaul. 1978. Effects of cadmium, nickel, copper, and zinc on nitrogen fixation by soybeans. Water, Air, Soil Pollut. 9:413-422.

Waksman, S. A. 1967. The Actinomycetes. A Summary of Current Knowledge. Ronald Press, New York, 280 pp.

11

Forest Metabolism: Influence of Air Contaminants on Photosynthesis and Respiration

Photosynthesis is the most important metabolic process of forest ecosystems. In simple outline the process amounts to the reduction of carbon dioxide to CH_2O and the oxidation of water to molecular oxygen and results in the derivation of approximately 95% of the dry weight of plants. Photosynthesis is the process primarily responsible for forest productivity.

The few published reports of rates of net photosynthesis (gross carbon dioxide fixation less the losses due to respiration in the light and dark) of intact leaves of mature trees suggest that such rates are in general greater than 10 and less than 200 mg of carbon dioxide taken up per gram dry weight per day (Botkin, 1968; Kozlowski and Keller, 1966). Reported daily rates are generally greater for decid-uous trees than for conifers on a leaf weight basis. Conifers, on the other hand, may be photosynthetically active for many months, even in northern latitudes. Linder and Troeng (1977) have indicated that Scotch pine may perform photo-synthesis in mid-Sweden for approximately 10 months per year when supplied with adequate moisture. Polster (1950) reported carbon dioxide uptake per gram fresh weight for European birch to be 67 mg carbon dioxide g^{-1} day^{-1}, beech 53 mg carbon dioxide g^{-1} day^{-1} and Norway spruce 14 mg carbon dioxide g^{-1} day^{-1}. Other reports show similar differences. Helms (1965) reports rates for Douglas fir on the order of 5 to 10 mg carbon dioxide g^{-1} day^{-1}. Fritts (1966) reported rates for ponderosa pine generally less than 20 mg carbon dioxide g^{-1} day^{-1}. In contrast, net photosynthesis for American white oak and scarlet oak are observed to range from 40 to 110 mg carbon dioxide g^{-1} day^{-1} (Botkin, 1968).

Differences among tree species are suggested in these reports but little data exist enabling direct comparative studies of species differences. Genetic differences in rates of net photosynthesis have been found under certain environmental conditions for seedlings of white pine (Bourdeau, 1963). Clonal differences have been observed for poplar by Huber and Polster (1955) and for larch by Polster and Weise (1962). Krueger and Ferrell (1965) observed differences for ecotypes of seedlings of Douglas fir from coastal and inland locations. Species differences have been found for seedlings of tree species under laboratory conditions. Kramer and Decker (1944) observed rates for seedlings of two oak species to be twice those of dogwood on a leaf area basis. Other similar reports are reviewed in Kramer and Kozlowski (1960), Kozlowski and Keller (1966), and Kramer and Kozlowski (1979), but it must be emphasized that too little is known to extrapolate from laboratory observations of seedlings to physiological reactions of mature trees under natural conditions.

From these studies, one would expect that differences exist among mature individuals of different tree species. Species differences have been found many times for herbaceous plants, as reported in Blackman and Black (1959), Blackman and Wilson (1951) and Hesketh and Moss (1963).

The situation for trees is complicated because differences have also been observed for leaves in different locations on the same tree. For example, shade leaves of European beech showed higher rates of net photosynthesis at low light intensities than sun leaves (Pisek and Tranquillini, 1954).

Rates of net photosynthesis for any tree are a function of time as well as location. Seasonal patterns have been studied particularly in conifers, and such differences observed for ponderosa pine (Fritts 1966) and Douglas fir (Hodges, 1962) among others. There are many reports of diurnal patterns in net photosynthesis, particularly in regard to a midday dip. These have been reported for coconut palm (McLean, 1920), ponderosa pine (Fritts, 1966), and other plants of the prairie and desert (Stocker, 1960), as well as for European beech and birch (Polster, 1950). The cause of this dip is not known and the subject of controversy, attributed by some to environmental conditions, such as changes in carbon dioxide or water vapor concentrations in the air (Pisek and Tranquillini, 1954), to internal factors, such as the accumulation of net photosynthates (Thomas and Hill, 1949) and to methods of measurement themselves (Bosian, 1965).

A. Constraints on Photosynthesis

Climate exerts profound control over photosynthetic rates (Zelitch, 1975; Govindjee, 1975). In addition to the obvious importance of solar radiation and ambient carbon dioxide (Anderson, 1973), moisture, temperature, and nutrition are critically important environmental variables. Water deficit decreases photosynthetic carbon dioxide uptake by restricting the transport of this gas to the chloroplasts in both the gaseous and liquid phase and by limiting the metabolic function of the chloroplasts themselves (Slavik, 1973). Carbon dioxide uptake generally increases with increasing temperature to some maximum and then decreases. Low temperature, chilling, and frost stress reduce photosynthetic rates (Bauer et al., 1973). Similar relationships between temperature and photosynthesis have been observed repeatedly under laboratory conditions, for example, between leaf temperature and carbon dioxide uptake of excised twigs of mature trees of Norway spruce and Swiss stone pine (Pisek and Winkler, 1958); between air temperature and carbon dioxide uptake by entire shoots of 35-day-old Douglas fir seedlings (Krueger and Ferrell, 1965); between air temperature and carbon dioxide uptake by four species of woody desert perennials, *Larrea divaricata, Hymenudea salsola, Encelia farinosa,* and *Chilopsis linearis* (Strain and Chase, 1966); and between air temperature and carbon dioxide uptake of excised apple leaves (Waugh, 1939). A second degree polynomial gave a significant regression of the logarithm of carbon dioxide uptake on air temperature for laboratory observations of seedlings of sand pine (Pharis and Woods, 1960). In a nonhomogeneous and mixed species forest, highly significant relationships were obtained between sunlight intensity and leaf temperature and rates of photosynthesis of individual oak leaves (Botkin, 1968). Sunlight and temperature accounted for more than 70% of the variation in rates of photosynthesis under these conditions.

There is abundant evidence demonstrating that the rate of photosynthesis may be decreased by a deficiency in any one of the essential nutrient elements (Nátr, 1973). Keller (1968, 1970) has shown a dramatic effect of nitrogen deficiency on the rate of photosynthesis of forest trees.

In addition to the restrictions imposed by these important environmental variables, it has become increasingly apparent that numerous materials introduced by human beings into their environment can reduce net photosynthesis of plants. These materials include radioisotopes (Nazirov, 1966), salt (Nellen, 1966), and numerous herbicides (Sasaki and Kozlowski, 1966a,b) and insecticides (Heinicke and Foott, 1966) among others. What is the evidence for adding air contaminants to this list?

B. Photosynthetic Suppression: Nonwoody Species

Many excellent reviews are available on the relationship between specific air contaminants and plant photosynthetic processes, for example; sulfur dioxide (Hällgren, 1978), ozone (Verkroost, 1974), and fluoride (McCune and Weinstein, 1971).

1. Sulfur Dioxide

The influence that sulfur dioxide exerts on plant photosynthesis has received more research attention than any other air pollutant. It has a research history extending over more than 40 years as Thomas and Hill reported that short fumigations with high concentrations of sulfur dioxide caused large reductions in the photosynthetic rate of alfalfa in 1937. Two hour exposure of alfalfa to 25 pphm (655 μg m^{-3}) sulfur dioxide produced a 2-3% inhibition of photosynthetic rate as measured by carbon dioxide uptake (White et al., 1974). Similar exposure to 25 pphm (470 μg m^{-3}) nitrogen dioxide failed to depress photosynthesis. Exposure to both gases simultaneously each at 25 pphm, however, caused a 9-15% inhibition. At 15 pphm, the mixture precipitated a 7% decrease. The inhibitory action of the two gases decreased as the concentrations increased and at 50 pphm each gas there was no synergistic response. Pinto bean photosynthesis was reduced approximately 15, 70, and 90% by 1-hr exposure to 1 (2620 μg m^{-3}), 3 (7860 μg m^{-3}), or 5 (13.1 $\times$ 10^3 μg m^{-3}) ppm sulfur dioxide (Sij and Swanson, 1974). When Pinto bean were subjected to 10 pphm (262 μg m^{-3}) sulfur dioxide and 10 pphm (188 μg m^{-3}) nitrogen dioxide singly and in combination for 20 hours, the whole plant transpiration rate was decreased only by the combination, indicating synergism (Ashenden, 1979).

The threshold of photosynthetic reduction by either nitrogen dioxide or nitric oxide alone appeared to be approximately 0.6 ppm for 2 hr for oaks and alfalfa (Hill and Bennett, 1970).

Joseph E. Miller and co-workers of the Radiological and Environmental Research Division, Argonne National Laboratory, Argonne, Illinois, have employed an elaborate field design (Zonal Air Pollution System) to monitor the influence of sulfur dioxide on soybean photosynthesis under natural conditions. Fumigations of 706 (1850 μg m^{-3}) ppb, 236 (618 μg m^{-3}) ppb, and 92 (241 μg m^{-3}) ppb sulfur dioxide for periods averaging 4.5 hr were applied on 24 separate occasions throughout the growing season. During fumigation photosynthesis was reduced to 37-63% of the control in the high-dose plot and 68-83% in the medium-dose plot. The low-dose plot did not consistently depress photosynthesis. The investigators judged that stomatal closure by soybean was not a causal factor in the photosynthetic depression (Muller et al., 1978).

2. Ozone

Photosynthetic response to ozone also has a rather lengthly research history. Early efforts employing ozonated hexene suggested photosynthetic reduction in duckweed (Erikson and Wedding, 1956), Pinto bean (Todd, 1958), and bean, tomato, and coleus (Todd and Propst, 1963). A variety of plants has been added to this early list of species exhibiting reduced carbon assimilation caused by ozone, for example, tobacco (Macdowall, 1965) and *Euglena gracilis* (De Koning and Jegier, 1968). Coyne and Bingham (1978) have examined the interaction of ozone and hydrogen sulfide on snap bean photosynthesis. Plants were exposed to 7.2 pphm (14.1 $\times$ 10^3 μg m^{-3}) ozone and/or 0.74 (1029 μg m^{-3}), 3.25 (4518 μg m^{-3}), and 5.03 (6992 μg m^{-3}) ppm hydrogen sulfide for 4 hr per day for 18 days. Ozone plus hydrogen sulfide depressed stomatal and photosynthetic response more than hydrogen sulfide alone. At the highest hydrogen sulfide concentration plus ozone, stomatal conductance was 41% and apparent photosynthesis was 52% of the control.

For a variety of agricultural species, Hill and Littlefield (1969) have reported a 40-70 percent reduction in carbon dioxide assimilation following exposure to

40-70 pphm (787-1372 μg m^{-3}) ozone for 30 to 90 min (Table 11-1). The effect of ozone influence appeared reversible as plants frequently recovered within a few hours.

3. Heavy Metals

Bazzaz et al. (1974a,b) have monitored photosynthesis of detached leaves of corn and sunflower exposed to various concentrations of cadmium chloride. In sunflower, net photosynthesis was completely inhibited within 45 minutes after the introduction of 18 mM cadmium. Within two hr photosynthesis was reduced to 40% and 70% of maximum after exposure to 9 and 4.5 mM cadmium, respectively. Response of corn was similar to sunflower except that inhibition was more pronounced at all treatment levels. Continuation of their studies employing detached sunflower leaves revealed that several heavy metals apparently interfere with stomatal function and that photosynthesis was reduced by 50% of maximum when leaf tissue concentrations were 63 μg g^{-1} thallium, 96 μg g^{-1} cadmium, 193 μg g^{-1} lead, or 79 μg g^{-1} nickel. When whole corn and sunflower

Table 11-1. Effect of Ozone on Apparent Photosynthesis of Several Agricultural Plants

Species	Age (days)	O$_3$ (pphm)	Treatment time (min)	Photosynthesis (% of control)	Leaf injury (%)
Oat	50	40	30	67	1
	48	45	30	60	1
	47	50	60	55	1
	50	60	60	69	2
Barley	36	62	30	42	0
	47	60	60	62	0
Wheat	56	70	60	50	4
Tobacco	99	40	90	22	0
	110	45	75	38	1
	69	50	45	67	4
	70	48	35	62	2
	118	48	80	50	1
Pinto bean	56	45	90	50	10
Lima bean	26	45	120	52	1
	39	70	45	36	5
Bush bean	51	50	80	65	3
Chard	78	60	70	55	3
Corn	49	50	90	68	3
Cauliflower	77	75	50	38	0
Sugar beet	74	65	90	80	4
	80	90	90	49	7
Potato	39	60	60	50	3
Tomato	72	60	60	57	1

Source: Hill and Littlefield (1969).

plants were grown in hydroponic culture and subjected to various levels of these heavy metals salts, thallium was found to be the most toxic to net photosynthesis followed in order by cadmium, nickel, and lead (Carlson et al., 1975). Thallium caused a 50% reduction in net photosynthesis at a leaf content of less than 82 μg g^{-1} for both sunflower and corn. Cadmium caused this same level of reduction at 160 μg g^{-1} in corn and 340 μg g^{-1} in sunflower. Sunflower and corn treated with nickel exhibited a 50% reduction at leaf concentrations of 290-350 μg g^{-1}. Lead content of treated plant leaves was not significantly different from the control.

C. Photosynthetic Suppression: Forest Tree Seedlings

Carbon dioxide uptake is a favored technique for monitoring photosynthetic rates. Since this measurement along with pollutant fumigation is most conveniently performed in a relatively small chamber, seedling size plants have been the preferred research material for investigators interested in forest trees.

1. Sulfur Dioxide

Roberts et al. (1971) exposed 5-month-old red maple seedlings to 1 ppm (2620 μg m^{-3}) sulfur dioxide for 4 to 6 hr and observed a slight stimulation in carbon dioxide exchange relative to unfumigated trees. At 6 ppm (15.7×10^3 μg m^{-3}) sulfur dioxide depressed photosynthesis but the suppression was due to gross foliar necrosis. Quaking aspen and white ash seedlings were exposed to 0.2 (524 μg m^{-3}), 0.5 (1310 μg m^{-3}), 1.0 (2620 μg m^{-3}), and 4.0 (10.5×10^3 μg m^{-3}) ppm sulfur dioxide for 2-4 hr by Jensen and Kozlowski (1974). No influence on photosynthetic rate was detected at 0.2 or 0.5 ppm. One ppm caused a slight reduction and 4 ppm caused a substantial decrease in photosynthetic rate.

Theodore Keller of the Swiss Forest Research Institute in Birmensdorf, Switzerland, has done considerable research on the relationship between sulfur dioxide and forest tree seedling metabolism. Exposure of Scotch pine to 0.05-0.10 (131-262 μg m^{-3}) ppm sulfur dioxide for many weeks caused a reduction in carbon dioxide assimilation of very approximately 40% of control rates (Keller, 1977a). Keller (1977b, 1978a,b) subsequently exposed genetically uniform 3-year-old grafts of white fir, Norway spruce, and Scotch pine to 0.05 (131 μg m^{-3}), 0.1 (262 μg m^{-3}), and 0.2 (524 μg m^{-3}) ppm sulfur dioxide for periods of up to 70 days. Prolonged exposure to 0.2 ppm caused a dramatic decrease in the photosynthesis of all three species, especially the fir (Table 11-2). Visible symptoms of foliar injury occurred at the end of the experiment only in Norway spruce subject to the 0.2 ppm fumigation. Seasonal observations with the latter species have revealed significant photosynthetic suppression at low dose during the spring and has led Keller (1978a) to observe that reduced annual increment may be substantial because of this. Since suppressed root development may also be associated with high dose (0.1 and 0.2 ppm) exposure to sulfur dioxide, Keller (1979) has judged that the impact of this gas on spruce may be substantial growth reduction.

Table 11-2. Photosynthesis of 3-Year-Old Grafts of Tree Seedlings Subject to Continuous Exposure to Varying Concentrations of Sulfur Dioxide under Controlled Environmental Conditions[a]

Season	Fumig. period (days)	Relative photosynthesis (%) at SO_2 conc.			
		Control	0.05 ppm	0.1 ppm	0.2 ppm
Silver fir					
Spring	1-15	40	28	27	10
	16-30	64	58	51	34
	31-60	93	74	49	24
Summer	1-15	109	108	114	100
	16-30	109	104	104	90
	31-60	96	87	77	58
Fall	1-15	122	94	91	87
	16-30	106	83	83	80
	31-60	100	72	70	62
Norway spruce					
Spring	1-15	56	52	58	53
	16-30	104	88	107	94
	31-60	110	89	88	72
Summer	1-15	105	113	108	95
	16-30	131	125	113	96
	31-60	127	120	101	80
Fall	1-15	113	134	114	96
	16-30	116	133	110	88
	31-60	100	100	84	68
Scotch pine					
Spring	1-15	75	77	70	38
	16-30	71	49	80	57
	31-60	86	64	83	49
Summer	1-15	130	132	124	111
	16-30	149	140	114	102
	31-60	136	116	83	44
Fall	1-15	110	113	118	116
	16-30	109	101	109	97
	31-60	100	80	82	72

Source: Keller (1977b).
[a] Carbon dioxide uptake rates are relative to uptake rate of control at bud break.

2. Ozone

In 1961, Taylor et al. presented evidence that the rate of carbon dioxide fixation by potted lime seedlings was significantly inhibited by exposure to ozone at 0.6 ppm (1176 μg m^{-3}) for 1 hr without the development of leaf symptoms.

The first observations on the relationship between forest tree photosynthesis and oxidants was presented by Paul R. Miller in 1966. This investigator subjected

3-year-old ponderosa pine, grown from seed collected in the San Bernardino Mountains, California, to ozone fumigation 9-hr daily in a controlled environment chamber (Miller et al., 1969). A 60-day fumigation with 0.15 ppm (294 μg m^{-3}) resulted in a final 25% decrease in apparent photosynthesis and a 30-day fumigation with 0.30 ppm (588 μg m^{-3}) revealed a 67% depression (Figure 11-1). Soluble sugars and polysaccharides were observed to be significantly decreased in 1-year-old needles following a 33-day exposure to 0.30 ppm ozone.

Barnes (1972) has examined the response of greenhouse-grown seedlings of four pine species to ozone fumigation. Seedlings of various ages, 2 to 8 months, were exposed to continuous fumigation with ozone in growth chambers at concentrations of 5 or 15 pphm (98-294 μg m^{-3}). Barnes recorded substantial variation in foliar response depending on seedling age. He also noted a consistent stimulation of respiration, in one instance 90%, at the 15 pphm fumigation dose. In younger seedlings of eastern white pine, which bore only primary needles, ozone had little influence on photosynthetic rate. In older seedlings with secondary needles, photosynthesis was slightly depressed. With seedlings of slash, pond, and loblolly pines, ozone exposure at 15 pphm (294 μg m^{-3}) had a relatively consistent depressing influence on photosynthesis of all species. At 5 pphm (98 μg m^{-3}), however, ozone appeared to have a stimulating influence on older secondary needles and a depressing affect on younger secondary needles (Table 11-3).

3. Fluoride

Keller (1973) has evaluated the influence of particulate fluoride compounds, sodium fluoride, calcium fluoride, and natural and synthetic cryolite, on the photosynthetic rate of 3-year-old Scotch pine seedlings. 4-year-old Douglas fir seedlings, and 1-year-old birch grafts. Foliage was coated using fungicide-applicators to a very thin whitish film under high and normal relative humidity in a greenhouse and in specially designed glass covers. Only treatment with sodium fluoride induced visible symptoms. In birch all dusts caused a significant depression of photosynthesis. In conifers the photosynthetic reduction was not significantly different from the controls in the absence of visible symptoms.

Seedlings of twelve forest tree species were potted and placed by Keller (1977c) at varying distances from an aluminum shelter. Photosynthesis was monitored for several months through summer and fall. Pine seedlings were observed to be particularly sensitive as their photosynthetic rates dropped to below 60% of controls when foliar concentrations reached approximately 30 μg g^{-1} (or less) fluoride on a dry weight basis. Photosynthetic suppression decreased with both increasing distance from the fluoride source and increasing foliar tissue maturity. Simultaneous exposure of young spruce cuttings to sulfur dioxide and rooting substrate treated with sodium fluoride has revealed a synergistic suppression of photosynthesis (Keller, 1979, personal communication).

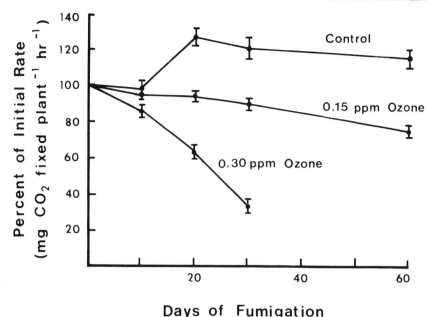

Days of Fumigation

Figure 11-1. The influence of ozone on the apparent photosynthesis of ponderosa pine seedlings. (From Miller et al., 1969.)

4. Heavy Metal Particulates

While the trace metal content was not determined, Auclair (1976) evaluated the influence of artificially applied coal (and cement dust) on 2-year-old Norway spruce seedlings. At all experimental light intensities employed, coal dust was found to significantly reduce photosynthesis. Whether this inhibition was due to stomatal blockage, attenuated light reception, or heavy metal toxicity was not

Table 11-3. Influence of Ozone on Photosynthesis and Respiration of Three Yellow Pine Seedlings

Concentration (pphm)	CO_2 exchange (mg g^{-1} dry wt hr^{-1})								
	Time of exposure (days)								
	36			77			84		
	Slash	Pond	Loblolly	Slash	Pond	Loblolly	Slash	Pond	Loblolly
Photosynthesis									
0	2-6	3-5	3-7	0-9	0-9	1-8	1-9	3-4	4-3
5	3-1	4-2	4-0	1-6	1-8	1-8	1-7	3-8	4-0
15	2-0	3-3	3-2	0-9	1-6	1-7	1-4	3-3	4-1
Respiration									
0	0-7	1-0	0-7	0-4	0-6	0-6	0-5	1-0	1-0
5	0-6	0-9	0-7	0-4	0-5	0-6	0-5	1-2	1-0
15	1-3[a]	1-2	1-2[a]	0-6	0-5	0-5	0-8	1-2	1-2

Source: Barnes (1972).

[a] Significantly greater than control value ($P < 0.05$).

clear from the experiment. Subsequent experiments examining the photosyn-thetic response of Scotch pine and poplar to coal dust led the investigator to conclude that the observed reduction in this metabolic process was due to re-duced light (Auclair, 1977).

The influence of lead and cadmium on the photosynthesis of nursery grown 2- to 3-year-old seedlings of American sycamore has been studied by Carlson and Bazzaz (1977). These investigators planted seedlings in pots containing a silty clay loam field soil and treated the pots with 0, 50, 100, 250, 500, and 1000 μg g^{-1} lead as lead chloride; 0, 5, 10, 25, 50, and 100 μg g^{-1} cadmium as cadmium chloride; and an integrated treatment with each level of both metals combined. Figure 11-2 shows the relationship observed between lead and cadmium soil amendment and photosynthesis. While it can be seen that treatment with lead and cadmium both reduced photosynthetic rate, no synergism was observed when the metals were combined.

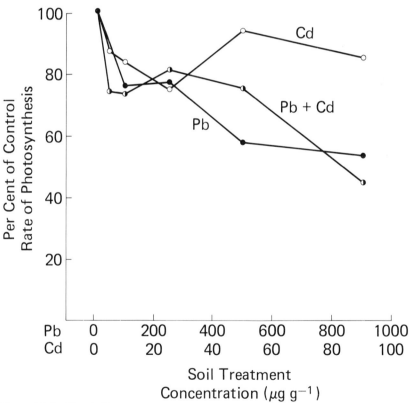

Figure 11-2. The influence of soil applied lead and cadmium on the photosyn-thesis of seedling American sycamore. (From Carlson and Bazzaz, 1977.)

D. Photosynthetic Suppression: Forest Tree Saplings

Investigations that have examined the response of photosynthetic rates of forest trees over the age of 5 years to air contaminant exposure have the advantage of avoiding metabolic pecularities that may be unique to the seedling stage of tree development. Because these studies are of necessity involved with relatively large plants, the monitoring of carbon dioxide flux and pollutant gas exposure is complicated. Experimental designs must monitor gas exchange in entire growth chambers, entire tree enclosures within growth chambers, in small enclosures placed on a portion of the foliage of experimental saplings, or deal with excised leaves. Reports from studies employing the latter two designs are summarized.

1. Excised Leaf Study

Silver maple leaves were obtained by detaching the distal 30 cm of twigs from 8-year-old silver maple saplings by Lamoreaux and Chaney (1978). Leaves of uniform size were excised from detached twigs in the laboratory. Some leaves were placed in solutions containing 0, 5, 10, or 20 ppm cadmium as cadmium chloride. Other leaves were treated with sulfur dioxide concentrations of 0, 1.0 (2620 μg m^{-3}), or 2.0 (5240 μg m^{-3}) ppm. Photosynthetic rates were determined after 45 hr of cadmium exposure and 30 min of sulfur dioxide treatment. Some leaves were exposed to the cadmium and sulfur dioxide treatments sequentially. Cadmium at 20 ppm, which resulted in a petiole concentration of 4 μg cadmium g^{-1} petiole tissue and 0 cadmium in leaf tissue, greatly reduced photosynthesis. The decrease caused by cadmium was greater when 1.0 or 2.0 ppm sulfur dioxide were present. In a peculiar deviation, treatment with 5 ppm cadmium precluded the synergistic influence of sulfur dioxide exposure.

2. Small Chamber Technique

The small chamber technique for the determination of photosynthetic rates was developed at Brookhaven National Laboratory, Upton, Long Island, by Woodwell and Whittaker (1968). It is thoroughly described in Woodwell and Botkin (1970). D. B. Botkin, several graduate students, and the author have employed the small chamber technique to assess the influence of gaseous contaminants on tree sapling photosynthetic rates.

In this method a leaf, or group of leaves or needles, is surrounded by a small transparent plastic (polyvinyl chloride) cover supported on a lightweight aluminum frame and forming a cylindrical chamber approximately 28 cm long and 15 cm in diameter. Air is supplied to and sampled from this chamber. Carbon dioxide uptake by leaf tissue is determined by maintaining an empty chamber in addition to chambers containing leaves and comparing the concentration of carbon dioxide in the air sampled from this chamber with that from a leaf-containing chamber. Gaseous pollutants, at monitored concentration, are introduced to the chambers with the supply air.

Since the small chamber technique completely encloses a leaf or group of leaves, it introduces some well appreciated artificialities. Air temperatures inside the chambers may become elevated in relation to ambient conditions outside the chambers. Temperatures within chambers are monitored with cooper-constantin thermocouples shielded at the entrance to the air sampling line. The effect of temperature distortion has been investigated and its influence on daily rates of carbon dioxide uptake by deciduous leaves in a coastal plain forest were found to be 15% or less (Botkin, 1968). Flow rates within the air of the chamber are unnatural in that it is impractical to simulate variable flow that mimics normal air movement within a forest canopy. Flow rates have a large effect on carbon dioxide uptake only when air flow is so slow that photosynthetic rate is restricted by carbon dioxide availability. Our experimental flow rates were maintained at 5 liters min^{-1} or better to avoid this problem.

A completely portable data monitoring system capable of recording photosynthetic rates, air pollutant levels, and pertinent environmental conditions was developed. The system recorded light intensity, temperature, and atmospheric concentrations of carbon dioxide, water vapor, and ozone or sulfur dioxide sequentially and continuously during experimental runs. Transducer measurements of all parameters, produced as a millivolt signal, were transmitted to a digital volt meter and through interfacing electronics to a paper tape punch, where each signal was recorded as a voltage. A schematic outline of the system is presented in Figure 11-3.

Our application of the small chamber technique to the study of eastern white pine photosynthesis as influenced by ozone has been presented in Botkin et al. (1971, 1972). All experimental trees were selected from suburban Connecticut and New York nurseries in locations subject to ozone pollution. Our monitoring system was applied to trees growing in large controlled environment facilities. Chamber temperature was maintained at $26°C$ daytime and $17°C$ nighttime with a photoperiod of approximately 13 hr. Lighting was mixed fluorescent and incandescent, giving a maximum of 0.23 g cal cm^{-2} min^{-1} at the top of the trees. Trees were exposed to ozone only during illumination.

The threshold of ozone suppression of white pine photosynthesis as indicated by this research was approximately 50 pphm (980 $\mu g\ m^{-3}$) for a minimum of 4 hr. Above this threshold, we distinguished three categories of ozone sensitivity: in sensitive trees dosages of 50-100 pphm (980-1960 $\mu g\ m^{-3}$) for 10 hr reduced net photosynthesis to essentially zero (Figure 11-4); in intermediate trees dosages within a similar range reduced net photosynthesis by approximately 50% (Figure 11-5); and in resistant trees these dosages had no effect on photosynthesis (Figure 11-6). In the intermediate trees, ozone-induced photosynthetic suppression was reversible if an ozone-free recovery period was made possible. Visible symptom expression (chlorosis or distal necrosis) on the foliage of the current year was not a good index of the timing or the severity of photosynthetic suppression of the needles following ozone exposure. Visible symptoms were induced by ozone exposure in some trees. Where both occurred, photosynthetic suppression preceded visible indication of injury.

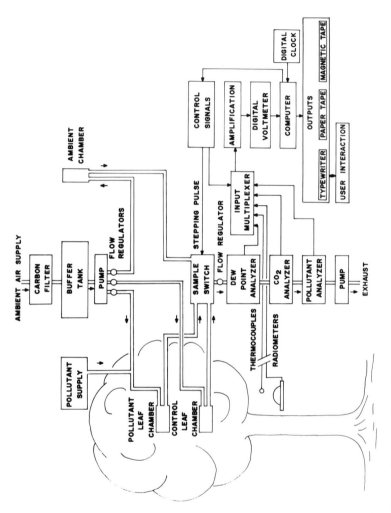

Figure 11-3. Outline of the data acquisition and recording system employed in the Botkin et al. (1971, 1972) small chamber technique for evaluating the influence of gaseous air pollutants on tree photosynthesis.

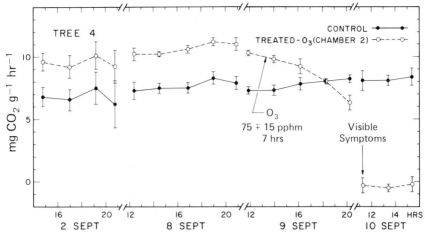

Figure 11-4. Net photosynthetic rates of a sensitive white pine exposed to ozone. Points represent means for 2-hr periods with 95% confidence intervals. Where not shown, confidence intervals were too small to graph. (From Botkin et al., 1972.)

Carlson (1979) employed the small chamber technique as described to evaluate the influence of sulfur dioxide, ozone and combinations of both gases on 8- to 15-year-old saplings of black oak, sugar maple, and white ash collected from natural Connecticut forests and grown in containers filled with natural forest soil profiles. Treatments included fumigation at low (0.2-0.4), intermediate (0.5-0.6), and high (1.3-1.5 cal cm^{-2} min^{-1}) light intensity and low (22-43%) and high (55-92%) relative humidity. Following 1 week of fumigation at low humidity and low light intensity the rate of photosynthesis was 52, 46, and 80%

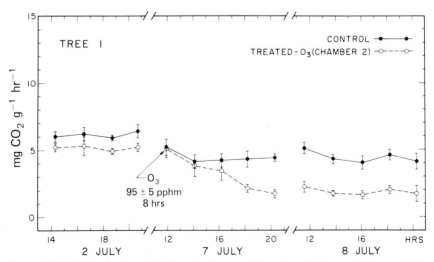

Figure 11-5. Net photosynthetic rates for a white pine of intermediate sensitivity to ozone. (From Botkin et al., 1972.)

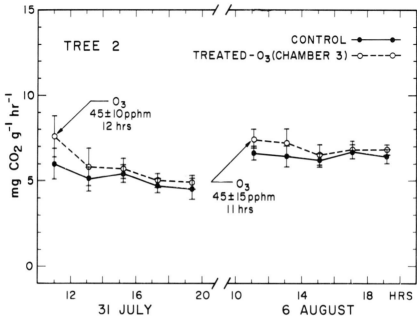

Figure 11-6. Net photosynthetic rates for a white pine resistant to ozone. (From Botkin et al., 1972.)

of control for treatment with 50 pphm (1400 μg m^{-3}) sulfur dioxide; 52, 73, and 100% of control for treatment with 50 pphm (1100 μg m^{-3}) ozone; and 56, 59, and 62% of control for treatment with 50 pphm sulfur dioxide plus 50 pphm ozone, respectively, for black oak, sugar maple, and white ash. After 3 weeks of treatment, sulfur dioxide reduced photosynthesis to 26, 57, and 93% of control while ozone reduced carbon dioxide uptake to 57, 45, and 94% of control, respectively, in black oak, sugar maple, and white ash. These results are summarized in Table 11-4. Simultaneous exposure to both sulfur dioxide and ozone caused a greater than additive reduction in net photosynthesis during the first 2 days of treatment in the case of sugar maple and white ash.

E. Photosynthetic Suppression: Large Forest Trees

The influence of air contaminants on photosynthesis of large trees growing under field conditions is very poorly understood. Experimental designs necessary for this research are complicated and expensive. Soil moisture, relative humidity, temperature, and light characteristics all influence photosynthetic rates under natural conditions and all must be continuously and carefully monitored. Access to mid- and upper-crown areas of large trees is generally by elaborate scaffolding which is awkward, risky, and expensive to erect.

Table 11-4. Influence of Sulfur Dioxide, Ozone, and a Mixture of Both Gases on the Net Photosynthesis of Forest Saplings following 1 to 2 days, 1 week, and 3 weeks of Intermittent Fumigation[a]

Species	Treatment (pphm)	% expected rate of net photosynthesis ± 1 SD		
		1-2 days	1 week	3 weeks
Black oak	50 SO_2	60 ± 15	52 ± 11	26 ± 7
	50 O_3	70 ± 10	52 ± 5	57 ± 5
	50 SO_2 + 50 O_3	55 ± 8	56 ± 9	−
Sugar maple	50 SO_2	78 ± 5	46 ± 9	57 ± 10
	50 O_3	79 ± 10	73 ± 16	45 ± 11
	50 SO_2 + 50 O_3	26 ± 10	59 ± 10	−
White ash	50 SO_2	73 ± 13	80 ± 12	93 ± 5
	50 O_3	100 ± 18	100 ± 15	94 ± 12
	50 SO_2 + 50 O_3	54 ± 8	62 ± 13	−

Source: Carlson (1979).
[a]The small chamber technique was employed along with environmental conditions of varying temperature, humidity, and light conditions.

Legge et al. (1977) have conducted field studies of photosynthesis in lodgepole pine-jack pine hybrids, white spruce, and aspen potentially influenced by sulfur dioxide and hydrogen sulfide from a natural gas processing plant near Whitecourt, Alberta, Canada. Small assimilation chambers, or cuvettes, were placed around portions of tree branches and used to monitor carbon dioxide flux. A 15-m high scaffold was used to gain access to midcrown foliage of the pine hybrids (20 m ht.). The spruce (3 m ht.) and aspen (2 m ht.) were reached from the ground. Efforts to correlate trends in photosynthetic rates of these trees with episodic release of sulfur gases from the processing plant were frustrated by reductions in light quantity coincident with peak sulfur gas exposure and the fact that the measurements were made in September and the authors judged that foliar senescence may have been a factor in the metabolic responses observed. Nevertheless, ambient sulfur dioxide concentrations ranging from 5-10 pphm (131-262 μg m^{-3}) and persisting for several hours in the study forest canopy were recorded. It was the authors' judgment that all the sampled trees exhibited less than expected photosynthetic rates. The maximum rates for monitored trees were pine, 3.28 mg dm^{-2} hr^{-1}; spruce, 2.3 mg dm^{-2} hr^{-1}; and aspen, 3 mg dm^{-2} hr^{-1}. Additional data would be required to more specifically relate ambient sulfur gas concentrations to photosynthetic performance of the forest surrounding the Windfall Gas Plant.

We have made an effort to monitor the photosynthetic response of mature deciduous species growing in natural Connecticut forests to artificially applied ozone by employing the portable small chamber technique described in Section D (Figure 11-7). Preliminary and unpublished results indicated that thresholds of suppression in red maple and white ash were comparable to the threshold determined for young white pine.

Figure 11-7. Field installation employed to obtain information on the photosynthetic response of mature Connecticut deciduous trees to artificially applied ozone using the small chamber technique. Scaffolding employed to gain upper crown access is visible to the left. The trailer to the right housed all data monitoring and recording equipment while the polyethylene "balloon" was used to stabilize air to be supplied to the chambers against short-term fluctuations in carbon dioxide concentration.

F. Photosynthetic Response to Air Contaminants: Mechanisms of Suppression

As is evident from this review, the typical method for monitoring the photosynthetic rate of forest species is carbon dioxide flux. When used to evaluate the influence of an air contaminant, this method tells us little about the physiological and biochemical mechanism(s) of alterations in rates that may be observed. Fortunately this topic has been approached by numerous investigators, commonly employing nonwoody species, and several hypotheses have been offered. While a thorough discussion of these is considered to be beyond the scope of this book, the most important suggestions have been as follows. Evidence has been provided to suggest that air pollutants may alter stomatal opening, interfere with chloroplast membranes, influence chlorophyll concentrations, affect pH, alter redox reactions, electron flow, and phosphorylation, or attack critical proteins or enzymes involved in photosynthesis. Reviews addressing the relationship between photosynthetic metabolism and specific air pollutants are available for sulfur dioxide (Hällgren, 1978; Ziegler, 1975; Keller and Schwager, 1977), ozone (Evans and Ting, 1974; Verkroost, 1974), fluoride (McCune and Weinstein, 1971), and heavy metals (Arndt, 1974; Höll and Hampp, 1975; Flückiger et al., 1978).

G. Summary

Photosynthesis is the most fundamental metabolic process of forest ecosystems and is the primary determinant of growth and biomass accumulation. The rate of net photosynthesis of mature trees frequently is within the range of 10-200 mg of carbon dioxide taken up per gram of dry weight per day. The rate is extremely variable, however, and influenced by genetic, clonal, and provenance differences, season of the year, time of day, position within the crown of the tree, age of foliage, climate, and edaphic factors.

Studies with a wide variety of agricultural and herbaceous species, under controlled environmental conditions, have indicated that air contaminants must be added to the list of environmental variables that can potentially alter the rate of photosynthesis.

Because of ease of handling and experimental design, investigators studying the relationship between air pollutants and tree photosynthesis have primarily employed tree seedlings for research material and controlled environmental facilities for growth. Evidence has been provided, under the above circumstances, for photosynthetic suppression caused by sulfur dioxide, ozone, fluoride, heavy metals, and coal dust. The thresholds of photosynthetic toxicity for tree seedlings vary with individual species and individual pollutants. For several seedlings the threshold of sulfur dioxide photosynthetic influence may approximate 1 ppm (2620 μg m^{-3}) or less for an exposure of several hours. For ozone, the threshold of photosynthetic response may approximate 0.5 ppm (980 μg m^{-3}) or less for an exposure of several days (Table 11-5).

Considerable risk is associated with extrapolation of seedling photosynthetic data accumulated in controlled environmental facilities to older trees in natural forests. Excised leaf and small chamber techniques, therefore, have been employed to assess the air pollutant influence on photosynthetic rates of trees 5 years old and older. The use of sapling-age experimental material avoids the unique characteristics of seedling metabolism. Evidence for forest tree sapling photosynthetic suppression has been presented for sulfur dioxide, ozone, and cadmium. For sulfur dioxide and ozone exposure, the sapling evidence suggests that the threshold of photosynthetic reduction may approximate 0.5 to 1 ppm for 5-10 hr for 1 or 2 days (Table 11-6).

Much of the seedling and sapling evidence suggests that the photosynthetic inhibition caused by sulfur dioxide and ozone is reversible if the pollutant stress is removed. Under the circumstance of variable pollutant concentration in ambient atmospheres, photosynthetic recovery might be common. Synergism, or

Table 11-5. Threshold Dose for Photosynthetic Suppression of Selected Forest Tree Seedlings by Air Contaminants

Pollutant	Concentration		Time	Experiment duration	Species	Reference
SO_2	6 ppm	$(15.7 \times 10^3 \ \mu g \ m^{-3})$	4-6 hr	Single treatment	Red maple	Roberts et al. (1971)
	1 ppm	$(2620 \ \mu g \ m^{-3})$	2-4 hr	Single treatment	Quaking aspen White ash	Jensen and Kozlowski (1974)
	0.1 ppm	$(262 \ \mu g \ m^{-3})$	Continuous	2 weeks	White fir	Keller (1977b)
	0.2 ppm	$(524 \ \mu g \ m^{-3})$	Continuous	2 weeks	Norway spruce Scotch pine	Keller (1977b)
O_3	0.30 ppm	$(588 \ \mu g \ m^{-3})$	9 hr day^{-1}	10 days	Ponderosa pine	Miller et al. (1969)
	0.15 ppm	$(294 \ \mu g \ m^{-3})$	Continuous	19 days	White pine	Barnes (1972)
	0.15 ppm	$(294 \ \mu g \ m^{-3})$	Continuous	84 days	Slash pine Pond pine Loblolly pine	Barnes (1972)
F	$30 \ \mu g \ g^{-1}$ d.w. basis foliar tissue				Pine (various)	Keller (1977c)
Pb	$<10 \ \mu g \ g^{-1}$ d.w. basis foliar tissue				American sycamore	Carlson and Bazzaz (1977)
Cd	$<10 \ \mu g \ g^{-1}$ d.w. basis foliar tissue				American sycamore	Carlson and Bazzaz (1977)

Table 11-6. Threshold Dose for Photosynthetic Suppression of Selected Forest Tree Saplings by Air Contaminants

Pollutant	Concentration	Time	Experiment duration	Species	Reference
SO$_2$	1 ppm (2620 μg m^{-3})	30 min	Single treatment	Silver maple (excised leaves)	Lamoreaux and Chaney (1978)
	0.5 ppm (1310 μg m^{-3})	7-11 hr	1-2 days	Black oak	Carlson (1979)
	0.5 ppm (1310 μg m^{-3})	7-11 hr	1-2 days	Sugar maple	
	0.5 ppm (1310 μg m^{-3})	7-11 hr	1-2 days	White ash	
O$_3$	0.5 ppm (980 μg m^{-3})	4 hr	Single treatment	White pine	Botkin et al. (1972)
O$_3$	0.5 ppm (980 μg m^{-3})	7-11 hr	1-2 days	Black oak	Carlson (1979)
	0.5 ppm (980 μg m^{-3})	7-11 hr	1-2 days	Sugar maple	Carlson (1979)
Cd	$\approx$ 100 μg g^{-1} (?)	45 hr	Single treatment	Silver maple (excised leaves)	Lamoreaux and Chaney (1978)

greater stress resulting from simultaneous pollutant exposure relative to either pollutant alone, appears frequently in the seedling and sapling literature. Evidence for synergistic photosynthetic suppression by sulfur dioxide and ozone and fluoride and cadmium has been presented. Almost all the studies report photosynthetic depression in the absence of, or at least prior to, the appearance of visible foliar symptoms.

The evidence for air pollution induced photosynthetic suppression in large trees in natural environments is extremely meager and fragile. The seedling-sapling evidence, however, demonstrates a threshold of effect that approaches ambient concentrations in numerous temperate environments. Because of the profound importance of the photosynthetic process and the potential for suppression by widespread air contaminants, appropriate field studies must be conducted in spite of their difficulty and cost. The opportunity to examine the impact of contaminants on respiration and transpiration should also be included in experimental designs. Inclusion of one or both of these physiologic processes in seedling-sapling studies has revealed some indication for significant alteration. Increased respiration coupled with reduced photosynthesis could exacerbate growth consequences.

References

Anderson, M. C. 1973. Solar radiation and carbon dioxide in plant communities —conclusions. *In*: J. P. Cooper (Ed.), Photosynthesis and Productivity in Different Environments. Cambridge Univ. Press, New York, pp. 245-354.

Arndt, V. 1974. The Kaulsky-effect: A method for the investigation of the actions of air pollutants in chloroplasts. Environ. Pollut. 6:181-194.

Ashenden, T. W. 1979. Effects of SO_2 and NO_2 pollution on transpiration in *Phaseolus vulgaris*. Environ. Pollut. 18:45-50.

Auclair, D. 1976. Effects of dust on photosynthesis. I. Effects of cement and coal dust on photosynthesis of spruce. Ann. Sci. For. 33:247-255.

Auclair, D. 1977. Effects of dust on photosynthesis. II. Effects of particulate matter on photosynthesis of Scots pine and poplar. Ann. Sci. For. 34:47-57.

Barnes, R. L. 1972. Effects of chronic exposure to ozone on photosynthesis and respiration of pines. Environ. Pollut. 3:133-138.

Bauer, H., W. Larcher, and R. B. Walker. 1973. Influence of temperature stress on CO_2-gas exchange. *In*: J. D. Cooper (Ed.), Photosynthesis and Productivity in Different Environments. Cambridge Univ. Press, New York, pp. 557-586.

Bazzaz, F. A., R. W. Carlson, and G. L. Rolfe. 1974a. The effect of heavy metals on plants: Part 1. Inhibition of gas exchange in sunflower by Pb, Cd, Ni and Tl. Environ. Pollut. 7:241-246.

Bazzaz, F. A., G. L. Rolfe, and R. W. Carlson. 1974b. Effect of cadmium on photosynthesis and transpiration of excised leaves of corn and sunflower. Physiol. Plant. 34:373-376.

Blackman, G. E., and J. N. Black. 1959. Physiological and ecological studies in the analysis of plant environment. XII. The role of the light factor in limiting growth. Ann. Bot. N. S. 23:131-145.

Blackman, G. E., and G. L. Wilson. 1951. Physiological and ecological studies in the analysis of plant environment. VI. The constancy for different species of a logarithmic relationship between the net assimilation rate and light intensity and its ecological significance. Ann. Bot. N. S. 15:63-94.

Bosian, G. 1965. The controlled climate in the plant chamber and its influence upon assimilation and transpiration. *In*: Methodology of Plant Ecophysiology. Proc. Montpellier Symp., UNESCO, New York, pp. 225-232.

Botkin, D. B. 1968. Observed and predicted rates of carbon dioxide uptake for oak leaves in a coastal plain forest. Ph.D. Thesis. Rutgers Univ., New Brunswick, New Jersey, 171 pp.

Botkin, D. B., W. H. Smith, R. W. Carlson. 1971. Ozone suppression of white pine net photosynthesis. J. Air Pollut. Control Assoc. 21:778-780.

Botkin, D. B., W. H. Smith, R. W. Carlson, and T. L. Smith. 1972. Effects of ozone on white pine saplings: Variation in inhibition and recovery of net photosynthesis. Environ. Pollut. 3:273-289.

Bourdeau, P. F. 1963. Photosynthesis and respiration of *Pinus strobus* seedlings in relation to provenance and treatment. Ecology 44:710-716.

Carlson, R. W. 1979. Reduction in the photosynthetic rate of *Acer, Quercus* and *Fraxinus* species caused by sulphur dioxide and ozone. Environ. Pollut. 18:159-170.

Carlson, R. W., and F. A. Bazzaz. 1977. Growth reduction in American sycamore (*Plantanus occidentalis* L.) caused by Pb-Cd interaction. Environ. Pollut. 12:243-253.

Carlson, R. W., F. A. Bazzaz, and G. L. Rolfe. 1975. The effect of heavy metals on plants. II. Net photosynthesis and transpiration of whole corn and sunflower plants treated with Pb, Cd, Ni and Tl. Environ. Res. 10:113-120.

Coyne, P. I., and G. E. Bingham. 1978. Photosynthesis and stomatal light responses in snap beans exposed to hydrogen sulfide and ozone. J. Air Pollut. Control Assoc. 28:1119-1123.

De Koning, H. W., and Z. Jegier. 1968. A study of the effects of ozone and sulfur dioxide on the photosynthesis and respiration of *Euglena gracilis*. Atmos. Environ. 2:321-326.

Erickson, L. C., and R. T. Wedding. 1956. Effects of ozonated hexene on photosynthesis and respiration of *Lemma minor*. Am. J. Bot. 43:32-36.

Evans, L. S., and I. P. Ting. 1974. Ozone sensitivity of leaves: Relationship to leaf water content, gas transfer resistance, and anatomical characteristics. Am. J. Bot. 61:592-597.

Flückiger, W., H. Flückiger-Keller, and J. J. Oertli. 1978. Biochemische Veränderungen in jungen Birken im Nahbereich einer Autobahn. Eur. J. For. Pathol. 8:154-163.

Fritts, H. C. 1966. Growth-rings of trees: Their correlation with climate. Science 154:973-979.

Govindjee. 1975. Bioenergetics of Photosynthesis. Academic Press, New York, 678 pp.

Hällgren, J. E. 1978. Physiological and biochemical effects of sulfur dioxide on plants. *In*: J. E. Nriagu (Ed.), Sulfur in the Environment. Part II. Ecological Impacts. Wiley, New York, pp. 163-209.

Heinicke, D. R., and J. W. Foott. 1966. The effect of several phosphate insecticides on photosynthesis of red delicious apple leaves. Can. J. Plant Sci. 46:589-591.

Helms, J. A. 1965. Diurnal and seasonal patterns of net assimilation in Douglas-fir, *Pseudotsuga menziesii* (Mirb.) Franco, as influenced by environment. Ecology 46:698-708.

Hesketh, J. D., and D. N. Moss. 1963. Variation in the response of photosynthesis to light. Crop Sci. 3:107-110.

Hill, A. C., and J. H. Bennett. 1970. Inhibition of apparent photosynthesis by nitrogen oxides. Atmos. Environ. 4:341-348.

Hill, A. C., and N. Littlefield. 1969. Ozone. Effect on apparent photosynthesis, rate of transpiration, and stomatal closure in plants. Environ. Sci. Technol. 3:52-56.

Hodges, J. D. 1962. Photosynthetic efficiency and patterns of photosynthesis of seven different conifers under different natural environmental conditions. M. F. Thesis, Univ. Washington, Seattle, Washington, 99 pp.

Höll, W., and R. Hampp. 1975. Lead and plants. Residue Rev. 54:79-111.

Huber, B., and H. Polster. 1955. Zur Frage der physiologischen Ursachen der unterschiedlichen Stofferzeugung von Pappelklonen. Bio. Zentralbl. 74:370-420.

Jensen, K. F., and T. T. Kozlowski. 1974. Effect of SO_2 on photosynthesis of quaking aspen and white oak seedlings. North Amer. For. Biol. Workshop Proc. 3:359.

Keller, T. 1968. The influence of mineral nutrition on gaseous exchange by forest trees. *In*: Phosphorus in Agriculture. Inter. Superphosphate Manufact. Assoc., Ltd., Agricul. Comm. Bull. No. 50, June 1968, Paris, pp. 1-11.

Keller, T. 1970. Gaseous exchange—A good indicator of nutritional status and fertilizer response of forest trees. Proc. 6th Internat. Colloq. Plant Analysis and Fertilizer Problems. ISHS, Tel Aviv, pp. 669-678.

Keller, T. 1973. On the phytotoxicity of dust-like fluoride compounds. Staub Reinhal. Luft 33:379-381.

Keller, T. 1977a. Definition and importance of latent injury by air pollution. Allg. Forst-u. Jagdztg. 148:115-120.

Keller, T. 1977b. The effect of long term low SO_2 concentrations upon photosynthesis of conifers. 4th Internat. Clean Air Congress, pp. 81-83.

Keller, T. 1977c. The influence of air pollution by fluorides on photosynthesis of forest tree species. *In*: W. Bosshard (Ed.). Mitt. schweiz. Anst. forstl. Ver'wes 53:163-198.

Keller, T. 1978a. Influence of low SO_2 concentrations upon CO_2 uptake of fir and spruce. Photosynthetica 12:316-322.

Keller, T. 1978b. The influence of SO_2 treatment at different seasons on the spruce trees intake of CO_2 and its yearly growth pattern. Schweizer. Zeitsch. Forstwesen 129:381-393.

Keller, T. 1979. The influence of SO_2 gasing on the growth of spruce tree roots. Schweizer. Zeitsch. Forstwesen 130:429-435.

Keller, T., and H. Schwager. 1977. Air pollution and ascorbic acid. Eur. J. For. Pathol. 7:338-350.

Kozlowski, T. T., and T. Keller. 1966. Food relations of woody plants. Bot. Rev. 32:293-382.

Kramer, P. A., and T. T. Kozlowski. 1960. Physiology of Trees. McGraw-Hill, New York, 642 pp.

Kramer, P. J., and J. P. Decker. 1944. Relation between light intensity and rate of photosynthesis of loblolly pine and certain hardwoods. Plant Physiol. 19: 350-358.

Kramer, P. J., and T. T. Kozlowski. 1979. Physiology of Woody Plants. Academic Press, New York, 811 pp.

Krueger, K. W., and W. K. Ferrell. 1965. Comparative photosynthetic and respiratory responses to temperature and light by *Pseudotsuga menziesii* var. *menziesii* and var. *glauca* seedlings. Ecology 46:797-801.

Lamoreaux, R. J., and W. R. Chaney. 1978. Photosynthesis and transpiration of excised silver maple leaves exposed to cadmium and sulphur dioxide. Environ. Pollut. 17:259-268.

Legge, A. H., and D. R. Jaques. 1977. Field studies of pine, spruce and aspen periodically subjected to sulfur gas emissions. Water, Soil, Air Pollut. 8:105-129.

Linder, S., and E. Troeng. 1977. The seasonal course of net photosynthesis and stem respiration in a 20-year-old stand of Scots pine (*Pinus silvestris* L.) V. K. Sci. Comm., 4th Internat. Congr. Photosynthesis, London, p. 221.

Macdowall, F. D. H. 1965. Stages of ozone damage to respiration of tobacco leaves. Can. J. Bot. 43:419-427.

McCune, D. C., and L. H. Weinstein. 1971. Metabolic effects of atmospheric fluorides on plants. Environ. Pollut. 1:169-174.

McLean, F. T. 1920. Field studies of the carbon dioxide absorption of coconut leaves. Ann. Bot. 34:367-389.

Miller, P. R. 1966. The relationship of ozone to suppression of photosynthesis and to the cause of chlorotic decline of ponderosa pine. Diss. Abstr. 26:3574-3575.

Miller, P. R., J. R. Parmeter, Jr., B. H. Flick, and C. W. Martinez. 1969. Ozone dosage response of ponderosa pine seedlings. J. Air Pollut. Control Associ. 19:435-438.

Muller, R. N., J. E. Miller, and D. G. Sprugel. 1978. Photosynthetic response of field-grown soybeans to fumigations with sulfur dioxide. Argonne National Laboratory, Environ. Res. Contr. No. 78-18. Argonne, Illinois, 21 pp.

Nátr, L. 1973. Influence of mineral nutrition on photosynthesis and the use of assimilates. *In*: J. P. Cooper (Ed.), Photosynthesis and Productivity in Different Environments. Cambridge Univ. Press, New York, pp. 537-555.

Nazirov, N. N. 1966. Deistvie ioniziruy-uschei radiatsu na fatozintez u razlichnykk po radioustoichivasti sortov khlopchatnika. Uzbeksku Biol. Zh. 10: 3-8.

Nellen, V. R. 1966. Über den Einfluss des Salzgehaltes auf die photosynthetische Leistung verschiedener Standardformen von *Delesseria sanguinea* und *Fucus serratus.* Helgoländer Wiss. Meeresunters 13:288-313.

Pharis, R. P., and F. W. Woods. 1960. Effects of temperature upon photosynthesis and respiration of Choctawatchee sand pine. Ecology 41:797-799.

Pisek, A., and W. Tranquillini. 1954. Assimilation und Kohlenstoffhaushalt in der Krone von Fichten (*Picea excelsa* Link) und Rotbuchenbäumen (*Fagus silvatica* L.) Flora (Jena) 141:237-270.

Pisek, A., and E. Winkler. 1958. Assimilationsvermögen und Respiration der Fichte (*Picea excelsa* Link) in verschiedener Höhenlage und der Zirbe (*Pinus cembra* L.) der alpinen Waldgrenze. Planta 51:518-543.

Polster, H. 1950. Die physiologischen Grundlagen der Stofferzeugung im Walde. Untersuchungen über Assimilation, Respiration und Transpiration unserer Hauptholzarten. Bayrischer Landwirtschaftsverlag, G.m.b.H. München, 96 pp.

Polster, H., and G. Weise. 1962. Vergleichende Assimilation-suntersuchungen an Klonen verschiedener Lärchenherkünfte (*Larix decidua* and *Larix leptolepis*) unter Freiland und Klimaraumbedingungen. Zuchter 32:103-110.

Roberts, B. R., A. M. Townsend, and L. S. Dochinger. 1971. Photosynthetic response to SO_2 fumigation in red maple. Plant Physiol. 47:30.

Sasaki, S., and T. T. Kozlowski. 1966a. Variable photosynthetic responses of *Pinus resinosa* seedlings to herbicides. Nature 209:1042-1043.

Sasaki, S., and T. T. Kozlowski. 1966b. Effects of herbicides on carbon dioxide uptake by pine seedlings. Can. J. Bot. 45:961-971.

Sij, J. W., and C. A. Swanson. 1974. Short-term kinetic studies on the inhibition of photosynthesis by sulfur dioxide. J. Environ. Qual. 3:103-107.

Slavik, B. 1973. Water stress, photosynthesis and the use of photosynthates. *In*: J. P. Cooper (Ed.), Photosynthesis and Productivity in Different Environments. Cambridge Univ. Press, New York, pp. 511-536.

Stocker, O. 1960. Die photosynthetischen Leistungen der Steppen und Wüstenpflanzen. *In*: W. Ruhland (Ed.), Handbuch der Pflanzenphysiologie 5:460-491.

Strain, B. R., and V. C. Chase. 1966. Effect of past and prevailing temperatures on the carbon dioxide exchange capacities of some woody desert perennials. Ecology 47:1043-1045.

Taylor, O. C., W. M. Dugger, Jr., M. D. Thomas, and C. R. Thompson. 1961. Effect of atmospheric oxidants on apparent photosynthesis in citrus trees. Plant Physiol. (Suppl.) 36:xxvi.

Thomas, M. D., and G. R. Hill. 1937. Relation of sulphur dioxide in the atmosphere to photosynthesis and respiration of alfalfa. Plant Physiol. 12:309-383.

Thomas, M. D., and G. R. Hill. 1949. Photosynthesis under field conditions. *In*: J. Frank and W. F. Loomis (Ed.), Photosynthesis in Plants. Iowa State Univ. Press, Ames, Iowa, pp. 19-52.

Todd, G. W. 1958. Effect of ozone and ozonated 1-hexene on respiration and photosynthesis of leaves. Plant Physiol. 33:416-420.

Todd, G. W., and B. Propst. 1963. Changes in transpiration and photosynthetic rates of various leaves during treatment with ozonated hexene. Physiol. Plant. 16:57-65.

Verkroost, M. 1974. The effect of ozone on photosynthesis and respiration of *Scendesmus obtusiusculus* Chod., with a general discussion of effects of air pollutants in plants. Mededelingen Landbouwhogeshool Wageningen 19:1-78.

Waugh, J. G. 1939. Some investigations on the assimilation of apple leaves. Plant Physiol. 14:436-477.

White, K. L., A. C. Hill, and J. H. Bennett. 1974. Synergistic inhibition of apparent photosynthetic rate of alfalfa by combinations of sulfur dioxide and nitrogen dioxide. Environ. Sci. Technol. 8:575-576.

Woodwell, G. M., and D. B. Botkin. 1970. Metabolism of terrestrial ecosystems by gas exchange techniques: The Brookhaven approach. *In*: D. E. Reichle (Ed.), Analysis of Temperate Forest Ecosystems. Springer-Verlag, New York, pp. 73-85.

Woodwell, G. M., and R. H. Whittaker. 1968. Primary production in terrestrial ecosystems. Amer. Zoolog. 8:19-30.

Zelitch, I. 1975. Environmental and biological control of photosynthesis: General assessment. *In*: R. Marcelle (Ed.), Environmental and Biological Control of Photosynthesis. Dr. W. Junk, The Hague, pp. 251-262.

Ziegler, I. 1975. The effect of SO_2 pollution on plant metabolism. Residue Rev. 56:79-105.

12

Forest Stress: Influence of Air Pollutants on Phytophagous Forest Insects

Arthropods have roles of enormous importance in the structure and function of terrestrial ecosystems. Forest ecosystems, in particular, typically have large and diverse arthropod populations. The importance of pollinating (Chapter 7) and litter metabolizing (Chapter 8) species has already been introduced. The damaging influence of high population densities of certain insects can be very visible and cause widespread forest destruction; witness contemporary North American situations involving the Douglas fir tussock moth, the gypsy moth, the eastern spruce budworm, and the southern pine bark beetle. It is critically important, however, to keep in perspective that there is substantial evidence to support the notion that forest insects, even those that cause massive destruction in the short run, may play essential and beneficial roles in forest ecosystems in the long run. These roles may involve regulation of tree species competition, species composition and succession, primary production, and nutrient cycling (Huffaker, 1974; Mattson and Addy, 1975).

There is increasing indication that a variety of particularly damaging forest insects detect and respond to stress induced alterations in host tree physiology. The stresses are variable and may include microbial infection, climatic extremes, edaphic factors, and age. Massive insect infestations are characteristically initiated in middle-aged to mature forests typified by reduced productivity rates. Localized and scattered insect outbreaks are associated with forests of all ages, but are generally associated with the least vigorous trees with slow growth rates. Some investigators judge that insect population growth is inversely related to host plant vigor (Mattson and Addy, 1975).

It is essential, therefore, to appreciate the interactions between air pollutants and forest insects because of the critical importance of these animals to forest ecosystem structure and function and because air contaminants may be an additional environmental stress factor capable of predisposing forest tree species to detrimental arthropod influence.

Air pollutants may directly affect insects by influencing growth rates, mutation rates, dispersal, fecundity, mate finding, host finding, and mortality. Indirect effects may occur through changes in host age structure, distribution, and acceptance. Research dealing with these possible interactions, however, is not extensive despite the fact that insect-air pollution has a research history that extends over fifty years. European literature dealing with this topic is substantially larger than North American literature.

A variety of studies has presented data indicating that species composition or population densities of insect groups are altered in areas of high air pollution stress, for example, roadside (Przybylski, 1979) or industrial (Lebrun, 1976; Novakova, 1969; Sierpinski, 1967) environments. Specific information is further available on the general influence of polluted atmospheres on population characteristics of forest insects (Charles and Villemant, 1977; Boullard, 1973; Hay, 1975; Schnaider and Sierpinski, 1967; Sierpinski, 1970, 1971, 1972a,b; Sierpinski and Chlodny, 1977; Templin, 1962; Wiackowski and Dochinger, 1973). Johnson (1969) has reviewed much of the literature dealing with air pollutants and insect pests of conifers. One of the most comprehensive literature reviews available concerning forest insects and air contaminants has been presented by Villemant (1979).

A. Sulfur Dioxide

While primarily concerned with honeybees, Hillman (1972) reviews several references concerned with the interaction of insects with sulfur dioxide. Additional references are contained in Ginevan and Lane (1978). As is generally true with insects from a variety of ecosystems, forest insect population densities appear to be both increased and decreased by exposure to sulfur dioxide depending on species. In an inventory of eastern white pine stems for white pine weevil deformity, Linzon (1966) recorded fewer deformed stems near to, relative to far from, sulfur dioxide sources. These results suggest an adverse impact on the weevil. Several European researchers, on the other hand, have associated increased population densities of the European pine shoot moth, a serious pest of red, Scotch, and Austrian pine, with increased ambient levels of sulfur dioxide and smoke (Sierpinski and Chlodny, 1977).

O. L. Gilbert of the University of Sheffield, England has been primarily concerned with lichen distribution as influenced by sulfur dioxide in urban and industrial areas. In one of his surveys of the Newcastle upon Tyne area he recorded arthropod numbers occurring 1-2 m above the ground on the bark of European ash trees. Numbers of herbivorous insects were significantly reduced in regions with high sulfur dioxide concentrations, while carnivorous insects did not show significant correlation with sulfur dioxide levels (Gilbert, 1971).

A few studies have suggested that sulfur dioxide may be involved in predisposition to insect infestation (Struble and Johnson, 1964). Anderson (1970) has reported an association between abnormal Christmas tree growth and an eriophytid mite on white pine subject to sulfur dioxide exposure from a power plant in a West Virginia-Maryland site. Saunders (1972) has also recorded increased mite infestations in Christmas tree plantations damaged by power plant effluent.

B. Ozone

There has been surprisingly little research concerned with the direct influence of ozone on insects. Concentrations of 10 pphm (196 μg m^{-3}) ozone for prolonged exposures have been found to cause mortality of adult house flies (Beard, 1965). Beard also observed reduced egg laying by females following high ozone exposure. Low levels of ozone appeared to have a favorable influence on adult flies. Levy et al. (1972) exposed three Diptera species to high ozone levels and found slightly reduced egg hatch in two species. Adult response, however, included a dramatic stimulation of oviposition with subsequent increase in adult populations. In subsequent trials with ozone exposure of cockroach species and the red imported fire ant, Levy et al. (1974) observed that several days exposure to 30 pphm (588 μg m^{-3}) did not produce any obvious deleterious effect to adult or immature stages. These species, in fact, appeared even more tolerant than house flies. Comparable research for important forest insects is not available.

1. Ozone Predisposition to Bark Beetle Infestation

Bark beetles are the most damaging and economically significant insect pests of commercially important conifers in the United States. Beetle outbreaks in western forests are associated with host weakening caused by microbial infection; for example, root disease initiated by *Fomitopsis annosa* or *Verticicladiella wagenerii* fungi in ponderosa pine (Stark and Cobb, 1969); and insect defoliation, for example, pine looper stripping of ponderosa pine (Dewey et al., 1974); or various climatic stresses including drought and windthrow (Rudinsky, 1962). In the latter 1960s, California investigators added ozone to the list of biotic and environmental stresses that predispose ponderosa pine to bark beetle infestation. This is perhaps the most completely documented example of insect damage enhancement caused by air pollution in North America (Miller and Elderman, 1977).

During the summer of 1966 a survey of ponderosa pines was carried out in the San Bernadino Mountains of California. These forests are subject to elevated atmospheric oxidants from the Los Angeles urban complex to the west. Over 1000 trees were examined for degree of ozone damage and infestation and mortality from either the western pine beetle or mountain pine beetle or both. Trees with the greatest pollution injury were found to be most commonly supporting populations of one or both bark beetle species (Figure 12-1). As the degree of oxidant damage increased, live crown ratio decreased and the occurrence of bark beetle infestation increased (Stark et al., 1968).

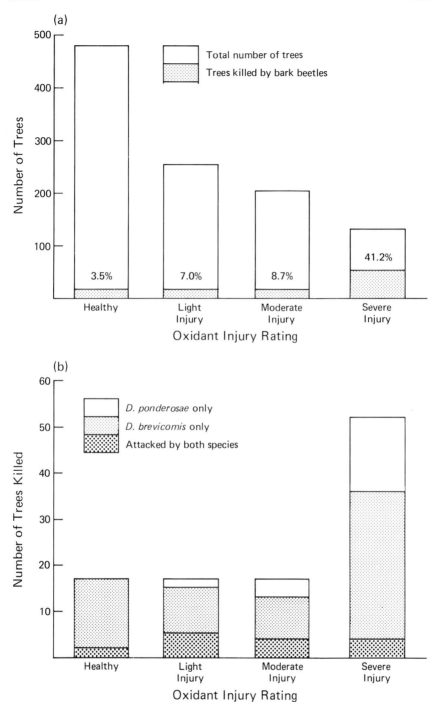

Figure 12-1. Relationship between the degree of photochemical air pollution injury and bark beetle attack (a) and mortality (b) in the San Bernardino Mountains of California. (From Stark and Cobb, 1969.)

In an effort to appreciate ponderosa pine resistance mechanisms for bark beetle infestation, the investigators embarked on an ambitious attempt to examine resin and sapwood characteristics of this species. The goal was to elucidate specific hypotheses regarding the apparent predisposition to insect attack by oxidants. A variety of field and laboratory tests was performed on 65- to 85-year-old ponderosa pine growing in an all-age, second growth stand. Several physiological and chemical differences were recorded between trees exhibiting various degrees of oxidant symptoms (Table 12-1). In addition to the examination of terpene fractions of stem xylem, Cobb et al. (1972) have also examined the essential oils from foliage of stressed and healthy ponderosa pine. No significant differences were observed in the monoterpenoids of foliage, as in the xylem observations, but methyl chavicol was found to be much lower in the injured trees.

Cobb et al. (1968b), based on their various observations, advanced the hypothesis that reduction of oleoresin exudation pressure, quantity, rate of flow, and increased propensity of oleoresin to crystallize, and a reduction in phloem and sapwood moisture content, all correlated with photochemical atmospheric pollution in ponderosa pine, increased the damage caused by the western pine beetle and mountain pine beetle in California. The investigators concluded that the changes induced in resin and sapwood characteristics facilitated bark beetle activity particularly in the concentration and establishment phases. Air pollution stressed trees were judged to be less suitable for beetle breeding than nonstressed trees. As a result, symptomatic trees may act to "trap" beetles. The authors observed that the latter may have accounted for their failure to observe obvious increases in bark beetle infestations in their study sites.

In the southern United States longleaf, loblolly, shortleaf, and slash pines are all subject to economically significant damage by the southern pine beetle. Loblolly and shortleaf are especially susceptible. Hodges et al. (1979) have recently completed measurements on physical and chemical parameters of the resin of 50 trees each of the four southern pine species. As in the California study, susceptibility was strongly correlated with physical properties of the

Table 12-1. Comparison of Resin and Sapwood Characteristics of Ponderosa Pine with and without Symptoms of Oxidant Stress

Characteristic	Without symptoms	With symptoms
Oleoresin exudation pressure	Greater	Lesser
Yield/rate of resin flow	Greater	Lesser
Crystallization rate	Lesser	Greater
Sapwood moisture content	Greater	Lesser
Phloem thickness	Greater	Lesser
Oleoresin chemistry	Same	Same
Resin acid chemistry	Same	Same
Soluble sugars	Greater	Lesser
Reserve polysaccharides	Greater	Lesser
Phloem pH	Same	Same

Source: Cobb et al. (1968a); Miller et al. (1968).

resin. If these properties are influenced by any southern air contaminants, an atmospheric predisposition factor for beetle infestation may occur in the Southeast as well as the West. Fortunately ambient oxidant levels are generally less in the Southeast relative to Southern California or the Northeast.

C. Fluoride

The tendency for fluoride accumulation in the biota has precipitated several studies concerned with fluoride uptake and metabolic influence in insects. Bees have received the greatest research attention (Lezovic, 1969; Caparrini, 1957; Guilhon et al., 1962; Marier, 1968; Maurizio and Staub, 1956), but other species including the Mexican bean beetle (Boyce Thompson Institute, 1971), desert locust, yellow meal worm (Outram, 1970), and others (Weismann and Svatarakova, 1974) have been investigated. A limited number of studies dealing with fluoride and forest insects are available.

1. Glacier National Park and Flathead National Forest Studies

Clinton E. Carlson and co-workers of the Forest Insect and Disease Branch, U.S.D.A. Forest Service, Missoula, Montana, have been involved in a comprehensive study of the influence of fluorides on the forest ecosystems surrounding an aluminum reduction plant in Columbia Falls, Montana (Carlson and Dewey, 1971). Specific attention has been given to the level and consequences of fluoride contamination of trees and insects. Aluminum is produced by the electrolytic reduction of alumina. The process releases to the atmosphere approximately equal amounts of fluoride in the particulate and gaseous forms. Particulates consist of sodium fluoride (NaF) and aluminum fluoride (AlF_3), while gaseous emissions include hydrogen fluoride (HF) along with small amounts of carbon tetrafluoride (CF_4). Symptoms of fluoride injury occur on trees and shrubs throughout large areas of the Glacier National Park and Flathead National Forest in the vicinity of Columbia Falls. High but asymptomatic levels of fluoride are associated with other portions of the forests.

A wide variety of insects was collected within approximately 1 km of the aluminum plant and analyzed for fluoride (Dewey, 1973). Control insects collected from areas not subject to plant emissions had fluoride body burdens ranging from 3.5-16.5 $\mu g\ g^{-1}$ (dry weight basis). Analysis of forest insects within the 1-km zone revealed the following ranges for major insect groups: 58.0-585.0 $\mu g\ g^{-1}$ pollinators, 6.1-170.0 $\mu g\ g^{-1}$ predators, 21.3-255 $\mu g\ g^{-1}$ foliar feeders, and 8.5-52.5 $\mu g\ g^{-1}$ cambial region feeders. Among the phytophagous species, the foliage feeders had the highest mean fluoride concentration. This is presumably due to the injestion of particulate fluoride associated with leaf surfaces as well as the fluoride contained within the leaves. The relatively high body burdens of the predatory ostomids, dragonflies, and damselflies suggested to Dewey that these insects may obtain some fluoride via respiratory processes or food chain accumulation. Dewey also observed that the elevated fluoride in the bark beetles examined may indicate fluoride contamination of host tree vascular tissue as well as foliage.

Lodgepole pine occupies a position of prominence in the forest ecosystems surrounding the aluminum reduction plant. It forms nearly pure stands over roughly 50% of the study area under the influence of fluoride deposition. Over 50 years ago, Keen and Evenden (1929) observed that defoliators, bark beetles, and wood borers of lodgepole pine may have contributed to the mortality of numerous trees which were severely defoliated by fume damge in the vicinity of a smelter at Northport, Washington. Since much of the lodgepole pine in the vicinity of the Columbia Falls aluminum plant was dying or symptomatic, Carlson et al. (1974) initiated efforts to correlate foliar and ambient fluoride concentrations with insect infestation. Stepwise multiple regression techniques were employed to statistically analyze the relationships between damage caused by the pine needle sheath miner, a needler miner, sugar pine tortrix, and the fluoride parameters mentioned. Foliar fluoride concentration was significantly related to needle miner damage and to damage caused by the pine needle sheath miner. The authors concluded that the data strongly suggest that fluoride contamination may be a contributing factor in predisposing lodgepole pines to damage by these insects. While not specifically given by the authors, it may be judged that the foliar fluoride threshold level predisposing to needle miner and pine needle sheath miner may approximate $30 \ \mu g \ g^{-1}$.

2. Other Forest Insect-Fluoride Studies

Foliar fluoride concentrations can become substantial in woody vegetation. McClenahen and Weidensaul (1977) have mapped the distribution of fluorides in black locust leaves in the vicinity of a point source in southeast Ohio (Fig. 12-2).

Elevated fluoride concentrations have been found associated with the European pine shoot moth, in the vicinity of an aluminum production facility (Mankovska, 1975). An infestation of black pine leaf scale on ponderosa pine near Spokane, Washington, from 1948 to 1950, originally thought associated with fluoride released from an aluminum facility, was ultimately judged to be correlated with cement and silicon dust (Johnson, 1950).

D. Trace Metals

In consideration of the potential non-soil-inhabiting, phytophagous insects have for accumulation of heavy metals from hosts growing in roadside, industrial, or urban environments, it is surprising to find relatively little published data.

Giles et al. (1973) have examined lead burdens of several insect species in the roadside environment of Interstate 83, 4 km north of Baltimore, Maryland, in an area with an average daily traffic of 13,000 vehicles. Japanese beetles did not show significant increases in lead relative to nonroadside control samples. Damselflies did not exhibit elevated lead burdens consistently over the sampling period. All collections of the European mantids, however, did exhibit higher concentrations of lead than control samples, but the levels were not excessive and averaged approximately 8 μg g^{-1}.

Price et al. (1974) have also sampled insects in the roadside environment from locations having average daily traffic values ranging from < 10 to 12,900 vehicles in the vicinity of Urbana, Illinois. In areas of high lead emission, insect species that suck plant juices, chew plant parts, or prey on other insects had approximately 10, 16, and 25 μg g^{-1} lead, respectively. In low lead emission areas the same feeding categories had 5, 3, and 3 μg g^{-1}, respectively.

E. Other Air Pollutants

Acid precipitation in western and central Europe may predispose Scotch pine to increased infestation by the pine bud moth (Hagvar et al., 1976). Baker and Wright (1977) have observed that the seven-spot ladybird and stick insects are capable of tolerating high concentrations of carbon monoxide for weeks under laboratory conditions.

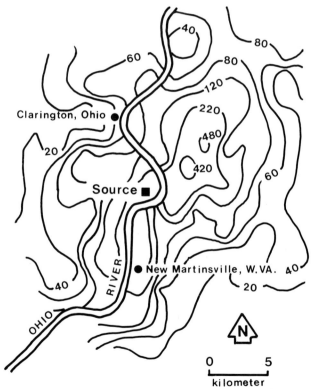

Figure 12.2. Isopleths of black locust foliar fluoride (μg g^{-1} dry wt basis) surrounding an Ohio point source (from McClenahen and Weidensaul, 1977).

F. Summary

The literature addressing the relationship between air pollutants and temperate forest insects is very meager and extremely disproportionate to the importance of arthropods in forest ecosystems.

Much of the research deals with relationships in a superficial manner. Alteration of population densities has been related to general ambient air quality. The specific pollutants present are neither inventoried nor measured. The insect surveys are restricted to one or a few species and do not constitute comprehensive surveys. Very limited information is available on the interaction of specific air contaminants with specific behavioral and physiological functions of insects. Information on fecundity, stage specific morbidity and mortality, and influence on sex ratios, for example, are needed. Ginevan and Lane (1978) have provided an experimental protocol ideally suited for initial information needed for important forest insects. These investigators found that laboratory exposure of developing larvae of fruit flies to sulfur dioxide at 0.7 ppm ($1834\ \mu g\ m^{-3}$) for 10 days resulted in large increases in developmental time and decreased survival. Fecundity was not influenced by adult fumigation at 0.7 ppm for 4 days. Fumigation of prepupae and pupae at 0.4 ppm ($1048\ \mu g\ m^{-3}$) for 4 days resulted in decreased survival that averaged 17%. The authors observed that factors that cause approximately 5-10% mortality in the pupal stage have been judged to have potentially significant effects on natural insect population dynamics. While the doses employed were high, they are not unknown in certain ambient situations.

The limited data available do suggest that insect-air pollutant interactions are likely to be very variable. The same dose of sulfur dioxide may stimulate or inhibit arthropods depending on species. A given insect may be variously influenced by ozone depending on the gas concentration and time of exposure.

The capability that various air pollutants have to alter tree host susceptibility to insect influence is one of the most important relationships between the two potential stresses. Convincing evidence in the case of ponderosa pine has been presented for predisposition to bark beetle infesatation by oxidants. Additional evidence supports the potential importance of sulfur dioxide and fluoride on mite and lepidopterous infestation, respectively, in other pine species. The threshold doses required for significant predisposition under field conditions are not adequately described, but may preliminarily be judged relatively low in the ponderosa and lodgepole pine cases described.

The limited data available concerning insect body burdens of recalcitrant materials for example, fluoride and heavy metals, suggest that fluoride accumulation may be excessive only very close (within 0.5-1 km) to a primary source and that heavy metals may not accumulate to levels characteristic of floral components even in ecosystems subject to excessive loading, for example, roadside environments.

Priority research attention should be given to selected eastern United States insect species, most notably the eastern spruce budworm and gypsy moth. Air pollution interaction with insects predatory on important forest insects should be studied more intensively. Bark and wood boring insects, shoot and bud tunneling insects, and scales and aphids should receive more research attention as they frequently respond favorably to increased host stress and are insulated from the atmosphere for portions of their life cycle.

References

Anderson, R. F. 1970. Relation of insects and mites to the abnormal growth of Christmas trees in Mt. Storm, West Virginia-Gorman, Maryland vicinity. U.S. Environmental Protection Agency Report, Durham, North Carolina, 31 pp.

Baker, G. M., and E. A. Wright. 1977. Effects of carbon monoxide on insects. Bull. Environ. Contam. Toxicol. 17:98-104.

Beard, R. L. 1965. Observation on house flies on high-ozone environments. Ann. Entomol. Soc. Am. 58:404-405.

Boullard, B. 1973. Interactions entre les polluants atmosphériques et certains parasites des essences forestières (champignons et insects). For. Privée 94:31, 33, 35, 37.

Boyce Thompson Institute. 1971. Annual Report of the Boyce Thompson Institute for Plant Research. Yonkers, New York, pp. 4-7.

Caparrini, W. 1957. Fluorine poisoning in domestic animals (cattle) and bees. Zooprofilass 12:249-250.

Carlson, C. E., and J. E. Dewey. 1971. Environmental Pollution by Fluorides in Flathead National Forest and Glacier National Park. U.S.D.A. Forest Service. Forest Insect and Disease Branch, Missoula, Montana, 57 pp.

Carlson, C. E., W. E. Bousfield, and M. D. McGregor. 1974. The relationship of an insect infestation on lodgepole pine to fluorides emitted from a nearby aluminum plant in Montana. Report No. 74-14, U.S.D.A. Forest Service, Div. State Private For., Missoula, Montana, 21 pp.

Charles, P. J., and C. Villemant. 1977. Modifications des niveaux de population d'insectes dans les jeunes plantations de pins sylvestres de la fôret de Roumare (Seine–Maritime) soumises à la pollution atmosphérique. C. R. Acad. Agric. Fr. 63:502-510.

Cobb, F. W., Jr., D. L. Wood, R. W. Stark, and P. R. Miller. 1968a. II. Effect of injury upon physical properties of oleoresin, moisture content, and phloem thickness. Hilgardia 39:127-134.

Cobb, F. W., Jr., D. L. Wood, R. W. Stark, and J. R. Parmeter, Jr. 1968b. Theory on the relationship between oxidant injury and bark beetle infestation. Hilgardia 39:141-152.

Cobb, F. W., Jr., E. Zavarin, and J. Bergot. 1972. Effect of air pollution on the volatile oil from leaves of *Pinus ponderosa*. Phytochemistry 11:1815-1818.

Dewey, J. E. 1973. Accumulation of fluorides by insects near an emission source in western Montana. Environ. Entomol. 2:179-180.

Dewey, J. E., W. M. Ciesla, and H. E. Meyer. 1974. Insect defoliation as a predisposing agent to a bark beetle outbreak in eastern Montana. Environ. Entomol. 3:722.

Gilbert, O. L. 1971. Some indirect effects of air pollution on bark-living invertebrates. J. Appl. Ecol. 8:77-84.

Giles, F. E., S. G. Middleton, and J. G. Grau. 1973. Evidence for the accumu-
   lation of atmospheric lead by insects in areas of high traffic density. Environ.
   Entomol. 2:299-300.
Ginevan, M. E., and D. D. Lane. 1978. Effects of sulfur dioxide in air on the
   fruit fly, *Drosophila melanogaster*. Environ. Sci. Technol. 12:828-831.
Guilhon, J., R. Truhaut, and J. Bernuchon. 1962. Studies on the variations in
   fluorine levels in bees with respect to industrial atmospheric air pollution in a
   Pyrenean village. Acad. d'agr. de France, Compt. Rendt. 48:607-615.
Hagvar, S., G. Abrahamsen, and A. Bokhe. 1976. Attack by the pine bud moth
   in southern Norway: Possible effect of acid pollution. For. Abstr. 37:694.
Hay, C. J. 1975. Arthropod stress. *In*: W. H. Smith and L. S. Dochinger (Eds.),
   Air Pollution and Metropolitan Woody Vegetation. U.S.D.A. Forest Service,
   Publica. No. PIEFR-PA-1, Upper Darby, Pennsylvania, pp. 33-34.
Hillman, C. 1972. Biological effects of air pollution on insects emphasizing the
   reactions of the honey bee (*Apis mellifera* L.) to sulfur dioxide. Ph.D. Thesis.
   Pennsylvania State Univ., State College, Pennsylvania, 170 pp.
Hodges, J. D., W. W. Elam, W. F. Watson, and T. E. Nebeher. 1979. Oleoresin
   characteristics and susceptibility of four southern pines to southern pine
   beetle (Coleoptera: Scolytidae) attacks. Can. Entomol. 111:889-896.
Huffaker, C. B. 1974. Some implications of plant-arthropod and higher-level,
   arthropod-arthropod food links. Environ. Entomol. 3:1-9.
Johnson, P. C. 1950. Entomological aspects of the ponderosa pine blight study,
   Spokane, Washington. Unpubl. Report, U.S.D.A. Bur. Entomol. and Plant
   Quar., Forest Insect Laboratory, Coeur d'Alene, Idaho, 15 pp.
Johnson, P. C. 1969. Atmospheric pollution and coniferophagous invertebrates.
   Proc. 20th Ann. Western For. Insect Work Conf., Coeur d'Alene, Idaho.
Keen, F. P., and J. C. Evenden. 1929. The role of forest insects in respect to
   timber damage and smelter fume area near Northport, Washington. Unpubl.
   Report, U.S.D.A. Bur. Entomol., Stanford Univ., Stanford, California, 12 pp.
Lebrun, P. 1976. Effects écologiques de la pollution atmosphérique sur les popu-
   lations et communautés de microarthropodes corticoles (Acariens, Collem-
   boles et Ptérygotes). Bull. Soc. Ecol. 7:417-430.
Levy, R., Y. J. Chiu, and H. L. Cromroy. 1972. Effects of ozone on three species
   of Diptera. Environ. Entomol. 1:608-611.
Levy, R., D. P. Jouvenaz, and H. L. Cromroy. 1974. Tolerance of three species
   of insects to prolonged exposures to ozone. Environ. Entomol. 3:184-185.
Lezovic, J. 1969. The influence of fluorine compounds on the biological life
   near an aluminum factory. Fluoride Quart. Re. 2:1.
Linzon, S. N. 1966. Damage to eastern white pine by sulfur dioxide, semimature-
   tissue needle blight, and ozone. J. Air Pollut. Control Assoc. 16:140-144.
Mankovska, B. 1975. Influence of fluorine emissions from an aluminum factory
   plant on the content in different developmental stages of European pine
   shoot moth, *Rhyacionia buoliana* Schiff. Biologia (Bratislava) 30:355.
Marier, J. R. 1968. Fluoride research. Science 159:1494-1495.
Mattson, W. J., and N. D. Addy. 1975. Phytophagous insects as regulators of
   forest primary production. Science 190:515-522.
Maurizio, A., and M. Staub. 1956. Poisoning of bees with industrial gases containing fluorine in Switzerland. Schweiz. Bienen Ztg. 79:476-486.

McClenahen, J. R., and T. C. Weidensaul. 1977. Geographic Distribution of Airborne Fluorides Near a Point Source in Southeast Ohio. Ohio Agricultural Research and Development Center, Res. Bull. No. 1093, Wooster, Ohio, 29 pp.

Miller, P. R., and M. J. Elderman (Eds.). 1977. Photochemical Oxidant Air Pollutant Effects on a Mixed Conifer Forest Ecosystem. EPA-600/3-77-104. Environmental Protection Agency, Corvallis, Oregon, 338 pp.

Miller, P. R., F. W. Cobb, Jr., and E. Zavarin. 1968. III. Effect of injury upon oleoresin composition, phloem carbohydrates, and phloem pH. Hilgardia 39:135-140.

Novakova, E. 1969. Influence des pollutions industrielles sur les communautes animals et l'utilisation des animaux comme bioindicateurs. Proc. 1st Eur. Congr. Influence of Air Pollution on Plants and Animals, Wageningen, 1968, pp. 41-48.

Outram, I. 1970. Some effects of fumigant sulphryl fluoride on the gross metabolism of insect eggs. Fluoride Quart. Rep. 3:2.

Price, P. W., B. J. Rathcke, and D. A. Gentry. 1974. Lead in terrestrial arthropods: Evidence for biological concentration. Environ. Entomol. 3:370-372.

Przybylski, Z. 1979. The effects of automobile exhaust gases on the arthropods of cultivated plants, meadows and orchards. Environ. Pollut. 19:157-161.

Rudinsky, J. A. 1962. Ecology of Scolytidae. Annu. Rev. Entomol. 7:327-348.

Saunders, J. L. 1972. Disease and insect pests of Christmas trees. School for Christmas Tree Growers. College of Agriculture, Proc. Cornell Univ., Ithaca, New York, pp. 88-90.

Schnaider, Z., and Z. Sierpinski. 1967. Dangerous condition for some forest tree species from insects in the industrial region of Silesia. Prace Instytut Badawczy Tesnictwa (Warsaw) No. 316, pp. 113-150.

Sierpinski, Z. 1967. Influence of industrial air pollutants on the population dynamics of some primary pine pests. Proc. 14th Congr. Int. Union For. Res. Organiz. 5(24):518-531.

Sierpinski, Z. 1970. Economic significance of noxious insects in pine stands under the chronic impact of the industrial air pollution. Sylwan 114:59-71.

Sierpinski, Z. 1971. Secondary noxious insects of pine in stands growing in areas with industrial air pollution containing nitrogen compounds. Sylwan 115:11-18.

Sierpinski, Z. 1972a. The economic importance of secondary noxious insects of pine on territories with chronic influence of industrial air pollution. Mitt. Forstl. Bundesversuchsanst Wien 97:609-615.

Sierpinski, Z. 1972b. The occurrence of the spruce spider (*Paratetranychus* (*Oligonychus*) *ununquis* Jacoby) on Scotch pine in the range of the influence of industrial air pollution. In: Institute Badawczego Lesnictwa, Warsaw, Bull. No. 433-434, pp. 101-110.

Sierpinski, Z., and J. Chlodny. 1977. Entomofauna of forest plantations in the zone of disastrous industrial pollution. *In*: J. Woldk (Ed.), Relationship Between Increase in Air Pollution Toxicity and Elevation Above Ground. Institute Badawczego Lesnictwa, Warsaw, pp. 81-150.

Stark, R. W., and F. W. Cobb, Jr. 1969. Smog injury, root diseases and bark beetle damage in ponderosa pine. Califor. Agric., Sept., 1969:13-15.

Stark, R. W., P. R. Miller, F. W. Cobb, Jr., D. L. Wood, and J. R. Parmeter, Jr. 1968. I. Incidence of bark beetle infestation in injured trees. Hilgardia 39:121-126.

Struble, G. R., and P. C. Johnson, 1964. Black pine leaf scale. U.S.D.A. Forest Serv., Forest Pest Leaflet No. 91, Washington, D.C., 6 pp.

Templin, E. 1962. On the population dynamics of several pine pests in smoke-damaged forest stands. Wissenschafthche Zeitschrift der Technischen Universität, Dresden 113:631-637.

Villemant, C. 1979. Modifications de l'enlomocenose due pin sylvestre en liaison avec la pollution atmosphérique en fôret de Roumare (Seine-Maritime). Doctoral Dissertation. Pierre and Marie Curie University, Paris, 161 pp.

Weisman, L., and L. Svatarakova. 1974. Toxicity of sodium fluoride on some species of harmful insects. Biologia (Bratislava) 29:847-852.

Wiackowski, S. K., and L. S. Dochinger. 1973. Interactions between air pollution and insect pests in Poland. 2nd Inter. Congr. Plant Pathol., Univ. of Minnesota, Minneapolis, Minnesota. Abstr. No. 0736, p. 1.

13

Forest Stress: Influence of Air Pollutants on Disease Caused by Microbial Pathogens

Abnormal physiology, or disease, in woody plants follows infection and subsequent development of an extremely large number and diverse group of microorganisms internally or on the surface of tree parts. All stages of tree life cycles and all tree tissues and organs are subject, under appropriate environmental conditions, to impact by a heterogeneous group of microbial pathogens including viroids, viruses, mycoplasmas, bacteria, fungi, and nematodes. The influence of a specific disease on the health of an individual tree may range from innocuous to mild to severe. Over extended time periods, the interaction of native pathogens with natural forest ecosystems is significant, and frequently beneficial, in terms of ecosystem development and metabolism. As in the instance of insect interactions (Chapter 12), microbes, and the diseases they cause, play important roles in succession, species composition, density, competition, and productivity. In the short term, the effects of microbial pathogens may conflict with forest management objectives and assume a considerable economic or managerial as well as ecologic significance (Smith, 1970).

The interaction between air pollutants and microorganisms in general is highly variable and complex. Considerable attention has already been given to soil microorganisms (Chapter 8) and symbiotic microbes (Chapter 10). Babich and Stotzky (1974) have provided a comprehensive overview of the relationships between air contaminants and microorganisms. Microbes may serve as a source as well as a sink for air pollutants. A specific air pollutant, at a given dose, may be stimulatory, neutral, or inimical to the growth and development of a particular virus, bacterium, or fungus. In the latter, fruiting body formation, spore production, and spore germination may be stimulated or inhibited. Microorganisms that normally develop in plant surface habitats may be especially subject to air pollutant influence. These microbes have received considerable research attention and have been the subject of review (Saunders, 1971, 1973, 1975). The author has employed a strategy analogous to this volume in an attempt to summarize the interaction between air contaminants and plant-surface microbial ecosystems (Smith, 1976). Class I, II, and III interactions are identifiable for these ecosystems as well as for forest ecosystems. As in the latter case, variable physiologic response of individual elements of the biota translate into increased, no change, or decreased biomass and biological activity at the ecosystem level (Table 13-1).

Microorganisms that function as plant pathogens are, of course, no exception to Table 13-1 generalizations. As a consequence, it is of no surprise that the apparent influence of individual pollutants and combinations of pollutants on microbial plant parasites is to both increase and decrease their activities. The actual impact of air pollution stress on disease expression is especially complicated, however, as the air contaminants not only influence the metabolism and ecology of the microbe but also influence the physiology of the host plant. Even under "unpolluted atmospheric conditions" disease in plants is a complex integration of pathogen physiology, host plant physiology, and ambient environmental conditions. The addition of an air pollutant stress has the effect of adding an additional complexing variable to an already elaborate and complicated interaction. Numerous comprehensive reviews have summarized the interactions between air contaminants and plant diseases. In 1973, Allen S. Heagle, U.S.D.A., Agricultural Research Service, Raleigh, North Carolina, summarized nearly 100 references and found that sulfur dioxide, ozone, or fluoride had been reported to increase the incidence of 21 diseases and decrease the occurrence of 9 diseases in a variety of nonwoody and woody hosts. Michael Treshow of the Department of Biology, University of Utah, Salt Lake City, Utah, has provided a detailed review concerning the influence of sulfur dioxide, ozone, fluoride, and particulates on a variety of plant pathogens and the diseases they cause (Treshow, 1975). Treshow lamented the fact that most of the data available deal with in vitro or laboratory accounts of microbe-air pollutant interactions, while only a few investigations have been performed that have examined the influence of air pollutants on disease development under field conditions. In a review provided by William J. Manning, Department of Plant Pathology, University of Massachusetts, Amherst, Massachusetts, it was pointed out that most research attention has been directed to fungal pathogen-air pollutant interactions (Manning, 1975). Greater research perspective is needed concerning air pollution influence on viruses, bacteria, nematodes, and the diseases they cause. Macroscopic agents of disease, most importantly true- and dwarf-mistletoes, should also be examined relative to air pollution impact, especially in the western part of North America where the latter are extremely important agents of coniferous disease.

Table 13-1. Influence of Air Pollution on Plant-Surface Microbial Ecosystems

Air pollution dose	Response of microbe	Impact on microbial ecosystem	Reaction of host plant
Class I Low	1. Act as a source of air contaminants 2. Act as a sink for air contaminants	1. No effect or potentially some allelopathic influence 2. No or minimal physiological alteration or potentially some fertilization, stimulation	
Class II Intermediate	1. Abnormal metabolism altered pigmentation, morphology, enzyme activity 2. Reduced reproduction (reduced competitiveness) (a) lessened spore production or dispersal (b) reduced or delayed spore germination 3. Reduced growth (reduced productivity and competitiveness) (a) vegetative retardation (b) vegetative inhibition	1. No significant or very minor perturbation 2. Altered species composition and succession 3. Reduced microbial biomass, altered structure and function (energy flow, nutrient cycling, competition, succession)	Altered surface microflora, changed relationship with saprophytes, increased/decreased disease caused by parasites
Class III High	1. Stimulation of individual species 2. Acute morbidity of individual species 3. Mortality of individual species	1. Increased microbial biomass, altered structure and function 2. Reduced microbial biomass, altered structure and function 3. Simplification	Altered surface microflora, changed relationship with saprophytes, increased/decreased disease caused by parasites

Source: Smith (1976).

Specific Air Pollutants and Forest Tree Disease

Forest trees because of their large size, extended lifetimes and widespread geographic distribution are subject to multiple microbially induced diseases frequently acting concurrently or sequentially. The reviews of Heagle (1973), Treshow (1975), and Manning (1975) included consideration of a variety of pollutant-woody plant pathogen interactions, but were not specifically concerned with forest tree disease. In their review of the impact of air pollutants on fungal pathogens of forest trees of Poland, Grzywacz and Wazny (1973) referenced literature citations indicating that air pollution stimulated the activities of at least 12 fungal tree pathogens while restricting the activities of at least 10 others.

A. Sulfur Dioxide

Elemental sulfur has been long appreciated for its toxic influence on fungi. At extremely high concentrations 900-2500 ppm (24-66 $\times$ 10^5 μg m^{-3}) sulfur dioxide itself has been employed as a fungicide (Treshow, 1975). Even though ambient concentrations may only approximate 1% of this extreme fungicidal dose, it has been observed that sulfur dioxide appears to have the ability to adversely influence pathogens directly (Manning, 1975). The observation that numerous forest pathogens appear to be restricted in regions subject to high ambient concentrations from point sources supports this generalization. The pioneering research of Scheffer and Hedgcock (1955) provides a classic example. In their intensive observations of the forest ecosystems surrounding metal smelters in Washington and Montana, these pathologists included surveys of parasitic fungi. In those areas subject to elevated sulfur dioxide, a large number of fungi, particularly those parasitic on foliage, appeared to be suppressed. This observation appeared especially true for species of *Cronartium, Coleosporium, Melampsora, Peridermium, Pucciniastrum, Puccinia, Lophodermium, Hypoderma,* and *Hypodermella.* Gradients of rust infection, *Melampsora albertensis* on quaking aspen and *M. occidentalis* on black cottonwood, were observed coincident with ambient sulfur dioxide. These fungi were not found in the zone of greatest sulfur dioxide tree injury, were sparse in the zone of moderate injury, and were most abundant where the injury was least. Similar observations were made with *Pucciniastrum pustulatum* on grand and subalpine firs; *Coleosporium solidaginis* on lodgepole pine; *Cronartium harknessii, C. comandrae,* and *Lophodermium pinastri* on ponderosa and lodgepole pines. *Hypodermella laricis,* parasitic on larch needles, was absent from areas affected by sulfur dioxide, while more common in regions free of sulfur gas stress. *Cronartium ribicola,* causal agent of white pine blister rust, has been observed to be almost absent in forests to distances of 40 km (25 miles) northeast of the Sudbury, Ontario, smelters in the direction of the prevailing wind. With increasing distance from the Sudbury sulfur dioxide source, white pine blister rust incidence invariably increases (Linzon, 1978).

A variety of additional fungi that infect tree foliage has been shown to be variously impacted by sulfur dioxide. *Microsphaera alni,* causal agent of oak powdery mildew, has been observed to be absent from the vicinity of a paper mill in Hinterburg, Austria (Koeck, 1935). Althrough unable to implicate a specific pollutant, Hibben and Walker (1966) have determined that lilacs growing in New York City appear to have substantially less powdery mildew caused by *Microsphaera alni* than lilacs in nonurban areas. At 0.3-0.4 ppm (785-1048 μg m^{-3}) sulfur dioxide exposure for 24-72 hr, these investigators recorded decreased *M. alni* spore germination of 50-60% on leaf discs and decreased disease development beyond the appressorium (infection peg) stage (Hibben and Taylor, 1975). At a sulfur dioxide acute dose of 1 ppm (2620 μg m^{-3}) for 1, 2, 4, or 6 hr, no effect on detached spore metabolism was noted. Chronic exposure was apparently necessary for suppressive effect. Ham (1971) has indicated that *Scirrha acicola,* causal agent of brown spot disease of loblolly pine foliage, grew normally and produced viable spores when exposed in vitro to 1 ppm (2620 μg m^{-3}) sulfur dioxide for 4 hr. It is of interest to speculate on what the response of this fungus would have been if it had also been subjected to chronic, low dose exposure. Additional evidence for reduced foliar disease, in areas of high ambient sulfur dioxide, has been presented for the following fungal pathogens: *Lophodermium juniperi* and *Rhytisma acerinum* (Barkman et al., 1969), *Hysterium policore* (Skye, 1968), and *Venturia inaequalis* (Przybylski, 1967).

In contrast to these examples of suppression, the significance of some foliage infecting fungi has been shown to be enhanced under conditions of elevated ambient sulfur dioxide. In an effort to explain increased incidence of pine needle blight on Japanese red pine in central Japan caused by *Rhizosphaera kalkhoffii,* Chiba and Tanaka (1968) exposed inoculated and uninoculated seedlings to 2 ppm (5240 μg m^{-3}) sulfur dioxide for 1, 2, 3, and 4 hr fumigations. Generally disease was most severe on those seedlings receiving the greatest sulfur dioxide dose. Jancarik (1961) has recorded a higher incidence of *Lophodermium piceae* on spruce needles damaged by sulfur dioxide exposure.

Fungi that cause wood decay have also received some examination relative to interaction with sulfur dioxide stress. Jancarik (1961) conducted a survey of macroscopic fruiting bodies of wood decay producing Basidiomycete fungi in northern Czechoslovakia in areas with healthy and sulfur dioxide damaged conifers. Of 40 decay-producing species recorded, 12 were present in regions of slight injury but absent from areas of severe sulfur dioxide damage. Six decay fungi were recorded only on severely damaged trees. These included *Glocophyllum abietinum, Trametes serialis,* and *Trametes heteromorpha. Poria* sp., *Mycena* spp., *Schizophyllum commune,* and *Polyporus versicolor* were found exclusively in areas of minor sulfur dioxide injury. Scheffer and Hedgcock (1955), on the other hand, were unable to find any influence of sulfur dioxide on decay incidence in Montana conifers. Close to the Anaconda sulfur gas point source, approximately 7% of the mature lodgepole pine and 72% of the Douglas fir were infected in various degrees by *Polyporus schweinitsii* or by *Fomes pini.* These percentages were comparable to the incidence in surrounding forests not under the influence of elevated sulfur dioxide.

Fungi that induce root disease are among the most important pathogens of managed forest ecosystems. An example of enormous importance to temperate forest ecosystems is the ubiquitous *Amillariella mellea* (*Armillaria mellea*) (Figure 13-1). This fungus is very widespread, has an extremely broad host range and is especially significant in causing disease in trees under stress. Sinclair (1969) has pointed out that the relationship between *A. mellea* infection and air pollution stress may be a classic example of disease predisposition caused by air contamination. Scheffer and Hedgcock (1955) did indeed find that the association of *A. mellea* with pine roots was greatest inside the zone of sulfur dioxide pine damage. Additional evidence has been provided indicating predisposition to *A. mellea* infection by trees stressed by sulfur dioxide exposure (Donaubauer, 1968; Jancarik, 1961; Kudela and Novakova, 1962).

The information available frustrates attempts to generalize concerning the influence of elevated ambient sulfur dioxide on forest tree disease induced by biotic agents. Substantial observations of fungal disease incidence suggest that the activities of some pathogens are suppressed while others are enhanced. Much of this information, however, stems from relatively simple disease surveys in stressed and nonstressed environments. In some of the studies it is not obvious that appropriate attention has been given to factors other than sulfur dioxide that also could have accounted for altered disease incidence. Little attention has been given to specific mechanisms of sulfur dioxide-plant-pathogen inter-

Figure 13-1. Fruiting bodies of *Armillariella mellea* developing at the base of a yellow birch in Vermont. This fungus causes one of the most significant and common root diseases in temperate forests.

action that may account for increased or decreased disease. It has been hypothesized that increased stomatal aperture may facilitate foliar infection, but the evidence is not extensive (Unsworth et al., 1972; Williams et al., 1971). It is probable in natural forest ecosystems subject to sulfur dioxide stress that disease incidence may be altered by sulfur gas influence on the pathogen and the host depending on the relative susceptibilities of the organisms and the nature of the sulfur dioxide dose. Before and after fumigation of tomato and bean plants with less than 20 ppm (524 μg m^{-3}) sulfur dioxide for several days and inoculation with the fungal agents causing early blight and bean rust, respectively, revealed an influence on the latter disease only (Weinstein et al., 1975). In this case, however, the decreased incidence and severity of the disease was judged to have resulted from alterations in *both* the pathogen and the host. This study is representative of the considerable evidence indicating that spore production and germination, vegetative development of some microbial pathogens, and metabolism and physiology of foliar tissues of some hosts are proportional to sulfur dioxide dose (National Academy of Sciences, 1978).

The results of the fine study of air pollution impact on fungal pathogens of Polish forests conducted by Grzywacz and Wazny (1973) are consistent with this dose generalization. These investigators found substantial quantitative differences in the occurrence of *Armillariellaa mellea, Ophodermium pinastri, Fomitopsis annosa (Fomes annosus), Cronartium flaccidum, Melampsora pinitorqua, Phellinus pini, Cenangium abietis,* and *Microsphaera alphitoides* relative to air quality. In all these cases, high dose exposure to industrial sulfur dioxide acted to destroy or inhibit the growth of these fungi. Low dose exposure acted to stimulate their activities. Plots of disease incidence against distance from point sources revealed curves of similar shape for all fungi examined (for example, Figure 13-2).

B. Ozone

As in the instance of sulfur dioxide, ozone has a research history as a potential microbial pesticide. In the ozone case, its proposed use as a fungicide (Hartman, 1924) predates its recognition as an air pollutant by several decades. Also, as in the case of sulfur dioxide, an extensive literature is available concerning the interaction of this gas with microbial development under laboratory conditions. Unlike sulfur dioxide, however, extensive field correlations of plant disease incidence and ambient ozone concentration are lacking. Heagle (1973) has correctly indicated that this is primarily due to the absence of point sources of ozone and the lack of distinct gradients of ozone concentrations in natural environments.

Excellent reviews of the interactions between ozone, plant pathogens, and agricultural crops have been provided by Heagle (1973) and Treshow (1975). These reviews reveal considerable evidence from laboratory and greenhouse studies to indicate that ozone can decrease infection, invasion, and spore production of fungal pathogens and that this can inhibit parasitism. The majority of

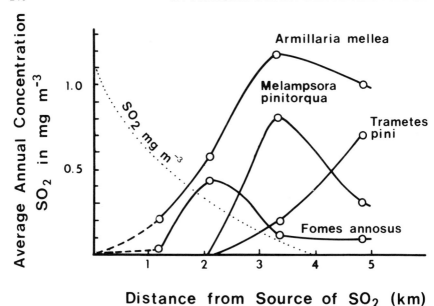

Distance from Source of SO₂ (km)

Figure 13-2. Average percentage of trees infected by various fungal pathogens in Polish forests at varying distances from sulfur dioxide point sources. (From Gryzywacz and Wazny, 1973.)

evidence indicates that ozone alters parasitism via effects on host plants. Ozone can also apparently stimulate the growth and development of microbes on plant surfaces.

Ozone dose required for direct impact on microbial metabolism may be quite high. The data of Hibben and Stotsky (1969) are illustrative. These investigators examined the response of detached spores of 14 fungi to 10-100 pphm (196-1960 μg m^{-3}) ozone for 1, 2, and 6 hr periods. The large pigmented spores of *Chaetomium* sp., *Stemphylium sarcinaeforme, S. loti,* and *Alternaria* sp. were uninfluenced by 100 pphm (1960 μg m^{-3}). Germination of *Trichoderma viride, Aspergillus terreus, A. niger, Penicillium egyptiacum, Botrytis allii,* and *Rhizopus strolonifera* spores were reduced by ozone exposure, but only by concentrations above 50 pphm (980 μg m^{-3}). The small colorless spores of *Fusarium oxysporum, Colletotrichum largenarium, Verticillium albo-atrum,* and *V. dahliae* had germination percentages reduced by 50 pphm (980 μg m^{-3}) and occasionally by doses of 25 pphm (490 μg m^{-3}) for 4 to 6 hr. Concentrations less than this latter dose stimulated spore germination in some cases.

The literature addressing the interaction of ozone with woody plant disease is very modest. The reduced incidence of powdery mildew disease on lilac documented by Hibben and Walker (1966) in urban areas could have been related to ozone. Laboratory exposure of conidia of *Microsphaera alni,* however, to 0.9-1.0 ppm (1760-1970 μg m^{-3}) ozone for 1, 2, 4, and 6 hr; 0.5 ppm (980 μg m^{-3}) ozone for 6 hr; 0.25 ppm (490 μg m^{-3}) ozone for 72 hr; or 0.1-0.15 ppm (200-

290 $\mu g \ m^{-3}$) ozone for 72 hr had essentially no influence on germination and early fungal development (Hibben and Taylor, 1975).

There is some indication that ozone may enhance disease development by pathogens that normally infect stressed or senescent plant parts or invade non-living woody plant tissues. *Lophodermia pinastri* and *Pullalaria pullulans* were most commonly associated with eastern white pine foliage injury when inoculated in conjunction with tree exposure to 7 pphm (137 $\mu g \ m^{-3}$) ozone for 4.5 hr (Costonis, 1968; Costonis and Sinclair, 1972).

Weidensaul and Darling (1979) inoculated Scotch pine seedlings with *Scirrhia acicola* 5 days before or 30 min following fumigation for 6 hr with 0.20 ppm (533 $\mu g \ m^{-3}$) sulfur dioxide, 0.20 ppm (332 $\mu g \ m^{-3}$) ozone, or both gases combined. Significantly more brown spot lesions were formed on seedlings fumigated with sulfur dioxide alone or combined with ozone than on controls when inoculation was done 5 days before fumigation. When inoculation was done 30 min after gas exposure, seedlings exposed to sulfur dioxide alone had more lesions than those exposed to ozone alone or combined with sulfur dioxide, but no significant differences were noted between treated seedlings and controls. The authors judged that ozone-induced stomatal closure may have been responsible for the latter observation.

Fomitopsis annosa is another Basidiomycete fungus capable of causing wide-spread and significant root disease and decay in a variety of coniferous hosts throughout temperate forests. A comprehensive examination of oxidant stress on California forest ecosystems (Chapter 15) has included a study of ozone influence on this fungus and the disease it causes in ponderosa and Jeffrey pines (Miller, 1977). Artificial root inoculation was conducted with trees exhibiting various degrees of oxidant stress. Pine seedlings were also artificially inoculated following fumigation with ozone. In light of the importance of freshly cut stump surfaces in the spread of this fungus, trees of various suceptibility classes were cut and their stump surfaces inoculated with *F. annosa*. Laboratory exposures of pure cultures of *F. annosa* to ozone were also performed. Preliminary results have indicated the following. Field inoculation of roots of both ponderosa and Jeffrey pines did not reveal any correlation with degree of oxidant damage. Stump inoculation, however, did suggest that air pollution injury may have increased the susceptibility of pine stumps to colonization by *F. annosa* (James et al., 1980). The percentage of infection of fumigated seedlings was also greater than that of nonfumigated seedlings (Table 13-2).

Heagle (1975) has recently determined that ozone, at doses comparable to ambient conditions in certain areas, was capable in inhibiting various phases of development of two rust fungi and a powdery mildew fungus of cereal crops. It would be of extreme interest to have comparable information on the fungi responsible for the large number of economically significant forest tree rust diseases.

In addition to fungal agents of disease, recent research attention has been given to other important microorganisms responsible for plant disease. Papers dealing with air contaminant interaction with bacterial disease agents (Laurence

Table 13-2. Infection of Ozone Fumigated and Unfumigated Jeffrey and Ponderosa Pine Seedlings by *Fomitopsis annosa*[a]

Pine species	Seedling number	Ozone fumigation concentration ($\mu g\ m^{-3}$)	Infection (%)
Jeffrey	32	0	53.1
Jeffrey	16	431.2	75.0
Jeffrey	16	882.0	81.0
Ponderosa	32	0	62.0
Ponderosa	16	431.2	81.0
Ponderosa	16	882.0	75.0

Source: Miller (1977).

[a] Seedlings at each concentration were exposed for a period ranging between 58 and 87 days.

and Wood, 1978a,b; Howell and Graham, 1977; Pell et al., 1977) and viral disease agents (Brennan, 1975; Bisessar and Temple, 1977; Davis and Smith, 1975, 1976; Moyer and Smith, 1975; Reinert and Gooding, 1978; Vargo et al., 1978) are available for nonwoody host species. Similar research on woody plant viruses and bacteria, and in addition on nematodes, would provide important perspective to those attempting to evaluate the importance of pathogen-plant-air pollutant interaction.

C. Fluoride

The influence of plant accumulated fluoride on disease development has received only very limited research attention. Some data have been provided concerning microbial response to elevated fluoride provided in laboratory media. As expected, various fungi respond differently to sodium fluoride incorporated into agar. *Pythium debaryanum* has a lower threshold of inhibition than *Verticilllium alboatrum* and *Helminthosporium sativum*. The growth of *Botrytis cinerea* and two *Colletotrichum* spp. was enhanced by low concentrations of fluoride (Heagle, 1973).

There is some evidence that foliar fluoride may reduce agricultural plant disease. Bean plants exposed to hydrogen fluoride at 7-10 $\mu g\ m^{-3}$ and with foliar fluoride concentrations approximating 400 $\mu g\ g^{-1}$ fluoride were less severely infected with powdery mildew than control plants (McCune et al., 1973). Threshow (1975) reported similar protection from powdery mildew for chrysanthemum plants that had foliar concentrations between 350-400 $\mu g\ g^{-1}$ following exposure to hydrogen fluoride at 2 $\mu g\ m^{-3}$ for 4 hr per day for several days. Elevated fluoride has further been correlated with reduced bacterial disease, but apparently tobacco mosaic virus symptoms of bean can be reduced or enhanced depending on fluoride concentration (Manning, 1975).

Unfortunately almost no information is available concerning fluoride and forest tree disease. Barkman et al. (1969) have recorded that *Melampsoridium betulinum*, the causal agent of a birch leaf rust and common in Norway, was absent from birches growing near a fluoride source.

D. Particulates

A large number of particulates, including coarse dust, trace metals and acid precipitation, have been implicated in alterations of plant disease.

1. Coarse Dust

Accumulation of coarse dust particles has been demonstrated to increase foliar disease. Lime dust significantly enhanced *Cercospora beticola* infection of sugar beet leaves following artificial application (Schönbeck, 1960). Manning (1971) has detailed the influence of dust on foliar disease in a forest ecosystem (Jefferson National Forest) surrounding quarries and limestone processing facilities in a mountain valley in southwestern Virginia. Grape and sassafras foliage with moderate dust deposits generally had more fungal infection than leaves lacking dust deposits. Dusty leaves also had increased numbers, but not kinds, of bacteria and fungi relative to clean leaves. On dusty hemlock leaves, bacterial numbers were greatly reduced while fungal incidence was increased (Table 13-3). Sassafras and grape were judged to be predisposed to leaf spot disease caused by the fungi *Guignardia bidwellii* and *Gloeosporium* sp. when dusty.

2. Trace Metals

The potential importance of trace metals associated with particles and microbial metabolism has been previously stressed in this volume (Chapter 8). In view of the considerable capacity of foliage to accumulate trace metal particles (Chapter 5), there is considerable interest to evaluate the interaction of metal cations with those microorganisms of the phyllosphere capable of causing disease or influencing those that do function as pathogens.

Gingell et al. (1976) positioned cabbage plants and 5- to 6-year-old Austrian pine saplings 0.6 km (0.4 mile) northeast of a smelter complex in Avonmouth, England, and monitored changes in foliar microbes at this site relative to plants located 7 km (4 miles) southeast of the industry. The zinc, lead, and cadmium levels of the cabbages and pine needles were much higher in the test plants located at 0.6 km. Isolations on Martin's rose-bengal streptomycin-agar and tryptic-soy-agar revealed significantly fewer microbes from the polluted relative to the control site for both pine needles and cabbage. The reduction in number and diversity of organisms on the contaminated cabbage was primarily due to a significantly lower population of bacteria and pigmented yeasts.

Specific information on the interaction of trace metals with foliar pathogens is not extensive. Leaves artificially contaminated with zinc, lead, and cadmium were shown to be less infected by *Botyritis cinerea* than uncontaminated leaves (Gingell, 1975).

In view of the excessive contamination of urban trees with trace metals in the roadside environment, the author's laboratory has attempted to explore the interactions between pollutant metals and leaf inhabiting fungi. Leaf washing and impression techniques were employed to isolate phylloplane fungi from the leaves

Table 13-3. Influence of Limestone Dust from Quarries and Rock Processing Plants on the Occurrence of Fungi on the Foliage of Woody Plants in the Jefferson National Forest, Virginia

	Leaf prints[a]						
	No. isolates per genus expressed as percent of total no. colonies isolated						
	Wild grape			Sassafras		Hemlock	
Genera of fungi isolated	Heavy crust	Moderate dust	No dust	Moderate dust	No dust	Heavy dust	No dust
Alternaria		6.3		2.8			28.0
Candida	20.0		15.6				
Cladosporium	26.6	2.3				26.9	19.4
Colletotrichum				31.8	21.6	3.5	
Cryptococcus		13.2	25.2	6.0	14.9	28.2	22.8
Curvularia							
Fusarium				8.1			
Penicillium		22.0	42.5				
Periconiella						12.8	
Pestalotia		9.0			5.5		12.3
Piricauda		2.5	9.2	12.7			
Rhodotorula	26.6	33.6	4.1		17.7		
Saccharomyces		5.8				17.8	
Black mycelium	6.8	5.3	4.2			6.8	
Brown mycelium	20.0			23.8	24.9		
Gray mycelium			3.3	15.4		6.8	17.5
White mycelium				8.0	2.7		
Total no. isolates	5	10	6	9	7	7	5
Total no colonies	15	534	237	362	179	78	57
	Dilution plates[b]						
Alternaria			4.2	8.1	4.0		
Candida	97.6		12.2			2.3	6.8
Cladosporium	1.2	16.4	18.5	32.5	8.7	11.7	13.6
Colletotrichum			8.4		6.2		
Cryptococcus						10.1	13.7
Curvularia					6.2	3.3	5.1
Fusarium					2.0		
Penicillium		2.7	4.2				
Periconiella				6.4	31.2		
Pestalotia							
Piricauda		4.1	1.7	4.2	6.2	3.3	4.0
Rhodotorula	0.6	56.1		12.1	12.6	69.3	56.8
Saccharomyces		2.7	34.6	28.4	16.6		

Particulates   253

Table 13-3 (continued)

	Dilution plates						
	No. isolates per genus expressed as percent of total no. colonies isolated						
	Wild grape			Sassafras		Hemlock	
Genera of fungi isolated	Heavy crust	Moderate dust	No dust	Moderate dust	No dust	Heavy dust	No dust
Black mycelium				4.2			
Brown mycelium	0.6						
Gray mycelium		18.0	16.2	4.1	6.2		
White mycelium							
Total no. isolates	4	6	8	8	9	7	6
Total no. colonies	21.5	365	49	123	48	295	58

Source: Manning (1971).
[a] Total no. colonies isolated (upper and lower surfaces) for leaf prints.
[b] No. colonies g^{-1} leaf tissue in 1000s for dilution plates.

of mature, roadside London plane growing in New Haven, Connecticut. Numerous fungi were consistently isolated from various crown positions and at different times during the growing season. Those existing primarily saprophytically included: *Aureobasidium pullulans, Chaetomium* sp., *Cladosporium* sp., *Epicoccum* sp., and *Phialophora verrucosa.* Those existing primarily parasitically included *Gnomonia platani, Pestalotiposis* sp., and *Pleurophomella* sp. The following cations were tested in vitro for their ability to influence the growth of these fungi: cadmium, copper, manganese, aluminum, chromium, nickel, iron, lead, sodium, and zinc. The results of this effort indicated variable fungal response with no correlation between saprophytic or parasitic activity and sensitivity to trace metals. Both linear extension and dry weight data indicated that the saprophytic *Chaetomium* sp. was very sensitive to numerous metals. *Aureobasidium pullulans, Epicoccum* sp., and especially *P. verrucosa,* on the other hand, appeared much more tolerant. Of the parasites *Gnominia platani* appeared more tolerant than *Pestalotiopsis* sp. and *Pleurophomella* sp. Metals exhibiting the broadest spectrum growth suppression were iron, aluminum, nickel, zinc, manganese, and lead (Smith, 1977).

Because of the important anthracnose disease caused by *Gnomonia platani* on *Platanus* species (Figure 13-3), we have been particularly interested in the tolerance of this organism to trace metals. In vitro linear extension of mycelial growth was significantly inhibited by aluminum, iron, and zinc. These three cations, in addition to cadmium, chronium, manganese, and nickel, also significantly suppressed spore formation (Figure 13-4). Dry weight determinations following

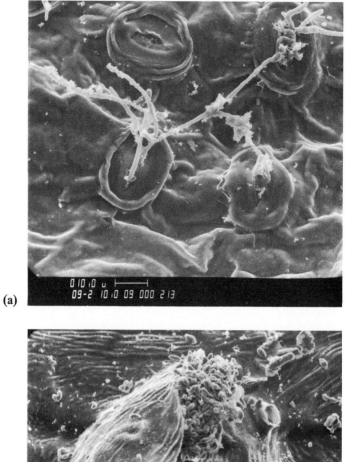

(a)

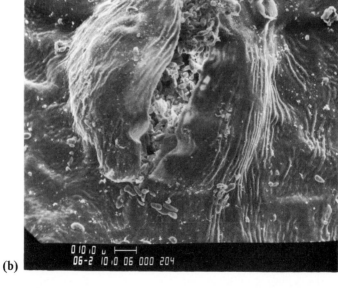

(b)

Figure 13-3. Scanning electron micrograph of *Gnomonia platani*, causal agent of anthracnose disease of *Platanus* species, growing on the leaf surface of American sycamore: (a) Mycelium, (b) Acervulus with conidia. Scale, 10 μm.

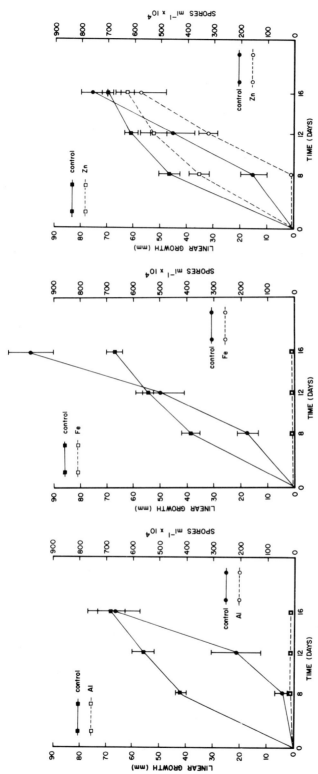

Figure 13-4. Linear growth (left axis) and conidial production (right axis) of *Gnomonia platani* on malt agar amended with aluminum, iron, and zinc concentrations equivalent to average *Platanus* leaf burdens in an urban area. Means are with 95% confidence intervals. (From Staskawicz and Smith, 1977.)

growth in liquid culture indicated that mycelial growth was significantly reduced by aluminum, iron, zinc, nickel, and copper. Amendment of shake cultures with lead, chromium, sodium, or manganese did not cause significant growth stimulation or inhibition (Figure 13-5). When condia were placed on a medium containing aluminum, iron, zinc, or nickel, spore germination was significantly suppressed (Figure 13-6). Reduction of mycelial growth, spore formation, and spore germination in nature would lessen the competitive capability of *Gnomonia plantani* and may lessen the ability of this fungus to cause foliar disease in *Platanus* (Staskawicz and Smith, 1977). Unfortunately it is difficult to extrapolate from our observations in vitro to the natural environment. The cation concentrations employed, while approximating measured field burdens, are arbitrary. It is most difficult to judge the actual concentration a specific fungus will encounter on a particular leaf surface. A dose-response test of metals and fungi discussed in this section, revealed that only zinc was toxic to *Chaetomium* under very low dose conditions (Smith et al., 1978). Our in vitro efforts have generally employed nitrate salts in order to supply a common anion and completely soluble compounds. In nature trace metals probably occur on leaf surfaces as less-soluble oxides, halides, sulfates, or phosphates (Koslow et al., 1977). The use of any natural product medium in in vitro efforts will presumably cause alteration of available metal concentrations due to binding by media components (Ko et al., 1976; Romamoorthy and Kushner, 1975). Since the metals were reacted with the fungi individually, it is possible that important antagonistic, additive, or

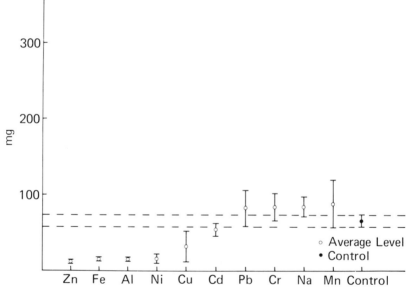

Figure 13-5. Dry weight of *Gnomonia platani* after seven days growth in malt broth amended with concentrations of trace metals equivalent to average leaf burdens in an urban area. Means with 95% confidence intervals. (From Staskawicz and Smith, 1977.)

synergistic interactions were overlooked. Because the phyllosphere has a complex microflora (Last and Deighton, 1965) with much interaction between parasitic and nonparasitic microbes (Fokkema and Lorbeer, 1974; Last and Warren, 1972), it is possible that trace element effect on other organisms that influence the fungi examined in our studies may be more significant in nature than the direct toxic effect on our test organisms. In spite of these limitations, we feel our data support the general suggestion that foliar fungi, including pathogens, respond differentially to foliar metal contamination and in this sense are very similar to the relationships recorded for other air contaminants and pathogenic microorganisms.

3. Acid Preparation

Fortunately, David S. Shriner, Environmental Sciences Division, Oak Ridge National Laboratory, Tennessee, has provided us with some very valuable perspective in this important but understudied area. Falling precipitation and the precipitation wetting of vegetative surfaces plays an enormously important role in the life cycles of a large number of plant pathogens. In recognition of this Shriner has examined the effects of simulated rain acidified with sulfuric acid on several host-parasite systems under greenhouse and field conditions (Shriner,

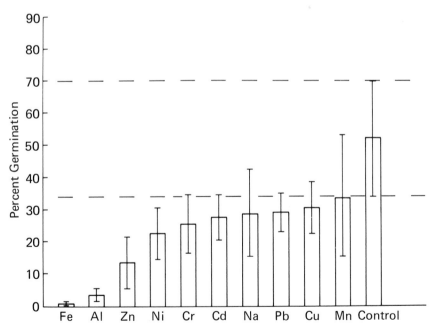

Figure 13-6. Percent germination of *Gnominia platani* conidia transferred to malt agar amended with concentrations of trace metals equivalent to average leaf burdens in an urban area. Means with 95% confidence intervals. (From Staskawicz and Smith, 1977.)

1974, 1975, 1977). The simulated rain he employed had a pH of 3.2 or 6.0 (representing extremes of natural precipitation).

The application of simulated rain of pH 3.2 resulted in (a) an 86% restriction of tilia production by *Cronortium fusiforme* (fungus) on willow oak; (b) a 66% inhibition of *Meloidogyne hapla* (root-knot nematode) on kidney bean; (c) a 29% decrease in percentage of leaf area of kidney bean affected by *Uromyces phaseoli* (fungus); and (d) either stimulated or inhibited development of halo blight of kidney bean caused by *Pseudomonas phaseolicola* (bacterium). In the latter case, the influence of acid precipitation varied and depended on the particular stage of the disease cycle when the exposure to acid precipitation occurred. Simulated sulfuric acid rain applied to plants prior to inoculation stimulated the halo blight disease by 42%. Suspension of inoculum in acid rain decreased inoculum potential by 100%, while acid rain applied to plants after infection had occurred inhibited disease development by 22% (Table 13-4).

Examination of the willow oak and bean leaves using the scanning electron microscope revealed distinct erosion of the leaf surface by rain of pH 3.2. This may suggest that altered disease incidence may be due to some change in the structure or function of the cuticle. Shriner has also proposed that the low pH rain may have increased the physiological age of exposed leaves. Shriner (1978) concluded his initial experiments by suggesting that he had not established threshold pH levels at which significant biological ramifications to pathogens occur from acid precipitation. He did suggest, however, that artificial precipitation of extremely low pH probably alters infection and disease development of a variety of microbial pathogens.

E. Other Air Contaminants

The sporulation of *Peronospora trifoliarum* has been shown to be inhibited by lower atmospheric chlorine concentrations then required to cause visible injury to host alfalfa plants (Fried and Stuteville, 1975).

In a novel experiment, uredospores of *Puccinia graminis tritici* and *P. striiformis* and conidia of *Pyricularia oryzae* and an *Alternaria* spp. failed to germinate on water agar when exposed to 6000 μliters of cigarette smoke liter^{-1} of air in an incubation chamber (Melching et al., 1974). Whatever component of the 1200 identified compounds known to occur in tobacco smoke was responsible for the inhibition, the authors judged that it was probably not nicotine, carbon monoxide, pyridine, phenol, or hydrogen cyanide acting alone.

Spore germination or mycelial growth of several fungi, including some forest tree pathogens, was recently shown to be reduced on cellophane previously exposed to smoke from burning pine needles. Prior exposure of Monterey pine seedlings to smoke reduced the amount of gall rust following inoculation (Parmeter and Uhrenholdt, 1975). These authors noted that forest burning, particularly wildfires, may result in dense clouds of smoke that drift for many kilometers through forest ecosystems (compare Figure 13-7). Smoke deposits on dead branches, stubs, exposed wounds, and other tree surfaces might reduce the activities of important forest fungi, including pathogens, if smoke deposits are toxic on plant surfaces as well as on cellophane.

F. Summary

Several conclusions are possible concerning the interaction between air pollutants and microbial pathogens of forest trees. The response of individual pathogens and the diseases they cause to atmospheric components is extremely variable and in a given host-parasite-environment system exposure to a particular pollutant may increase, decrease, or have no apparent influence on a given disease situation. In those instances in which air contaminants do alter disease occurrence or severity, the primary mechanism may be a direct influence on the causal microorganism, an indirect influence on the causal microorganism via a direct influence on an associated microbe, or an indirect influence on the pathogen via an alteration in host physiology or metabolism. Microorganisms that cause foliar disease, or that infect plants through the leaves, may be expected to be especially subject to influence by air pollutants. This is true for at least three reasons: (1) these microbes may grow and develop vegetatively and reproductively in environments with relatively high levels of ambient pollutant concentrations (for example, leaf surface); (2) foliar tissue is known to be the site of primary accumulation of recalcitrant materials from the atmosphere, for example fluoride and heavy metals; and (3) foliar tissue is the primary site of direct damage to the plant occasioned by air pollution and leaf tissue may be expected to be predisposed to infection if physical or metabolic resistance mechanisms are adversely influenced. Most of the evidence is consistent with these three observations. It is also true, however, that most investigators have concentrated their attention on foliar infecting groups such as rust fungi, powdery mildew fungi, and other bacteria and fungi that cause foliar symptoms. Unfortunately, there is less information available concerning the response to air pollution of microorganisms that infect trees via the root, stem, fruit, seed, or branch.

As in the instance of bark beetle infestation, it might be expected that those pathogens that normally infect woody plants and cause significant disease when the host is under stress would be facilitated in environments of poor air quality. Perhaps the best example of this is the situation involving *Armillariella mellea*. This fungus is known to be most severe under conditions of host predisposition by some insect, microbial, edaphic, or environmental stress. Evidence has been provided to indicate that air pollution may be an additional potential predisposing agent.

Unfortunately we are deficient in our appreciation of the interaction of air pollutants and important tree disease groups caused by viruses and mycoplasma, nematodes, dwarf mistletoes, and fungal species responsible for root disease, vascular disease, stem and branch disease, seed disease, and wood decay.

The pollutants that have received the greatest research effort include sulfur dioxide, ozone, fluoride, dust, and heavy metals. There is substantial circumstan-

Table 13-4. Effects of Simulated Rain Acidified with Sulfuric Acid on Selected Host-Parasite Interactions under Greenhouse and Field Conditions

Host-pathogen system	Acidity of simulated rain (pH)			Disease measure				
	Preinoculation	Inoculation	Postinoculation	Infected leaves/plant[a]	Telia/infected leaf[a]	Dead leaflets/plant[c,d] (no.)	Eggs/plant[f]	% root galled[f,g]
Greenhouse studies								
Willow oak	3.2	3.2	3.2	3.8[b]	15[b]			
Cronartium fusiforme oak-pine-rust of oak	6.0	6.0	6.0	6.5	115			
Phaseolus vulgaris-Pseudomonas phaseolicola halo blight of bean	3.2	3.2	3.2			0.0a[e]		
	3.2	3.2	6.0			0.0a		
	3.2	6.0	3.2			4.2d		
	3.2	6.0	6.0			4.8d		
	6.0	3.2	3.2			0.0a		
	6.0	3.2	6.0			0.0a		
	6.0	6.0	3.2			2.0b		
	6.0	6.0	6.0			3.2c		
			LSD 0.05			0.77		
Field studies								
Phaseolus vulgaris-Meloidogyne hapla root knot nematode on beans		3.2	3.2				74[b]	26
		6.0	6.0				217	50

			% leaf area affected at	
			Seven weeks	Nine weeks
Phaseolus vulgaris-	3.2	3.2	22[b]	48
Uromyces phaseoli	6.0	6.0	31	45
rust of beans				

Source: Shriner (1977).

[a] Each value is the mean of six plants/treatment.

[b] Analysis of variance (ANOVA) significant, $P = 0.05$.

[c] Number of inoculated leaflets (max. 6/plant) which were dead on plants which developed symptoms characteristic of halo blight.

[d] Each value is the mean of 5 plants.

[e] Values in a column followed by the same letter were not significantly different, $LSD_{0.05}$.

[f] Each value is the mean of 18 plants/treatment.

[g] % of root area galled by *Meloidogyne hapla*.

Figure 13-7. Excessive smoke resulting from the Poverty Flat wildfire, Payette National Forest, Idaho. (Photograph courtesy U.S.D.A. Forest Service.)

tial evidence to indicate that a variety of forest tree diseases appears to be less abundant in areas of grossly elevated ambient sulfur dioxide, for example, within a few kilometers of a major point source such as a smelter. Under these conditions it is presumed that sulfur compounds may be exerting a directly toxic influence on the microbial pathogen.

The direct influence of ozone on microbial pathogens may occur at substantially higher ambient concentrations than is the general case with sulfur dioxide. The influence of this gas on disease development, however, may be typically via an alteration in host metabolism, for example, through a change in some resistance mechanism. Accumulation of fluoride to levels approximately 300-400 μg g^{-1} of leaf tissue (dry wt. basis) appears, in several cases, to reduce the incidence of microbial disease. Lime dust contamination of leaves, on the other hand, appears to increase the incidence of foliar infection.

Trace metal pollutants have a high potential to interact with tree pathogens on vegetative surfaces. In vitro evidence indicates variable influence of metals on microbes depending on the cation and the specific pathogen. Generally, however, most fungi are probably not directly influenced unless the surface contamination on the plant is extremely high as might occur in the immediate vicinity of a smelter point source.

The influence of acid precipitation on tree pathogens is in need of significantly increased research attention. The large geographic area subject to this stress and the variety of ways that reduced pH could influence disease development justify this need. The influence of combustion products resulting from forest burning and deposited on plant and soil surfaces also requires research attention to evaluate the potential to influence microbial pathogens.

References

Babich, H., and G. Stotzky. 1974. Air pollution and microbial ecology. Crit. Rev. Environ. Cont. 4:353-420.

Barkman, J. J., F. Rose, and V. Westhoff. 1969. The effects of air pollution on non-vascular plants. Section 5 discussion. Proc. Eur. Congr. Influence Air Pollut. Plants. Wageningen, The Netherlands, pp. 237-241.

Bisessar, S., and P. J. Temple. 1977. Reduced ozone injury on virus-infected tobacco in the field. Plant Dis. Reptr. 61:961-963.

Brennan, E. 1975. On exclusion as the mechanism of ozone resistance in virus-infected plants. Phytopathology 65:1054-1055.

Chiba, O., and K. Tanaka. 1968. The effect of sulfur dioxide on the development of pine needle blight caused by *Rhizosphaera kalkhoffii*. Bubak. J. Jpn. For. Soc. 50:135-139.

Costonis, A. C. 1968. Relationship of ozone, *Lophodermium pinastri* and *Pullularia pullulans* to needle blight of eastern white pine. Ph.D. Thesis, Cornell Univ., Ithaca, New York, 176 pp.

Costonis, A. C., and W. A. Sinclair. 1972. Susceptibility of healthy and ozone-injured needles of *Pinus strobus* to invasion by *Lophodermium pinastri* and *Aureobasidium pullulans*. Eur. J. For. Path. 2:65-73.

Davis, D. D., and S. H. Smith. 1975. Bean common mosaic virus reduces ozone sensitivity of pinto bean. Environ. Pollut. 9:97-101.

Davis, D. D., and S. H. Smith. 1976. Reduction of ozone sensitivity of pinto bean by virus-induced local lesions. Plant Dis. Reptr. 60:31-34.

Donaubauer. E. 1968. Sekundärschäden in österreichischen Rauchschadensbebieten. Schwierigkeiten der Diagnose und Bewertung. *In*: Materialy VI Miedzynarodowej Konferencji, Katowice, Poland, Sept. 9-14, 1968. Polaska Akademia Nauk, pp. 277-284.

Fokhema, N. J., and J. W. Lorbeer. 1974. Interactions between *Alternaria porri* and the saprophytic mycoflora of onion leaves. Phytopathology 64:1128-1133.

Fried, P. M., and D. L. Stuteville. 1975. Effect of chlorine on *Peronospora trilfoliorum* sporangial production and germination. Phytopathology 65:929-930.

Gingell, S. M. 1975. The effect of heavy metal pollution on the leaf surface microflora. B. Sc. Thesis, Univ. of Bristol, England.

Gingell, S. M., R. Campbell, and M. H. Martin. 1976. The effect of zinc, lead and cadmium pollution on the leaf surface microflora. Environ. Pollut. 11:25-37.

Grzywacz, A., and J. Wazny. 1973. The impact of industrial air pollutants on the occurrence of several important pathogenic fungi of forest trees in Poland. Eur. J. For. Path. 3:129-141.

Ham, D. L. 1971. The biological interactions of sulfur dioxide and *Scirrhia acicola* on loblolly pine. Ph.D. Thesis. Duke University, Durham, North Carolina.

Hartman, F. E. 1924. The industrial application of ozone. J. Am. Soc. Heat. Vent. Engin. 30:711-727.

Heagle, A. S. 1973. Interactions between air pollutants and plant parasites. Annu. Rev. Phytopathol. 11:365-388.

Heagle, A. S. 1975. Response of three obligate parasites to ozone. Environ. Pollut. 9:91-95.

Hibben, C. R., and G. Stotsky. 1969. Effects of ozone on the germination of fungus spores. Can. J. Microbiol. 15:1187-1196.

Hibben, C. R., and M. P. Taylor. 1975. Ozone and sulphur dioxide effects on the lilac powdery mildew fungus. Environ. Pollut. 9:107-114.

Hibben, C. R., and J. T. Walker. 1966. A leaf roll-necrosis complex of lilacs in an urban environment. Proc. Am. Soc. Hort. Sci. 89:636-642.

Howell, R. K., and J. H. Graham. 1977. Interaction of ozone and bacterial leaf-spot of alfalfa. Plant Dis. Reptr. 61:565-567.

James, R. L., F. W. Cobb, Jr., W. W. Wilcox, and D. L. Rowney. 1980. Effects of photochemical oxidant injury of ponderosa and jeffrey pines on susceptibility of sapwood and freshly-cut stumps to *Fomes annosus*. Phytopathology 70: 704-708.

Jancarik, V. 1961. Vyskyt drevokaznych hub v kourem poskozovani oblasti Krusnych hor. Lesnictvi 7:667-692.

Ko, W. H., J. T. Kliejunas, and J. T. Shimooka. 1976. Effect of agar on inhibition of spore germination by chemicals. Phytopathology 66:363-366.

Koeck, G. 1935. Mildew on oak trees and flue-gas damage. Z. Pflanzenkr. Pflanzensch. 45:1-2.

Koslow, E. E., W. H. Smith, and B. J. Staskawicz. 1977. Lead-containing particles on urban leaf surfaces. Environ. Sci. Technol. 11:1019-1021.

Kudela, M., and E. Novakova. 1962. Lesni skudci a slpdu zvero v lesich poskozovanych Kourem. Lesnictvi 6:493-502.

Last, F. T., and F. C. Deighton. 1965. The nonparasitic micro-flora on the surfaces of living leaves. Trans. Brit. Mycol. Soc. 48:83-99.

Last, F. T., and R. C. Warren. 1972. Non-parasitic microbes colonizing green leaves: Their form and functions. Endeavour 31:143-150.

Laurence, J. A., and F. A. Wood. 1978a. Effects of ozone on infection of soybean by *Pseudomonas glycinea*. Phytopathology 68:441-445.

Laurence, J. A., and F. A. Wood. 1978b. Effect of ozone on infection of wild strawberry by *Xanthomonas fragariae*. Phytopathology 68:689-692.

Linzon, S. N. 1978. Effects of airborne sulfur pollutants on plants. *In*: J. O. Nriagu (Ed.), Sulfur in the Environment: Part II. Ecological Impacts. Wiley, New York, pp. 109-162.

Manning, W. J. 1971. Effects of limestone dust on leaf condition, foliar disease incidence, and leaf surface microflora of native plants. Environ. Pollut. 2:69-76.

Manning, W. J. 1975. Interactions between air pollutants and fungal, bacterial and viral plant pathogens. Environ. Pollut. 9:87-90.

McCune, D. C., L. H. Weinstein, J. F. Mancini, and P. Van Leuken. 1973. Effects of hydrogen fluoride on plant-pathogen interactions. Proc. Internat. Clean Air Congr., Dusseldorf, Germany.

Melching, J. S., J. R. Stanton, and D. L. Koogle. 1974. Deleterious effects of tobacco smoke on germination and infectivity of spores of *Puccinia graminis tritici* and on germination of spores of *Puccinia striiformis, Pyricularia oryzae,* and an *Alternaria* species. Phytopathology 64:1143-1147.

Miller, P. R. 1977. Photochemical Oxidant Air Pollutant Effects on a Mixed Conifer Ecosystem. A Progress Report. U.S. Environmental Protection Agency, Publica. Report No. EPA-600/3-77-104, Corvallis, Oregon, 338 pp.

Moyer, J. W., and S. H. Smith. 1975. Oxidant injury reduction on tobacco induced by tobacco etch virus infection. Environ. Pollut. 9:103-106.

National Academy of Sciences. 1978. Sulfur Oxides. NAS, Washington, D.C., pp. 80-129.

Parmeter, J. R., and B. Uhrenholdt. 1975. Some effects of pine-needle or grass smoke on fungi. Phytopathology 65:28-31.

Pell, E. J., F. J. Lukezic, R. G. Levine, and W. C. Weissberger. 1977. Response of soybean foliage to reciprocal challenges by ozone and a hypersensitive-response-inducing *Pseudomonad*. Phytopathology 67:1342-1345.

Przylbylski, Z. 1967. Results of observations of the effect of SO_2, SO_3 and H_2SO_4 on fruit trees, some harmful insects near the sulfur mine and sulfur processing plant at Machow near Tarnobrzeg. Postepy Nauk Roln 2:111-118.

Reinert, R. A., and G. V. Gooding, Jr. 1978. Effect of ozone and tobacco streak virus alone and in combination on *Nicotiana tabacum*. Phytopathology 68:15-17.

Romamoorthy, S., and D. J. Kushner. 1975. Binding of mercuric and other heavy metal ions by microbial growth media. Microb. Ecol. 2:162-176.

Saunders, P. J. W. 1971. Modification of the leaf surface and its environment by pollution. *In*: T. F. Preece and C. H. Dickinson (Eds.), Ecology of Leaf Surface Microorganisms. Academic Press, New York, pp. 81-89.

Saunders, P. J. W. 1973. Effects of atmospheric pollution on leaf surface microflora. Pestic. Sci. 4:589-595.

Saunders, P. J. W. 1975. Air pollutants, microorganisms and interaction phenomena. Environ. Pollut. 9:85.

Scheffer, T. C., and G. G. Hedgecock. 1955. Injury to Northwestern Forest Trees by Sulfur Dioxide from Smelters. U.S.D.A. Forest Service. Tech. Bull. No. 1117, Washington, D.C., 49 pp.

Schönbeck, H. 1960. Beobachtungen zur frage des Einflusses von industriellen Immissionen auf die Krankheitsbereitschaft der Pflanze. Ber. Landesanst. Bodennutzungsschutz 1:89-98.

Shriner, D. S. 1974. Effects of simulated rain acidified with sulfuric acid on host-parasite interactions. Ph.D. Thesis, North Carolina State Univ., Raleigh, North Carolina, 79 pp.

Shriner, D. S. 1975. Effects of simulated rain acidified with sulfuric acid on host-parasite interactions. *In*: L. S. Dochinger and T. A. Seliga (Eds.), The 1st Internat. Symp. on Acid Precipitation and the Forest Ecosystem. U.S.D.A. For. Serv. Genl. Tech. Rep. No. NE-23, Upper Darby, Pennsylvania, pp. 919-925.

Shriner, D. S. 1977. Effects of simulated rain acidified with sulfuric acid on host-parasite interactions. Water, Air, Soil Pollut. 8:9-14.

Shriner, D. S. 1978. Effects of simulated acidic rain on host-parasite interactions in plant diseases. Phytopathology 68:213-218.

Sinclair, W. A. 1969. Polluted air: Potent new selective force in forests. J. For. 67:305-309.

Skye, E. 1968. Lichens and air pollution. Acta Phytogeogr. Suec. 52:1-23.

Smith, W. H. 1970. Tree Pathology—A Short Introduction. Academic Press, New York, 309 pp.

Smith, W. H. 1976. Air pollution—effects on the structure and function of plant-surface microbial-ecosystems. In: C. H. Dickinson and T. F. Preece

(Eds.), Microbiology of Aerial Plant Surfaces. Academic Press, New York, pp. 75-105.

Smith, W. H. 1977. Influence of heavy metal leaf contaminants on the *in vitro* growth of urban tree phylloplane-fungi. Microb. Ecol. 3:231-239.

Smith, W. H., B. J. Staskawicz, and R. S. Harkov. 1978. Trace-metal pollutants and urban-tree leaf pathogens. Trans. Brit. Mycol. Soc. 70:29-33.

Staskawicz, B. J., and W. H. Smith. 1977. Trace-metal leaf-pollutants suppress *in vitro* development of *Gnomonia platani.* Eur. J. For. Pathol. 7:51-58.

Treshow, M. 1975. Interaction of air pollutants and plant diseases. *In*: J. B. Mudd and T. T. Kozlowski (Eds.), Responses of Plants to Air Pollution. Academic Press, New York, pp. 307-334.

Unsworth, M. H., P. V. Biscal, and H. R. Pinckey. 1972. Stomatal response to sulfur dioxide. Nature 239:458-459.

Vargo, R. H., E. J. Pell, and S. H. Smith. 1978. Induced resistance to ozone injury of soybean by tobacco ringspot virus. Phytopathology 68:715-719.

Weidensaul, T. C., and S. L. Darling. 1979. Effects of ozone and sulfur dioxide on the host-pathogen relationship of Scotch pine and *Scirrhia acicola.* Phytopathology 69:939-941.

Weinstein, L. H., D. C. McCune, A. L. Aluisio, and P. Van Leuken. 1975. The effect of sulfur dioxide on the incidence and severity of bean rust and early blight of tomato. Environ. Pollut. 9:145-155.

Williams, R. J. H., M. M. Lloyd, and G. R. Ricks. 1971. Effects of atmospheric pollution on deciduous woodland. I. Some effects on leaves of *Quercus petraea* (Mattuscha) Leibl. Environ. Pollut. 2:57-68.

14

Forest Stress: Symptomatic Foliar Damage Caused by Air Contaminants

Under conditions of sufficient dose, air pollutants directly cause visible injury to forest trees. The accumulation of particulate contaminants on leaf surfaces or the continued uptake of gaseous pollutants through leaf stomata will eventually result in cell and tissue damage that will be manifest in foliar symptoms obvious to the trained, but unaided eye. This direct induction of disease in trees by air pollutants is the most dramatic and obvious individual tree response of all Class II interactions. It is the only Class II interaction that can be detected in the field by causal observation. Unlike altered reproductive strategy, nutrient cycling, tree metabolism, or insect and disease relationships; the degree of foliar symptoms induced by air pollutants can be relatively easily observed, inventoried, and quantified. In the presence of sufficient dose, tree damage may be of sufficient severity to cause mortality. Tree death directly induced by ambient air pollution exposure is considered a Class III interaction and is treated in Chapter 16.

This chapter will consider each of the ten most important air contaminant groups capable of causing direct tree morbidity. For each pollutant summary perspective will be presented including (1) foliar symptoms, (2) physiological-biochemical mechanism(s) of toxicity, (3) threshold dose, and (4) relative woody plant susceptibility. There is, of course, an enormous literature on the above topics for a large variety of plants. It is well beyond the scope of this book to provide a comprehensive review of these areas. The author has found the following publications especially useful for general information and perspective: Darley and Middleton (1966), Heggestad (1968), Centre for Agricultural Publishing and Documentation (1969), Lacasse and Moroz (1969), Hindawi (1970), Jacobson and Hill (1970), Naegele (1973), Stern et al. (1973), Mudd and Kozlowski (1975), U.S. Environmental Protection Agency (1976), and Guderian (1977).

A. Limitations on Generalizations Concerning Direct Air Pollutant Influence on Trees

Before presenting the discussions of the influence of individual pollutants on forest species, it is enormously important to realize the limitations associated with these general comments. Readers are cautioned that published information concerning symptoms, toxicology, threshold doses, and susceptibility is not absolute but rather is quite arbitrary and variable. Substantial variations in the parameters discussed in this chapter occur for a variety of reasons. The most important reasons include variation in inherent characteristics and age of trees along with variation caused by environmental conditions, all of which have been the subject of review (Davis and Wood, 1973a,b; Heck, 1968; Heggestad and Heck, 1971).

1. Plant Factors

Evidence has been provided, some with woody plant species that vegetative response to air contaminants is genetically controlled (Townsend, 1975). Genetic considerations must be evaluated, therefore, when attempting to generalize about varietal or species reaction to a particular air pollutant. David F. Karnosky, Cary Arboretum, Millbrook, New York, has provided several reviews of this general topic (Karnosky, 1974, 1978a,b). Intraspecific variation in response to a variety of pollutants was recorded for eleven forest trees by Karnosky in 1974. Variable responses of 32 cultivars of eight forest tree species, widely employed in urban plantings, have been recorded for exposure to sulfur dioxide and ozone (Karnosky, 1978a) (Table 14-1). Henry D. Gerhold, Forest Resources Laboratory, Pennsylvania State University, University Park, Pennsylvania, has performed pioneering research on the genetic control of *Pinus* species response to air pollution in the United States. In his 1975 review, Gerhold tabulated the variable response of 20 Northeastern forest tree species to sulfur dioxide, nitrogen oxides, ozone, and fluoride. In a comprehensive review of North American and European literature, Gerhold (1977) cited references to intraspecific variation in eight conifers and seven deciduous species or hybrid groups within which significant genetic variation in response to one or more air contaminants has been found (Table 14-2). Additional evidence for genetically controlled variable response has recently been provided for pines and sulfur dioxide and ozone (Genys and Heggestad, 1978), trembling aspen and sulfur dioxide and ozone (Karnosky, 1977), and ash and ozone (Steiner and Davis, 1979).

Age is another plant factor that must be considered when reviewing tree response to air contaminants. Evidence from pine species indicates foliar age is important in tree reaction to sulfur dioxide (Berry, 1974; Smith and Davis, 1977) and ozone (Davis and Coppolino, 1974; Davis and Wood, 1973b). A unique age associated factor is encountered in air pollution studies involving trees. Laboratory, greenhouse, growth chamber, and field studies are most conveniently performed with seedling or sapling size plants. As a result, most acute response

Table 14-1. Response of 32 Forest Tree Cultivars, Widely Employed in United States Urban Plantings, to Ozone and Sulfur Dioxide, Alone and in Combination, as Determined by Chamber Tests and to Oxidants (Primarily Ozone) as Determined by Field Tests

Species	Cultivar	Ozone	Sulfur dioxide	Ozone plus sulfur dioxide	Oxidants
Norway maple	Cleveland	R	I	R	R
	Crimson King	R	R	R	–
	Crimson Sentry	R	I	R	R
	Columnar	I	R	S	R
	Emerald Queen	I	R	I	R
	Green Mountain	R	R	I	–
	Jade Glen	I	I	I	R
	Schwedler	I	I	I	R
	Summershade	R	R	R	–
Red maple	Autumn Flame	R	I	R	R
	Bowhall	I	R	S	–
	Red Sunset	R	R	R	R
	Tilford	I	S	I	–
Sugar maple	Goldspire	I	R	S	–
	Temple's Upright	R	R	R	–
European beech	Rotundifolia	R	R	R	–
White ash	Autumn Purple	S	S	S	I
European ash	Hessei	I	S	I	R
Green ash	Marshall's Seedless	I	S	S	I
	Summit	I	I	I	R
Ginko	Autumn Gold	R	I	I	–
	Fairmont	R	I	I	–
	Fastigiate	R	I	R	–
	Sentry	R	R	R	–
Honey locust	Emerald Lace	S	I	I	R
	Imperial	S	R	R	S
	Majestic	S	S	R	I
	Shademaster	S	R	I	I
	Skyline	S	S	R	R
	Sunburst	S	S	I	R
London plane	Bloodgood	S	S	S	S
English oak	Fastigiate	S	R	I	–

Source: Karnosky (1978a).

[a] S = sensitive, I = intermediate, R = tolerant.

data has been accumulated from studies using relatively young trees. The response of mature and overmature trees is invariably extrapolated from these studies. In awareness of differences in physiological processes of mature versus juvenile individuals, it should be recognized that risks are associated with this extrapolation.

Table 14-2. Forest Tree Species That Have Exhibited Genetically Variable Response to One or More Air Pollutants

Species	Differences among	Pollutant
Acacia	Populations	CO
Red maple	Populations	O_3
Sugar maple	Seedlings	O_3
European and Japanese larch	Families	SO_2
Norway spruce	Families	SO_2, HF
	Clones	
	Populations	
	Families	
Lodgepole pine	Populations	SO_2
	Populations	SO_2
	Populations	SO_2
Ponderosa pine	Populations	O_x, HF
	Populations	SO_2, O_3
White pine	Clones	O_3, SO_2
	Clones	HF, O_3, SO_2
Scotch pine	Clones	HF, SO_2
	Populations	SO_2
	Clones, families	SO_2
Platanus	Families	O_3, SO_2
Poplar	Clones	SO_2
	Clones	SO_2
	Clones	O_3
Trembling aspen	Clones	O_3, SO_2
Douglas fir	Populations	SO_2
American elm	Families	O_3, SO_2
	Clones	O_3, SO_2

Source: Gerhold (1977).

Tree health is a final factor that may influence the response of a tree to air pollution stress. Chapters 12 and 13 have clearly shown that insect and microbial stress may interact with air contaminants in a complex way. Reduced foliar gas exchange occasioned by microbial infection or arthropod feeding may reduce gaseous air pollutant uptake and reduce acute air pollution injury. Accumulation of persistent air contaminants, for example, heavy metals and fluoride, on the other hand, may exacerbate impaired tree health already abnormal by virtue of insect and disease impact.

2. Environmental Factors

Woody plant studies that have examined acute plant response to an air pollutant are typically performed under controlled environmental conditions. Most data have been produced from experiments conducted in growth chambers or greenhouses. A long list of environmental variables strongly mediate the response of a plant to air pollution exposure (Davis and Wood, 1973a; and Davis, 1975a). This realization has necessitated the use of controlled growth facilities in order to manage confusing and complexing environmental response. Controlled facilities themselves introduce well appreciate artificialities. Closed chambers, for example, typically have higher temperature and lower light than ambient conditions. Recent developments in open-top chamber design will hopefully partially address this problem (Heagle et al., 1979). Environmental variables of primary significance in plant response to pollutant stress include temperature (Davis, 1975a), light intensity and quality (Dunning and Heck, 1973), humidity (Leone and Brennan, 1969; Otto and Daines, 1969; Wilhour, 1970), and wind velocity (Brennan and Leone, 1968; Heagle et al., 1971).

In addition to the large number of above-ground environmental variables that influence plant response to air pollution, considerable evidence suggests that numerous below ground or edaphic variables are also important (Smith, 1975). The greatest amount of information relates to the influence of soil moisture and soil nutrient content on plant response. The general observation has been that plants are generally more resistant to damage from gaseous air contaminants when they are under moisture stress. Presumably the drought-induced reduced stomatal aperture permits less uptake of pollutants from the ambient atmosphere and therefore occasions less plant injury (cf. Fuhrer and Erismann, 1980). The influence of soil nutrient status on plant susceptibility to air contaminants, especially oxidants, has also been investigated. The nonwoody plant data are variable however, and generalizations are difficult (Heagle, 1979; Jager and Klein, 1976; Leone, 1976).

Some research consideration has been given to the response of trees to soil nutrients and air pollution damage. Bjorkman (1970) concluded that optimal soil nutrient concentrations reduced sulfur dioxide damage to Scotch pine. Applications of lime and fertilizer, however, failed to protect eastern white pine from air pollution damage (Dochinger, 1964; Dochinger and Seliskar, 1970). Following fertilization of field-grown white pine, Berry and Hepting (1964) did not detect any change in tree susceptibility to air pollution injury. Applications of N,P,K fertilizer to air pollution sensitive clones of eastern white pine, however, did increase their tolerance to air pollution stress (Cotrufo and Berry, 1970). Cotrufo (1974) has treated clonal ramets of eastern white pine with factorial combinations of N, P, and K and exposed the trees to ambient sulfur dioxide pollution. The addition of N at all levels of P and K significantly increased sulfur dioxide susceptibility. The addition of P at all levels of N and K reduced susceptibility, while the ramets did not respond significantly to the addition of K.

In addition to soil moisture and nutrient content, other soil variables may influence plant response to air pollution. Composition of the soil atmosphere, for example, oxygen or carbon dioxide tension, may be important. Soil temperature is another variable that may have some significance in controlling plant response.

3. Pollutant Interactions

In natural environments, trees may be exposed to more than one air pollutant concurrently or sequentially within a relatively short period of time. A large number of investigations cited in this volume and elsewhere (Reinhert, 1975) have indicated that contaminants may interact to produce a unique plant response. In general, three kinds of interaction have been recognized. When the influence of exposure to one pollutant is merely added onto the influence of another or several other pollutants, the interaction is termed additive. In this case an individual pollutant does not either increase nor decrease the influence of another contaminant. When exposure to two or more air pollutants simultaneously produces a plant response more severe than any induced by individual contaminant stress, then the interaction is termed synergistic. If two or more pollutants interact to reduce the impact of a given pollutant, then the interaction is termed antagonistic.

Investigations specifically designed to explore these interactions have been conducted and have demonstrated all of these responses in trees: additive (Matsushima and Brewer, 1972; Tingey and Reinert, 1975); synergistic (Dochinger et al., 1970; Krause and Kaiser, 1977); and antagonistic (Davis, 1977; Nielson et al., 1977). The size and longevity of forest tree species makes considerations of potential pollutant interactions especially important.

4. Dose-Response Considerations

Dose measurement of air pollutant exposure involves a measured concentration of a toxicant for a known duration of time (Weinstein, 1975). A large volume of literature reports only single dose exposures. It is frequently most difficult to predict tree response over a range of contaminant concentrations for varying time periods. Cultural treatments performed preceding or following pollutant exposure can modify dose-response data. The large variability in dose administration itself frequently makes comparison between laboratories and experiments difficult.

Ideally investigations should include doses that will induce effects that range from none to extreme and data that can be employed to estimate a median effective dose (compare ED_{50} or effective dose at which 50% of the test plants are influenced). The dose associated with the lowest probable threshold for the occurrence of an effect is also needed (Weinstein, 1975).

5. Summary

A very impressive set of variables influences the response of plants to air pollutant exposure (Figure 14-1). Because of the size and age of trees, research artificialities introduced by experimental designs are more severe for forest vegetation than other plant types. As a result severe limitations are placed on our ability to extrapolate much of the wealth of data available in the literature to natural forest environments. It is further important to recognize that the symptoms of acute foliar response to air pollutants are not unlike the symptoms in-

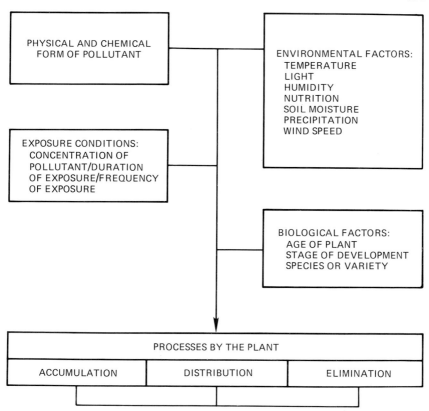

Figure 14-1. Factors affecting the response of plants to air pollutants. (From Weinstein and McCune, 1979.)

duced in tree foliage by a host of other abiotic and biotic stress factors. The diagnosis of specific tree response to air pollution stress will involve the same care, knowledge, and perspective that all woody plant diagnoses involve; that is, a good appreciation of all stress factors, and the symptoms they cause, for a given plant in a given location. It will further be necessary to have some appreciation of the historical stresses that have impacted the tree in order to make an accurate diagnosis.

B. Sulfur Dioxide

Due to the long history of awareness of the adverse impact of sulfur dioxide on vegetative health and the extensive global distribution of point sources of sulfur dioxide, an enormous literature has accumulated on the acute influence of this gas on plants. A variety of excellent reviews is available including those dealing with general vegetation (Linzon, 1978; National Academy of Sciences, 1978a; Tamm and Aronsson, 1972), agricultural crops (Heitschmidt and Altman, 1978), and forest trees (Forestry Commission, 1971; Davis and Wilhour, 1976).

1. Foliar Symptoms

In angiosperms, the formation of marginal and interveinal necrotic areas, dark green or dull in color, with a water-soaked appearance are the initial symptom. Eventually these necrotic zones dry and bleach to a light (ivory-white-hyaline color). Necrotic spots result as mesophyll cells collapse in those areas near stomata. Lamina portions immediately bordering veins rarely become necrotic as stomata are few and intercellular space is limited in these areas. Necrotic zones extend through the leaf and are visible on adaxial (upper) as well as abaxial (lower) surfaces. Leaves are most susceptible to damage when they have just reached their fully expanded size. Older leaves are less sensitive and very young expanding leaves are the most resistant. Gymnosperm needles develop a water-soaked appearance and typically turn reddish-brown in color. As in the case of angiosperms, recently fully expanded needles are the most sensitive (U.S. Environmental Protection Agency, 1976).

Excellent color photographs of sulfur dioxide symptoms are available in Jacobson and Hill (1970), Loomis and Padgett (1973), Hindawi (1970), U.S.D.A. Forest Service (1973), and U.S. Environmental Protection Agency (1976).

2. Mechanism of Toxicity

Following absorption of sulfur dioxide through stomatal pores the gas is presumed to be dissolved in moisture films on mesophyll cell walls. Sulfurous acid (H_2SO_3), and following oxidation, sulfuric acid (H_2SO_4) are eventually formed. Some of the toxic effects of sulfur dioxide may be due to its acidifying influence. Sulfite $(SO_3{}^{2-})$ and to a lesser extent sulfate $(SO_4{}^{2-})$ ions are toxic to a variety of biochemical processes. Sulfite adds to ketones and aldehydes to form α-hydroxysulfonates. It reacts with pyrimidine bases in DNA and RNA. Sulfite may interfere with a variety of enzymes by competing with phosphate and carbonate groups at binding sites, by reaction with disulfide bonds and altering structure, or by forming inhibiting compounds (Wallace and Spedding, 1976). Interference at the physiological level, resulting from biochemical alterations induced by sulfur compounds, may involve abnormalities in stomatal opening, amino acid metabolism, photosynthesis, or chlorophyll levels (Mudd, 1973, 1975).

3. Threshold Dose

The National Academy of Sciences (1978) has provided a comprehensive review of sulfur dioxide concentrations that cause acute foliar damage to plants under experimental laboratory and field conditions as well as following ambient exposures. The variation in threshold doses is extensive but not surprising in view of the experimental limitations previously discussed.

Linzon (1978) has concluded that acute injury to native vegetation does not occur below 0.70 ppm (1820 μg m^{-3}) of sulfur dioxide for 1 hr or 0.18 ppm (468 μg m^{-3}) for exposures of 8 hr. An intermediate dose approximating 0.25

ppm (650 μg m^{-3}) for several hours may injure some species. Mild foliar symptoms, including chlorosis, silvering and other discoloration, in forest ecosystems may occur from average concentrations of sulfur dioxide in the range of 0.008 ppm (21 μg m^{-3}) to 0.017 ppm (44 μg m^{-3}) over entire growing seasons, during which the sulfur dioxide episodes are of variable intensities.

4. Relative Susceptibility

A large number of forest tree rankings of relative susceptibility to acute sulfur dioxide injury are available. Specialty lists for Australia (O'Connor et al., 1974) and Europe (Tamm and Aronsson, 1972) have been provided. The most comprehensive and useful list for North American woody plants has been provided by Davis and Wilhour (1976) and is reproduced in Table 14-3.

C. Nitrogen Oxides

The most comprehensive and recent reviews of the influence of oxides of nitrogen on plants have been provided by Taylor et al. (1975) and the National Academy of Sciences (1977b). Nitric oxide is less phytotoxic than nitrogen dioxide and both are less damaging to plants than sulfur dioxide, gaseous fluoride, or oxidants.

1. Foliar Symptoms

Initial symptoms involve the development of diffuse discolored spots of gray-green or light brown color. These spots eventually weather and dry and become bleached as in the instance of sulfur dioxide injury. In angiosperms, the necrotic discolored spots that form in interveinal portions of the leaf frequently combine to form stripes. Marginal necrosis may appear, especially in locust, oak, and maple. High dose exposure may cause intense net necrosis and leave only "fingers of green" along the veins. Gymnosperm symptoms typically initially involve red-brown or fuchsia discoloration in the distal portion of the needle. The discoloration may eventually extend to the needle base. A distinct boundary between necrotic and healthy needle tissue may occur in pine and fir. A general chlorosis of older leaves may occur after long-term exposure to low concentration (National Academy of Sciences, 1977b). These symptoms are generally not distinctive, are infrequent, and field diagnosis based on symptoms alone would be extremely difficult.

2. Mechanism of Toxicity

Following absorption of nitrogen dioxide through the stomata, the gas will react with water on the moist surfaces of mesophyll cells and form nitrous (HNO_2) and nitric (HNO_3) acids. Toxicity may partially result from a pH decrease (Zeevaart, 1976). Deamination reactions may be induced in amino acids and

Table 14-3. Relative Sensitivity of North American Woody Plants to Acute Damage by Sulfur Dioxide

Sensitive	Intermediate	Tolerant
	Angiosperms	
Thinleaf alder	Mountain alder	Buck-brush
Large-toothed aspen	Chinese apricot	Buffalo-berry
Trembling aspen	Basswood	Redstem ceanothus
Green ash	Water birch	Forsythia
European birch	Box elder	Black hawthorn
Gray birch	Bitter cherry	Kinnikinnick
Western paper birch	Chokecherry	Littleleaf linden
White birch	Black cottonwood	Cure-leaf mountain mahogany
Yellow birch	Eastern cottonwood	Norway maple
Lowbush blueberry	Narrowleaf cottonwood	Silver maple
Bitter cherry	Sticky currant	Sugar maple
Chinese elm	Red osier dogwood	Gambel oak
Beaked hazel	Blueberry elder	Pin oak
California hazel	American elm	Northern red oak
Manitoba maple	Wild grape	Oregon grape
Rocky Mountain maple	Red hawthorne	London plane
Lewis mock-orange	Tatarian honeysuckle	Western poison ivy
Sitka mountain-ash	Witch hazel	Carolina poplar
Texas mulberry	Hydrangea	Squawbush
Pacific ninebark	Common lilac	Smooth sumac
Ocean-spray	Mountain mahogany	
Lombardy poplar	Douglas maple	
Creambush rockspirea	Rocky Mountain maple	
Low serviceberry	Red maple	
Saskatoon serviceberry	Coronarius mock-orange	
Utah serviceberry	Virginalis mock-orange	
Staghorn sumac	European mountain-ash	
Black willow	Western mountain-ash	
	Mountain laurel	
	White oak	
	Balsam poplar	
	Big sagebrush	
	Mountain snowberry	
	Columbia snowberry	
	Van Houts spirea	
	Shiny leaf spirea	
	Gymnosperms	
Western larch	Douglas fir	Arborvitae (white cedar)
Eastern white pine	Balsam fir	Western red cedar
Jack pine	Grand fir	Silver fir
Red pine	Western hemlock	White fir

Table 14-3 (continued)

Sensitive	Intermediate	Tolerant
	Gymnosperms	
	Austrian pine	Ginkgo
	Lodgepole pine	Common juniper
	Ponderosa pine	Rocky Mountain juniper
	Western white pine	Utah juniper
	Engleman spruce	Western juniper
	White spruce	Limber pine
		Pinyon pine
		Blue spruce
		Pacific yew

Source: Davis and Wilhour (1976).

nucleic acid bases. The acids may react with unsaturated compounds and cause isomerization and free radical formation. Excessive nitrate anions may react with amines to form nitrosamines. At the physiological level cellular pH is altered and pollutant interactions with cellular components lead to altered metabolism, for example, carbon dioxide fixation may be inhibited (nitrite is much more toxic than nitrate in this regard), acetate metabolism may be inhibited, and the final result is that growth may be suppressed (Mudd, 1973; Taylor et al., 1975; Zeevaart, 1976).

3. Threshold Dose

Nitrogen oxides are much less phytotoxic than ozone or peroxyacylnitrates. Vegetative injury would be anticipated only in localized regions immediately adjacent to excessive industrial sources. Leaf symptoms would be expected at doses approximating 1.6-2.6 ppm (3000-5000 μg m^{-3}) for periods of up to 48 hr. The threshold for leaf injury may require exposure to 20 ppm (38 $\times$ 10 μg m^{-3}) if the exposure is only for 1 hr, while a concentration of 1 ppm (1900 μg m^{-3}) might require up to 100 hr to produce symptoms (National Academy of Sciences, 1977b). Nitric oxide injury to trees in the field is unknown while nitrogen dioxide injury would be expected only in the vicinity of an excessive industrial source.

4. Relative Susceptibility

Information on relative susceptibility of forest trees to nitrogen dioxide is extremely limited and entirely based on experimental fumigations. A relative ranking provided by Davis and Wilhour (1976) is reproduced in Table 14-4.

Table 14-4. Relative Sensitivity of Woody Plants to Acute Damage by Artificial Exposure to Nitrogen Oxides, Primarily Nitrogen Dioxide

Sensitive	Intermediate	Tolerant
	Angiosperms	
Weeping birch	Norway maple	Black locust
Showy apple	Fan maple	Hornbeam
Wild pear	Winter lime	European beech
Rose	Summer lime	Common elder
Azalea	Rhododendron	Mountain elm
Pyracantha	Ligustrum	Common oak
	Gymnosperms	
European larch	Blue spruce	Yew
Japanese larch	White spruce	Black pine
	Lawson's cypress	Gingko
	Japanese fir	Shore juniper
	Common silver fir	

Source: Davis and Wilhour (1976).

D. Ozone

Since acute ozone damage to vegetation was first observed in the mid 1940s an enormous literature has developed. Reviews have been provided by Heath (1975), Treshow (1970), and the National Academy of Sciences (1977c). The latter is most comprehensive and contains in excess of 500 literature citations.

1. Foliar Symptoms

In angiosperms the classic symptoms are necrotic spots on the adaxial surface typically termed stipple or fleck. Unlike sulfur and nitrogen oxide impact on mesophyll cells, ozone symptoms result from palisade cell destruction. Eventually the entire upper leaf surface may assume a bleached appearance (coalescence of flecks). In other cases, palisade cells may accumulate dark alkaloid pigments and present the stipple symptom. Ultimately ozone influence may extend to the spongy cells and produce abaxial necrotic spots. High dose exposure may induce dark, water-soaked areas to form. In gymnosperms, eastern white pine elongating needles undergo distal necrosis, tip dieback and browning, when subject to ozone exposure. Chlorotic flecks and mottling may be associated with tip necrosis. In ponderosa pine, ozone exposure causes premature needle shed, dwarfed and chlorotic needles, and a general decline syndrome (National Academy of Sciences, 1977c).

Color photographs of ozone symptoms are available in Jacobson and Hill (1970), Loomis and Padgett (1973), Hindawi (1970), U.S.D.A. Forest Service (1973), and U.S. Environmental Protection Agency (1976).

2. Mechanism of Toxicity

Despite considerable research on this topic over the past ten years, a unified hypothesis for ozone toxicity has not been provided. Following ozone uptake through the stomata a large number of biochemical responses have been noted. Phenol, the activity of some enzymes, α-aminobutyric acid, reducing sugars, soluble sugars, and certain amino acids have been observed to increase. Other amino acids, other enzyme activities (nitrate reductase, carbohydrase), and sulfhydryl groups have been observed to decrease. Ozone may react with membrane lipid systems. A theory that ozone, or a product of chloroplast ozone oxidation, migrates through the cytoplasmic membrane into the chloropast interior before exerting an effect is consistent with much of the data available (National Academy of Sciences, 1977c). Unsaturated fatty acids of leaf membranes are susceptible to ozone attack. Enzymes are readily inactivated by ozone. Either or both of these ramifications could easily account for the abnormal physiology induced by ozone exposure. The relative importance and timing of specific effects, however, remains unclear (Mudd, 1973). At the cellular level, ozone exposure results in gross plasmolysis and cell death. Physiological ramifications of ozone contamination may include altered photosynthesis, respiration, membrane and stomatal function.

3. Threshold Dose

Several reports indicate that a variety of eastern deciduous species are injured by exposures to ozone of 0.20-0.30 ppm (392-588 μg m^{-3}) for 2-4 hr (National Academy of Sciences, 1977c). Eastern conifers may be somewhat more sensitive while western conifers may be somewhat more resistant. Several studies suggest that the threshold of visible injury of eastern white pine approximates 0.15 ppm (294 μg m^{-3}) for 5 hr (Costonis, 1976). Ponderosa pine in the San Bernardino Mountains of California has a threshold of symptomatic injury approximating 0.08 ppm (157 μg m^{-3}) for 12-13 hr (Taylor, 1973).

4. Relative Susceptibility

Davis and Wilhour (1976) have reviewed the considerable literature dealing with woody plant susceptibility and their summary listing is provided in Table 14-5.

E. Peroxyacetylnitrate

Peroxyacetylnitrate is one of a series of photochemically produced oxidant homologues. Its chemical formula is $CH_3 CO \cdot O_2 NO_2$ and the gas is commonly referenced using the abbreviation PAN. Peroxyacetylnitrate along with peroxypropionylnitrate (PPN) and peroxybutyrylnitrate (PBN) are phytotoxic. Excellent reviews of vegetative stress by peroxyacetylnitrate have been provided by Mudd (1975b) and the National Academy of Sciences (1977c).

Table 14-5. Relative Sensitivity of North American Woody Plants to Acute Damage by Ozone

Sensitive	Intermediate	Tolerant
	Angiosperms	
Tree-of-heaven (ailanthus)	Chinese apricot	Apricot
Green ash	Box elder	Arborvitae (white cedar)
White ash	Incense cedar	Chinese azalea
Quaking aspen	Lambert cherry	Avocado
Campfire azalea	Northern black currant	European beech
Hinodegiri azalea	Black bead elder	European birch
Korean azalea	Chinese elm	Japanese box
Snow azalea	Lynwood gold forsythia	Gray dogwood
Bridalwreath	Sweetgum	White dogwood
Bing cherry	Blue-leaf honeysuckle	Dwarf winged Euonymus
Rock cotoneaster	Common lilac	Laland's firethorne
Spreading cotoneaster	Sweet mock-orange	Black gum
Concord grape	Black oak	American holly
Honey locust	Pin oak	English holly
Chinese lilac	Scarlet oak	Hetz Japanese holly
European mountain-ash	Common privet	Mountain laurel
Gambel oak	Eastern redbud	American linden
White oak	Rhododendron	Little-leaf linden
Hybrid poplar	Vaccinioides snowberry	Black locust
Tulip poplar	Linden viburnum	Trailing mahonia
Londense privet	Tea viburnum	Bigtooth maple
Saskatoon serviceberry		Norway maple
Alba snowberry		Sugar maple
Fragrant sumac		Red maple
American sycamore		Bur oak
English walnut		English oak
		Northern red oak
		Shingle oak
		Pachiystima
		Japanese pagoda
		Barlett pear
		Japanese pieris
		Poison ivy
		Amur north privet
		Carolina rhododendron
		Rose woods
		Common sagebrush
		Korean spice viburnum
		Burkwoodii viburnum
		Black walnut
	Gymnosperms	
European larch	Japanese larch	Balsam fir
Austrian pine	Eastern white pine	Douglas fir

Table 14-5 (continued)

Sensitive	Intermediate	Tolerant
	Gymnosperms	
Coulter pine	Knobcone pine	White fir
Jack pine	Lodgepole pine	Eastern hemlock
Loblolly pine	Scotch pine	Digger pine
Monterey pine	Shortleaf pine	Red pine
Ponderosa pine	Slash pine	Redwood
Virginia pine	Sugar pine	Giant sequoia
	Torrey pine	Black Hills spruce
		Colorado blue spruce
		Norway spruce
		White spruce
		Dense yew
		Hatifield's yew

Source: Davis and Wilhour (1976).

1. Foliar Symptoms

Descriptions of ambient peroxyacetylnitrate symptoms come primarily (if not exclusively) from nonwoody species. Laboratory induced woody plant symptoms, however, are generally consistent with classic field symptoms resulting from exposure to this gas. Peroxyacetylnitrate preferentially influences the spongy mesophyll cells and deciduous species typically exhibit symptoms initially on the abaxial surface. The classic syndrome is a glaze appearance followed by bronzing of the lower leaf surface. High dose exposure may result in upper leaf surface injury. Premature senescence and leaf abscission may occur. Low dose exposure may cause chlorosis only. Symptoms associated with peroxypropionylnitrate and peroxbutyrylnitrate are similar. Young expanding leaves are normally most sensitive to injury (National Academy of Sciences, 1977c). Adequate differentiation of peroxyacetylnitrate symptoms in gymnosperms has not been provided.

2. Mechanism of Toxicity

Following stomatal uptake of peroxyacetylnitrate, the pollutant may be involved in a variety of biochemical effects. Several enzymes are inhibited by the gas. Peroxyacetylnitrate may react with sulfhydryl groups and may inhibit fatty acid synthesis. The gas is capable of oxidizing a large number of sulfur compounds. Eventually, as in the case of ozone, leaf cell plasmolysis occurs and cell death results. The specific sequence and relative importance of various biochemical events remain unclear. Physiological abnormalities may be caused in carbohydrate or hormone metabolism and photosynthesis (Mudd, 1975b; National Academy of Sciences, 1977c).

3. Threshold Dose

Peroxyacetylnitrates are the most phytotoxic of the various photochemical oxidants described. Insufficient data are available to develop reliable dose-response curves for peroxyacetylnitrate and homologues, especially for woody plants. Laboratory evidence with agricultural species indicate a sensitive plant threshold dose of approximately 0.01-0.02 ppm (49-99 μg m^{-3}) for 4 hr (U.S. Environmental Protection Agency, 1976). Davis (1975b) could not injure cotyledons or primary needles of ponderosa pine with peroxyacetylnitrate concentrations of 0.08, 0.20, or 0.40 ppm (396, 989, or 1979 μg m^{-3}) for 8-hr exposures. Variable symptoms were reported in maple, ash, oak, and honey locust when exposed to 0.20-0.30 ppm (989-1484 μg m^{-3}) for 8 hr (Drummond, 1971).

4. Relative Susceptibility

In their review, Davis and Wilhour (1976) do not record any North American woody plants sensitive or intermediate in resistance to damage by peroxyacetylnitrate. They list 36 species as tolerant of laboratory exposures (Table 14-6).

Table 14-6. North American Woody Plants Tolerant of Peroxyacetylnitrate Exposure under Laboratory Conditions

Angiosperms	Gymnosperms
Green ash	Arborvitae (white cedar)
White ash	Balsam fir
Basswood	Douglas fir
European white birch	White fir
White dogwood	Eastern hemlock
Sweet gum	European larch
Honey locust	Japanese larch
Common lilac	Austrian pine
Norway maple	Eastern white pine
Silver maple	Pitch pine
Sugar maple	Ponderosa pine
American mountain-ash	Red pine
English oak	Scotch pine
Northern red oak	Virginia pine
Pin oak	Hybrid poplar
White oak	Tulip poplar
	Black Hills spruce
	Blue spruce
	Norway spruce
	White spruce

Source: Davis and Wilhour (1976).

F. Fluoride

The impact of fluoride on vegetation has an extended research history and a very substantial literature. Numerous reviews of this research have been provided (Chang, 1975; Keller, 1975a,b; McCune and Weinstein, 1971; Weinstein and McCune, 1971). An extremely comprehensive and very useful review was recently provided by Leonard H. Weinstein, Boyce Thompson Institute for Plant Research, Ithaca, New York (Weinstein, 1977).

1. Foliar Symptoms

In angiosperms the initial symptom is leaf tip chlorosis. The chlorotic condition eventually expands around the leaf margin and inward along the midvein. With continued exposure chlorotic areas become necrotic. Necrotic tips and margins may ultimately fall from the leaf. Occasionally premature leaf abscission precedes development of intensive necrosis (Weinstein, 1977).

The typical response of gymnosperms to fluoride exposure is the development of distal necrosis or "tipburn." A well-defined boundary usually occurs between live and necrotic portions. The boundary zone may become darker brown than the necrotic tip which is typically some shade of reddish-brown (U.S. Environmental Protection Agency, 1976). Unfortunately these symptoms are not highly specific and may be induced by a large number of tree stresses. Color photographs of fluoride injury are provided in Weinstein (1977), U.S. Environmental Protection Agency (1976), Loomis and Padgett (1973), and Jacobson and Hill (1970).

2. Mechanisms of Toxicity

Trees may accumulate fluoride through stomata, the cuticle, or via other exterior surfaces. The specific biochemical ramifications of fluoride contamination remain incompletely appreciated. A variety of enzymes sensitive to fluoride stimulation or inhibition is known. Fluoride may alter cell nutrient status by complexing with a variety of metal cations. The interaction of fluoride with enzymes and nutrients is translated into effects on metabolite levels, metabolic reactions, for example, oxygen uptake, cell wall formation, starch synthesis, and growth, all thoroughly reviewed by Weinstein (1977).

3. Threshold Dose

Gaseous compounds containing fluoride are more toxic than particulate forms. The phytotoxicity of the latter is largely related to solubility. Ammonium fluoride, for example, is much more toxic to vegetation than low solubility cryolite (Na_3AlF_6). In general fluoride is considerably more phytotoxic than the previously discussed gaseous pollutants. Susceptible plants may be injured at ambient concentrations 10 to 1000 times lower than sulfur dioxide, ozone, peroxy-acetylnitrate, and nitrogen oxides. Since fluorides accumulate in foliar tissue, toxicities are generally expressed in terms of tissue concentrations on a dry weight basis. For most plants designated susceptible, this concentration is typically less than 100 μg g^{-1}. Plants classified as intermediate or tolerant can probably tolerate concentrations in excess of 200 μg g^{-1} without development of visible symptoms (Weinstein, 1977). Selected woody plants may accumulate in excess of 4000 μg g^{-1} without visible stress. For a large number of trees the fluoride dose for symptomatic injury ranges from approximately 0.75 to 50 μg m^{-3} fluoride for several hours to 10 or more days.

4. Relative Susceptibility

Plants exhibit a broad range of tolerances to fluoride. A comprehensive listing of relative susceptibility provided by Weinstein (1977) is reproduced in Table 14-7.

G. Trace Metals

A very large number of trace metals may be associated with higher plants. Many of these are heavy metals, that is, have a density greater than five. Some of these metals are required micronutrient elements, for example, iron, manganese, copper, and zinc. Other metals have no known metabolic function, for example, lead, cadmium, nickel, and tin. All these trace metals have the potential to be toxic to trees if present in sufficient concentrations. Trace metals identified as having especially high potential to acutely injury trees, because of widespread distribution or intensive local release as a result of anthropogenic activities, include cadmium, cobalt, chromium, copper, lead, mercury, nickel, thallium, vanadium, and zinc. These metals are the subject of an enormous literature. Comprehensive general reviews include Energy Research and Development Administration (1975), Kathny (1973), Oehme (1978), and Purves (1977). Particular element reviews include cadmium (Commission of the European Communities, 1978; Friberg et al., 1974), chromium (National Academy of Sciences, 1974), copper (National Academy of Sciences, 1977a; Nriagu, 1979), lead (Peterson, 1978), mercury (Friberg and Vostal, 1972; National Academy of Sciences, 1978b), nickel (National Academy of Sciences, 1975), and zinc (Nriagu, 1980).

1. Foliar Symptoms

Specific symptom descriptions for forest vegetation exposed to excessive trace metals are not abundant. The syndrome of a large number of plants acutely injured by trace metal accumulation, however, involves development of interveinal chlorosis, tip and margin necrosis, and premature leaf abscission. Mitchell and Fretz (1977) have described symptoms of cadmium and zinc toxicity to

Table 14-7. Relative Sensitivity of Woody Plants to Acute Damage by Fluoride

Sensitive	Intermediate	Tolerant
	Angiosperms	
Chinese apricot	Apple	Alder
Blueberry	Green ash	Japanese barberry
Box elder	Trembling aspen	Black birch
European grape	Azalea	White birch
Sheep laurel	Warty barberry	Cutleaf birch
Bradshaw plum	Wintergreen barberry	Blackberry
Italian prune	Common boxwood	Black-haw
	Bing cherry	Cotoneaster
	Royal Ann cherry	Currant
	Choke cherry	Flowering dogwood
	Flowering cherry	Red osier dogwood
	Concord grape	Elderberry
	Mountain laurel	American elm
	Lilac	Chinese elm
	American linden	Siberian elm
	Littleleaf linden	Winged euonymus
	Hedge maple	Pyracantha
	Norway maple	Forsythia
	Red maple	Honeysuckle
	Silver maple	Black locust
	Sugar maple	Honey locust
	European mountain ash	Oak
	Mulberry	Pear
	Peach	London plane
	Flowering plum	Balsam poplar
	Rhododendron	Carolina poplar
	Serviceberry	Lombardy poplar
	Smooth sumac	Silver-leaved poplar
	Staghorn sumac	Privet
	Double fill viburnum	Russian olive
	Black walnut	Spirea
	English walnut	Sweetgum
	Yew	Sycamore
		Tree-of-heaven
		Tuliptree
		Arrowwood viburnum
		Leatherleaf viburnum
		Scibold viburnum
		Goat willow
		Laurel leaf willow
		Weeping willow
	Gymnosperms	
Douglas fir	Balsam fir	Arborvitae (white cedar)
Western larch	Grand fir	Eastern red cedar

Table 14-7 (continued)

Sensitive	Intermediate	Tolerant
	Gymnosperms	
Eastern white pine	Ginkgo	Western red cedar
Mugho pine	Western white pine	Cypress
Loblolly pine	Jack pine	Andorra juniper
Lodgepole pine	Austrian pine	Creeping juniper
Ponderosa pine	Blue spruce	Pfitzen juniper
Scotch pine	Birds nest spruce	
	Black spruce	
	Engelman spruce	
	White spruce	

Source: Weinstein (1977).

forest tree seedlings grown in solution culture amended with heavy metals. Symptoms of cadmium toxicity appeared on the youngest foliage of red maple in the form of interveinal chlorosis. Foliage size was stunted in most cases. High cadmium levels resulted in loss of leaf turgor, wilting, and ultimately death. Initial symptoms of cadmium stress of white pine were inhibition of needle expansion, and on Norway spruce, chlorotic tips of new growth. Ultimately wilting progressed basipetally, followed by necrosis. Low zinc levels induced interveinal chlorosis in red maple. High zinc concentrations resulted in wilting of new growth and eventually necrosis of leaf and stem tissue. Zinc produced symptoms on white pine and Norway spruce similar to those described for cadmium.

2. Mechanisms of Toxicity

Specific mechanisms of toxicity of trace metals are presumed to be quite variable. It is probable, however, that trace metal injury to woody plants involves one or more of the following biochemical abnormalities: (1) metal interferes with enzyme function, (2) metal serves as an antimetabolite, (3) metal forms a stable precipitate or chelate with an essential metabolite, (4) metal catalyzes the decomposition of an essential metabolite, (5) metal alters the permeability of cell membranes or (6) the metal replaces important structural or electrochemically important elements in the cell (Bowen, 1966). Each of these abnormalities is capable of inducing substantial adverse impact on a variety of critical physiological functions.

3. Threshold Dose

It is extremely difficult to generalize concerning threshold doses required for acute injury to woody plants. Temple and Hill (1979) have presented threshold levels of contamination for vegetation for a variety of trace elements. Some of the trace metals are essential plant nutrients and are required by various species in differing amounts. Numerous plants have the capacity to evolve substantial tolerance to high environmental levels of certain heavy metals (Antonavics et al., 1971). Some elements, for example, cadmium, nickel, and thallium, are especially mobile in plants. Other elements have considerable potential to accumulate, for example, mercury and vanadium. Table 14-8 presents a general suggestion of baseline trace metal content for the ten metals judged to have especially high potential to injure trees.

Mitchell and Fretz (1977) have suggested that toxicity thresholds (foliar symptoms) for red maple, white pine, and Norway spruce seedlings were 23, 61, and 8 μg g^{-1} cadmium, respectively, in leaf tissue (dry weight basis). In the instance of zinc, foliar symptoms resulted when leaf concentrations reached 421 μg g^{-1} zinc in red maple, 1006 μg g^{-1} in white pine, and 596 μg g^{-1} in Norway spruce. There is insufficient information available to suggest threshold levels of these various trace metals required to cause acute injury to sensitive woody plants. These thresholds are surely very variable and most probably within the ranges contained in Table 14-8.

4. Relative Susceptibility

Relative susceptibility of forest trees to acute injury by trace metal contamination cannot be established with current information.

H. Other Air Contaminants

Under certain circumstances localized forest areas or ornamental woody plants may be subject to *acute* injury by a minor list of particulate and gaseous pollutants including acid rain, ammonia, chlorine, hydrocarbons, and hydrogen sulfide. Typically these injuries result following accidental or unusual industrial, commercial or transportation release, or in the presence of extremely atypical climatic conditions.

Droplets of rain with very low pH ($<$ 3.0) may cause necrotic spotting of tree leaves. Ammonia may induce the formation of marginal leaf spots and cause leaves to assume a dull and dark green color. Eventually leaves turn brown or black. In the presence of excessive chlorine, angiosperms typically develop interveinal leaf spots while gymnosperms generally exhibit distal necrosis. Exposure to excessive ethylene may cause trees to develop chlorosis and necrosis of older leaves followed by leaf abscission. Conifers may exhibit dwarfed needle growth and premature cone shed. Hydrogen sulfide typically causes interveinal white to tan discoloration.

The approximate threshold dose required for acute injury to sensitive plants is quite variable for miscellaneous pollutants. Agricultural species may be acutely injured by ammonia at 1000 ppm (70 $\times$ 10^4 μg m^{-3}) for 3 min or 55 ppm (38 $\times$ 10^3 μg m^{-3}) for 1 hr. For chlorine, injury thresholds approximate 0.5-1.5 ppm (1400 to 4350 μg m^{-3}) for 0.5 to 3 hr. Nonwoody vegetation susceptible to ethylene injury may become symptomatic at ambient concentrations as low as 0.04-0.1 ppm (4-115 μg m^{-3}) for 8 hr or 0.002 to 0.02 ppm (2.3-23 μg m^{-3}) for

Table 14-8. Baseline (Uncontaminated) Trace Metal Concentrations (Ash Weight Basis) for United States Forest Foliage

Element	Baseline concentration (mean and range) (ppm)	
Cadmium	7.0	(.05-60)
Chromium	8.0	($<$ 2-150)
Cobalt	6.2	($<$ 1-10,000)
Copper	128	($<$ 10-3000)
Lead	135	($<$ 10-3000)
Mercury	25	($<$ 25-50)
Nickel	37	($<$ 2-1300)
Thallium	4	(2-100)
Vanadium	7.7	($<$ 5-70)
Zinc	740	(100-7400)

Source: Calculated from Connor et al. (1975) and Shacklette et al. (1978).

24 hr. Hydrogen sulfide at 86 ppm (12×10^4 μg m^{-3}) will injure the most sensitive plant species in 5 hr (U.S. Environmental Protection Agency, 1976). Continuous fumigation of Douglas-fir seedlings with hydrogen sulfide caused slight symptoms at 100 ppb (139 μg m^{-3}), and extensive foliar damage at 300 ppb (417 μg m^{-3}) (Thompson and Kats, 1978).

The relative susceptibility of several forest tree species to chlorine is presented in Table 14-9.

Table 14-9. Relative Sensitivity of Woody Plants to Acute Damage by Chlorine

Sensitive	Intermediate	Tolerant
	Angiosperms	
Box elder		Azalea
Crab apple		
Horse chestnut		
Pin oak		
Sugar maple		
Sweetgum		
Tree-of-heaven		
	Gymnosperms	
		Balsam fir
		Pine

Source: U.S. Environmental Protection Agency (1976).

I. Summary

Acute foliar disease may be caused in forest vegetation by widespread air contaminants including sulfur dioxide, nitrogen oxides, ozone, peroxyacetylnitrates, fluoride, and several trace metals, and localized air contaminants including acid rain, ammonia, chlorine, hydrocarbons, and hydrogen sulfide. The response of woody plants to these atmospheric pollutants is extremely variable and dramatically controlled by genetic factors, plant age and health, and environmental conditions. Field symptoms of air pollution injury are not highly specific, are mimicked by a wide variety of other tree stress factors, and are useful only to experienced observers familiar with the range of edaphic, entomological, and pathological stress factors characteristic of a given flora in a given location. The dose required to produce acute injury varies widely with pollutant and vegetative type. There has been sufficient work done to enable a generalized ranking of relative forest tree sensitivity to the most important air pollutants. A summary treatment of general symptoms and injury thresholds for the major contaminants is contained in Table 14-10.

Table 14-10. Acute Foliar Disease of Forest Trees Caused by Major and Minor Air Pollutants

Pollutants	Symptoms	Threshold dose
A. Major		
1. Sulfur dioxide	Angiosperms: interveinal necrotic blotches Gymnosperms: red-brown dieback or banding	0.70 ppm (1820 μg m^{-3}) for 1 hr; 0.18 ppm (468 μg m^{-3}) for 8 hr; 0.008-0.017 ppm (21-44 μg m^{-3}) for growing season
2. Nitrogen dioxide	Angiosperms: interveinal necrotic blotches similar to SO$_2$ injury Gymnosperms: red-brown distal necrosis	20 ppm (38 $\times$ 10^3 μg m^{-3}) for 1 hr; 1.6-2.6 ppm (3000-5000 μg m^{-3}) for 48 hr; 1 ppm (1900 μg m^{-3}) for 100 hr
3. Ozone	Angiosperms: upper surface flecks Gymnosperms: distal necrosis, stunted needles	0.20-0.30 ppm (392-588 μg m^{-3}) for 2-4 hr; some conifers 0.08 ppm (157 μg m^{-3}) for 12-13 hr
4. Peroxyacetyl-nitrate	Angiosperms: lower surface bronzing Gymnosperms: chlorosis, early senescence	0.20-0.80 ppm (989-3958 μg m^{-3}) for 8 hr
5. Fluoride	Angiosperms: tip and margin necrosis Gymnosperms: distal necrosis	< 100 μg g^{-1} fluoride, dry wt. basis

Table 14-10 (continued)

Pollutants	Symptoms	Threshold dose
6. Trace metals	Angiosperms: interveinal chlorosis, tip and margin necrosis Gymnosperms: distal necrosis	Variable, undetermined
B. Minor		
1. Acid rain	Angiosperms: necrotic spots Gymnosperms: distal necrosis	$pH < 3.0$
2. Ammonia	Angiosperms: interveinal necrotic blotches similar to SO_2 injury Gymnosperms: distal necrosis	55 ppm (38,280 $\mu g\ m^{-3}$) for 1 hr
3. Chlorine	Angiosperms: chlorosis, upper surface fleck similar to O_3 Gymnosperms: distal necrosis	0.5-1.5 ppm (1400-4530 $\mu g\ m^{-3}$) for 0.5-3 hr
4. Ethylene	Angiosperms: chlorosis, necrosis, abscission Gymnosperms: dwarfing, premature defoliation	Variable, undetermined
5. Hydrogen sulfide	Angiosperms: interveinal necrotic blotches Gymnosperms: distal necrosis	100 ppm (14 $\times$ 10^4 $\mu g\ m^{-3}$) for 5 hr

[a]Symptoms and dose thresholds are for the most sensitive species.

References

Antonovics, J., A. D. Bradshaw, and R. G. Turner. 1971. Heavy metal tolerance in plants. Adv. Ecol. 7:1-85.

Berry, C. R. 1974. Age of pine seedlings with primary needles affects sensitivity to ozone and sulfur dioxide. Phytopathology 64:207-209.

Berry, C. R., and G. H. Hepting. 1964. Injury to eastern white pine by unidentified atmospheric constituents. For. Sci. 10:2-13.

Bjorkman, E. 1970. The effect of fertilization on sulfur dioxide damage to conifers in industrial and built-up areas. Stud. For. Suec. 78:1-48.

Bowen, H. J. M. 1966. Trace Elements in Biochemistry. Academic Press, New York, 241 pp.

Brennan, E., and I. A. Leone. 1968. The response of plants to sulfur dioxide or ozone-polluted air supplied at varying flow rates. Phytopathology 58:1661-1664.

Centre for Agricultural Publishing and Documentation. 1969. Air Pollution Proc. 1st European Congr. on the Influence of Air Pollution on Plants and Animals. Wageningen, The Netherlands, April 22-27, 1968, 415 pp.

Chang, C. W. 1975. Fluorides. In: J. B. Mudd and T. T. Kozlowski (Eds.), Responses of Plants to Air Pollution. Academic Press, New York, pp. 57-95.

Commission of the European Communities. 1978. Criteria Dose-Effect Relationships for Cadmium. Pergamon Press, New York, 202 pp.

Connor, J. J., H. T. Shacklette, R. J. Ebens, J. A. Erdman, A. T. Miesch, R. R. Tidball, and H. A. Bourtelot. 1975. Background Geochemistry of Some Rocks, Soils, Plants and Vegetables in the Conterminous United States. U.S. Geological Survey, Professional Paper No. 574-F, Washington, D.C., 168 pp.

Costonis, A. C. 1976. Criteria for evaluating air pollution injury to forest trees. IUFRO Congress, Oslo, Norway, June 21-26, 1976.

Cotrufo, C. 1974. The sensitivity of a white pine clone to air pollution as affected by N, P, and K. U.S.D.A. Forest Service, Research Note No. SE-198, Southeastern Forest Experiment Station, Asheville, North Carolina, 4 pp.

Cotrufo, C., and C. R. Berry. 1970. Some effects of a soluble NPK fertilizer on sensitivity of eastern white pine to injury from SO_2 air pollution. For. Sci. 16:72-73.

Darley, E. F., and J. T. Middleton. 1966. Problems of air pollution in plant pathology. Annu. Rev. Phytopath. 4:103-118.

Davis, D. D. 1975a. Variable tree response due to environmental factors-climate. In: W. H. Smith and L. S. Dochinger (Eds.), Air Pollution and Metropolitan Woody Vegetation. U.S.D.A. Forest Service. PIEFR-PA-1, Upper Darby, Pennsylvania, pp. 14-16.

Davis, D. D. 1975b. Resistance of young ponderosa pine seedlings to acute doses of PAN. Plant Dis. Reptr. 59:183-184.

Davis, D. D. 1977. Response of ponderosa pine primary needles to separate and simultaneous ozone and PAN exposures. Plant Dis. Reptr. 61:640-644.

Davis, D. D., and J. B. Coppolino. 1974. Relationship between age and ozone sensitivity of current needles of ponderosa pine. Plant Dis. Reptr. 58:660-663.

Davis, D. D., and R. G. Wilhour. 1976. Susceptibility of Woody Plants to Sulfur Dioxide and Photochemical Oxidants. U.S. Environmental Protection Agency Publica. No. EPA-600/3-76-102, Corvallis, Oregon, 71 pp.

Davis, D. D., and F. A. Wood. 1973a. The influence of environmental factors on the sensitivity of Virginia pine to ozone. Phytopathology 63:371-376.

Davis, D. D., and F. A. Wood. 1973b. The influence of plant age on the sensitivity of Virginia pine to ozone. Phytopathology 63:381-388.

Dochinger, L. S. 1964. Effects of nutrition on the chlorotic dwarf disease of eastern white pine. Plant Dis. Reptr. 48:107-109.

Dochinger, L. S., and D. E. Seliskar. 1970. Air pollution and the chlorotic dwarf disease of eastern white pine. For. Sci. 16:46-55.

Dochinger, L. S., F. W. Bender, F. L. Fox, and W. E. Heck. 1970. Chlorotic dwarf of eastern white pine caused by an ozone and sulphur dioxide interaction. Nature 225:476.

Drummond, D. B. 1971. Influence of high concentrations of peroxyacetylnitrate on woody plants. Phytopathology 61:178.

Dunning, J. A., and W. W. Heck. 1973. Response of pinto bean and tobacco to ozone as conditioned by light intensity and/or humidity. Environ. Sci. Technol. 7:824-826.

Energy Research and Development Administration. 1975. Biological Implications of Metals in the Environment. ERDA Symposium Series No. 42, Washington, D.C., 682 pp.

Forestry Commission (England). 1971. Fume Damage to Forests. Research and Development Paper No. 82, London, 50 pp.

Friberg, L., M. Piscator, G. Nordberg, and T. Kjellstrom. 1974. Cadmium in the Environment. Chemical Rubber Co. Press, Cleveland, Ohio, 248 pp.

Friberg, L., and J. Vostal. 1972. Mercury in the Environment. Chemical Rubber Co. Press, Cleveland, Ohio, 215 pp.

Fuhrer, J., and K. H. Erismann. 1980. Uptake of NO_2 by plants grown at different salinity levels. Experientia 36:409-410.

Genys, J. B., and H. E. Heggestad. 1978. Susceptibility of different species, clones and strains of pines to acute injury caused by ozone and sulfur dioxide. Plant Dis. Reptr. 62:687-691.

Gerhold, H. D. 1975. Resistant varieties. In: W. H. Smith and L. S. Dochinger (Eds.), Air Pollution and Metropolitan Woody Vegetation. U.S.D.A. Forest Service. PIERR-PA-1, Upper Darby, Pennsylvania, pp. 45-49.

Gerhold, H. D. 1977. Effect of Air Pollution on Pinus strobus L. and Genetic Resistance. U.S. Environmental Protection Agency, Publica. No. EPA-600/3-77-002. Corvallis, Oregon, 45 pp.

Guderian, R. 1977. Air Pollution Phytoxicity of Acidic Gases and Its Significance in Air Pollution Control. Ecological Studies No. 22. Springer-Verlag, New York, 122 pp.

Heagle, A. S. 1979. Effects of growth media, fertilizer rate and hour and season of exposure on sensitivity of four soybean cultivars to ozone. Environ. Pollut. 18:313-322.

Heagle, A. S., W. W. Heck, and D. Body. 1971. Ozone injury to plants as influenced by air velocity during exposure. Phytopathology 61:1209-1212.

Heagle, A. S., R. B. Philbeck, H. H. Rogers, and M. B. Letchworth. 1979. Dispensing and monitoring ozone in opentop field chambers for plant-effects studies. Phytopathology 69:15-20.

Heath, R. L. 1975. Ozone. In: J. B. Mudd and T. T. Kozlowski (Eds.), Responses of Plants to Air Pollution. Academic Press, New York, pp. 23-55.

Heck, W. W. 1968. Factors influencing expression of oxidant damage to plants. Annu. Rev. Phytopathol. 6:165-188.

Heggestad, H. E. 1968. Diseases of crops and ornamental plants incited by air pollutants. Phytopathology 58:1089-1097.

Heggestad, H. E., and W. W. Heck. 1971. Nature, extent, and variation of plant response to air pollutants. Adv. Agronomy 23:111-145.

Heitschmidt, R. K., and J. Altman. 1978. Probable Effects of SO_2 on Agricultural Crops. Experiment Station, Colorado State University, Tech. Bull. No. 133, Fort Collins, Colorado, 7 pp.

Hindawi, I. J. 1970. Air Pollution Injury to Vegetation. U.S. Dept. Health, Education and Welfare, National Air Pollution Control Administration, Raleigh, North Carolina, 44 pp.

Jacobson, J. S., and A. C. Hill. 1970. Recognition of Air Pollution Injury to Vegetation: A Pictorial Atlas. Air Pollution Control Association, Pittsburgh, Pennsylvania.

Jäger, H. J., and H. Klein. 1976. Studies on the influence of nutrition on the susceptibility of plants to SO_2. Eur. J. For. Pathol. 6:347-353.

Karnosky, D. F. 1974. Implications of genetic variation in host resistance to air pollutants. Proc. 9th Central States Forest Tree Improvement Conference, Ames, Iowa, pp. 7-20.

Karnosky, D. F. 1977. Evidence for genetic control of response to sulfur dioxide and ozone in *Populus tremuloides*. Can. J. For. Res. 7:437-440.

Karnosky, D. F. 1978a. Selection and testing programs for developing air pollution tolerant trees for urban areas. Proc. IUFRO Air Pollution Meeting, Sept. 18-23, 1978, Ljubljana, Yugoslavia.

Karnosky, D. F. 1978b. Genetics of air pollution tolerance of trees in the Northeastern United States. Proc. 26th Northeastern Forest Tree Improvement Conf., July 25-27, 1978, Pennsylvania State Univ. State College. Pennsylvania, pp. 161-178.

Kathny, E. L., Ed. 1973. Trace Elements in the Environment. American Chemical Soc., Adv. in Chem. Series No. 123, Washington, D.C., 149 pp.

Keller, T. 1975a. On the phytotoxicity of fluoride immissions for woody plants. Mitt. eidg. Anst. forstl. Vers'wes 51:303-331.

Keller, T. 1975b. On the translocation of fluoride in forest trees. Mitt. eidg. Anst. forstl. Vers'wes 51:335-356.

Krause, G. H., and H. Kaiser. 1977. Plant response to heavy metals and sulphur dioxide. Environ. Pollut. 12:63-71.

Lacasse, N. L., and W. J. Moroz. 1969. Handbook of Effects Assessment—Vegetation Damage. Center for Air Environment Studies, Pennsylvania State University, University Park, Pennsylvania.

Leone, I. A. 1976. Response of potassium deficient tomato plants to atmospheric ozone. Phytopathology 66:734-736.

Leone, I. A., and E. Brennan. 1969. The importance of moisture in ozone phytotoxicity. Atmos. Environ. 3:399-406.

Linzon, S. N. 1978. Effects of airborne sulfur pollutants on plants. *In*: J. O. Nriagu (Ed.), Sulfur in the Environment: Part II, Ecological Impacts. Wiley, New York, pp. 109-162.

Loomis, R. C., and W. H. Padgett. 1973. Air Pollution and Trees in the East. U.S. D.A. Forest Service, State and Private Forestry, Atlanta, Georgia, 28 pp.

Matsushima, J., and R. F. Brewer. 1972. Influence of sulfur dioxide and hydrogen fluoride as a mix or reciprocal exposure on citrus growth and development. J. Air Pollut. Assoc. 22:710-713.

McCune, D. C., and L. H. Weinstein. 1971. Metabolic effects of atmospheric fluorides on plants. Environ. Pollut. 1:169-174.

Mitchell, C. D., and T. A. Fretz. 1977. Cadmium and zinc toxicity in white pine, red maple, and Norway spruce. J. Am. Soc. Hort. Sci. 102:81-84.

Mudd, J. B. 1973. Biochemical effects of some air pollutants on plants. *In*: J. A. Naegele (Ed.), Air Pollution Damage to Vegetation. Adv. Chem. Series No. 122, Amer. Chem. Soc., Washington, D.C., pp. 31-47.

Mudd, J. B. 1975a. Sulfur dioxide. *In*: J. B. Mudd and T. T. Kozlowski (Eds.), Responses of Plants to Air Pollution. Academic Press, New York, pp. 9-12.

Mudd, J. B. 1975b. Peroxyacetyl nitrates. *In*: J. B. Mudd and T. T. Kozlowski (Eds.), Responses of Plants to Air Pollution. Academic Press, New York, pp. 97-119.

Mudd, J. B., and T. T. Kozlowski. 1975. Responses of Plants to Air Pollution. Academic Press, New York, 383 pp.

Naegele, J. A. 1973. Air Pollution Damage to Vegetation. Advances in Chemistry Series No. 122, Amer. Chem. Soc., Washington, D.C., 137 pp.

National Academy of Sciences. 1974. Chromium. NAS, Washington, D.C., 155 pp.

National Academy of Sciences. 1975. Nickel. NAS, Washington, D.C., 277 pp.

National Academy of Sciences. 1977a. Copper, NAS, Washington, D.C., 115 pp.

National Academy of Sciences. 1977b. Effects of nitrogen oxides on vegetation. *In*: Nitrogen Oxides. NAS, Washington, D.C., pp. 147-158.

National Academy of Sciences. 1977c. Ozone and Other Photochemical Oxidants. NAS, Washington, D.C., 789 pp.

National Academy of Sciences. 1978a. Effects of atmospheric sulfur oxides and related compounds on vegetation. *In*: Sulfur Oxides. NAS, Washington, D.C., pp. 80-129.

National Academy of Sciences. 1978b. An Assessment of Mercury in the Environment. NAS, Washington, D.C., 192 pp.

Nielsen, D. G., L. E. Terrell, and T. C. Weidensaul. 1977. Phytotoxicity of ozone and sulfur dioxide to laboratory fumigated Scotch pine. Plant Dis. Reptr. 61: 699-703.

Nriagu, J. O., Ed. 1980. Zinc in the Environment. Part I. Ecological Cycling. Wiley-Interscience, Somerset, New Jersey, 464 pp.

Nriagu, J. O., Ed. 1980. Zinc in the Environment. Part. Ecological Cycling. Wiley-Interscience, Somerset, New Jersey, 464 pp.

O'Connor, J. A., D. G. Parbery, and W. Strauss. 1974. The effects of phytotoxic gases on native Austrialian plant species. Part I. Acute effects of sulphur dioxide. Environ. Pollut. 7:7-23.

Oehme, F. W., Ed. 1978. Toxicity of Heavy Metals in the Environment. Dekker, New York, Part I, 515 pp.; Part II, 970 pp.

Otto, H. W., and R. H. Daines. 1969. Plant injury by air pollutants. Influence of humidity on stomatal apertures and plant response to ozone. Science 163: 1209-1210.

Peterson, P. J. 1978. Lead and vegetation. *In*: J. O. Nriagu (Ed.), The Biochemistry of Lead in the Environment. Part B. Biological Effects. Elsevier/North-Holland Biomedical Press, New York, pp. 355-384.

Purves, D. 1977. Trace Element Contamination of the Environment. Elsevier, New York, 260 pp.

Reinert, R. A. 1975. Pollutant interactions and their effects on plants. Environ. Pollut. 9:115-116.

Shacklette, H. T., J. A. Erdman, T. F. Harms, and C. S. E. Pupp. 1978. Trace elements in plant foodstuffs. *In*: F. W. Oehme (Ed.), Toxicity of Heavy Metals in the Environment. Dekker, New York, pp. 25-68.

Smith, H. J., and D. D. Davis. 1977. The influence of needle age on sensitivity of Scotch pine to acute doses of SO_2. Plant Dis. Reptr. 61:870-874.

Smith, W. H. 1975. Variable tree response due to environmental factors-edaphic. *In*: W. H. Smith and L. S. Dochinger (Eds.), Air Pollution and Metropolitan

Woody Vegetation, U.S.D.A. Forest Service. PIEFR-PA-1, Upper Darby, Pennsylvania, pp. 17-18.

Steiner, K. C., and D. D. Davis. 1979. Variation among *Fraxinus* families in foliar response to ozone. Can. J. For. Res. 9:106-109.

Stern, A. C., H. C. Wohlers, R. W. Boubel, and W. P. Lowry. 1973. Fundamentals of Air Pollution. Academic Press, New York, 492 pp.

Tamm, C. O., and A. Aronson. 1972. Plant Growth as affected by Sulphur Compounds in Polluted Atmosphere. A Literature Survey. Royal College of Forestry, Dept. Forest Ecology and Forest Soils, Research Note No. 12, Stockholm, Sweden, 53 pp.

Taylor, O. C. 1973. Oxidant Air Pollutant Effects on a Western Coniferous Forest Ecosystem. Task C Report No. EP-R3-73-043B. Statewide Air Pollut. Res. Center, Riverside, California, 189 pp.

Taylor, O. C., C. R. Thompson, D. T. Tingey, and R. A. Reinert. 1975. Oxides of nitrogen. *In*: J. B. Mudd and T. T. Kozlowski (Eds.), Responses of Plants to Air Pollution. Academic Press, New York, pp. 121-139.

Temple, P. J., and R. Wills. 1979. Sampling and analysis of plants and soils. *In*: W. W. Heck, S. V. Krupa, and S. N. Linzon (Eds.), Methodology for the Assessment of Air Pollution Effects on Vegetation. Air Pollution Control Assoc., Pittsburgh, Pennsylvania, Chap. 13, pp. 1-23.

Thompson, C. R., and G. Kats. 1978. Effects of continuous H_2S fumigation on crop and forest plants. Environ. Sci. Technol. 12:550-553.

Tingey, D. T., and R. A. Reinert. 1975. The effect of ozone and sulphur dioxide singly and in combination on plant growth. Environ. Pollut. 9:117-125.

Townsend, A. M. 1975. Variable tree response due to genetic factors. *In*: W. H. Smith and L. S. Dochinger (Eds.), Air Pollution and Metropolitan Woody Vegetation. U.S.D.A. Forest Service. PIEFR-PA-1, Upper Darby, Pennsylvania, pp. 18-19.

Treshow, M. 1970. Ozone damage to plants. Environ. Pollut. 1:155-161.

U.S.D.A. Forest Service. 1973. Air Pollution Damages Trees. State and Private Forestry, Upper Darby, Pennsylvania, 32 pp.

U.S. Environmental Protection Agency. 1976. Diagnosing Vegetation Injury Caused by Air Pollution. U.S.E.P.A., Washington, D.C.

Wallace, R. G., and D. J. Spedding. 1976. The biochemical basis of plant damage by atmospheric sulphur dioxide. Clean Air 10:61-64.

Weinstein, L. 1975. Dose-response relationships. *In*: W. H. Smith and L. S. Dochinger (Eds.), Air Pollution and Metropolitan Woody Vegetation. U.S. D.A. Forest Service. PIEFR-PA-1, Upper Darby, Pennsylvania, pp. 11-13.

Weinstein, L. H. 1977. Fluoride and plant life. J. Occupa. Med. 19:49-78.

Weinstein, L. H., and D. C. McCune. 1971. Effects of fluoride on agriculture. J. Air Pollut. Control Assoc. 21:410-413.

Weinstein, L. H., and D. C. McCune. 1979. Air pollution stress. *In*: H. Mussell and R. Staples (Eds.), Stress Physiology in Crop Plants. Wiley, New York, pp. 328-341.

Wilhour, R. G. 1970. The influence of temperature and relative humidity on the response of white ash to ozone. Phytopathology 70:579.

Zeevaart, A. J. 1976. Some effects of fumigating plants for short periods with NO_2. Environ. Pollut. 11:97-108.

15

Class II Summary: Forest Responds by Exhibiting Alterations in Growth, Biomass, Species Composition, Disease, and Insect Outbreaks

In the presence of a sufficient dose of an air pollutant, forest trees will be adversely impacted. When this occurs the threshold between Class I and Class II interactions is crossed. At intermediate dose, the specific contaminant concentration and time of exposure varying greatly with specific pollutant and forest situation, the influence on individual forest components may range from extremely subtle to visibly dramatic. Chapters 7 through 14 have reviewed the evidence available to support the hypotheses that intermediate air pollution loads may alter or inhibit forest tree reproduction, alter forest nutrient cycling, alter tree metabolism, or change forest stress conditions by influencing insect pests, microbial pathogens, and by directly damaging foliar tissue. All but the latter of these impacts would be extremely subtle, visibly asymptomatic, and detectable only by very careful forest monitoring.

The primary response of a forest ecosystem to sustained intermediate dose and Class II interaction would be reduced growth and consequently reduced biomass. Reduced essential element availability, decreased photosynthesis, increased respiration, increased insect and disease stress, and decreased foliar tissue would all contribute to a reduction in tree growth rates and ultimately to lessened forest biomass. Alterations in the reproductive strategies of individual tree species or differential response of these species to reduced nutrition, altered metabolism and pest stress, and to direct foliar injury may cause changes in competitive ability and ultimately lead to alterations in tree succession and species composition. These relationships are outlined in Table 15-1. Recent reviews of Class II vegetative responses to air pollutants include Heck et al. (1977), Jensen et al. (1976), and Weinstein and McCune (1979).

Table 15-1. Interaction of Air Pollution and Temperate Forest Ecosystems under Conditions of Intermediate Air Contaminant Load, Designated Class II Interaction

Forest soil and vegetation: activity and response	Ecosystem consequence and impact
1. Forest tree reproduction, alteration or inhibition	1. Altered species composition

2. Forest nutrient cycling, alteration
  a. Reduced litter decomposition
  b. Increased plant leaching, soil leaching, and soil weathering
  c. Disturbance of microbial symbioses
	2. Reduced growth, less biomass

3. Forest metabolism, alteration
  a. Decreased photosynthesis
  b. Increased respiration
	3. Reduced growth, less biomass

4. Forest stress, alteration
  a. Phytophagous insects, increased or decreased activity
  b. Microbial pathogens, increased or decreased activity
  c. Foliar damage increased by direct air pollution influence
	4. Altered ecosystem stress
	Increased or decreased insect infestations;
	Increased or decreased disease epidemics;
	Reduced growth, less biomass, altered species composition

What is the evidence available to support the hypotheses that forest ecosystem exposure to intermediate doses of air pollutants result in reduced growth and biomass, altered species composition, and increased forest pest impact?

A. Forest Growth Reduction Caused by Air Pollution

Forest growth is complex in concept and measurement. Addition of woody tissue is the dominant feature of forest growth. The accumulation of woody biomass (living weight) represents gross photosynthetic production less respiratory losses. Individuals interested in the commercial value of forest systems will measure growth in terms of annual wood increment. The latter is calculated from measurements of tree bole diameter and height and is modified by the amount of bole taper. Typically annual increment is expressed as volume of wood per unit of land area. It may also be expressed as weight of wood per unit of land area. Site index is another indicator of forest growth. It represents the average height of the dominant trees in a given forest system at some arbitrary age, typically 50 years. Temperate forest trees conveniently add a readily discernable layer of wood every year to their main stem. The annual increment of radial growth is easily measured by examining ring width (Nash et al., 1975). Sensitive measurement of cambial growth may be obtained by the use of dendrometers. These devices quantify growth by measuring increases in stem circumference or single stem radii (Fritts, 1976; Kramer and Kozlowski, 1979). Measurement of basal area, or stem cross-sectional area in square centimeters at an arbitrary height above the ground, is yet another measure used to quantify forest tree growth.

The most fundamental characteristic of an ecosystem is its productivity. Forest productivity is high relative to other ecosystems and net productivity of 1200 dry g m^{-2} year^{-1} for trees and shrubs together is quite typical for temperate forests (Whittaker, 1975). Productivity is strongly controlled, however, by a variety of variables including system age and environmental parameters. The most important of the latter include nutrient availability, water availability, and temperature. Because of the variety of Class II interactions identified, air quality may also influence forest productivity in certain environments.

1. Sulfur Dioxide

The capacity of sulfur dioxide to reduce plant productivity is well established. The ability of this gas to reduce yield of commercially important plants is of special concern to those investigating agricultural ecosystems (Davis, 1972; Jones et al., 1977; Sprugel et al., 1979). Chamber fumigations have provided evidence that sulfur dioxide can reduce the growth of forest seedlings (for example, Keller, 1978; Kress and Skelly, 1980) as reflected in height growth and radial wood increment (Figure 15-1). The most impressive evidence for sulfur dioxide suppression of forest growth stems from investigations carried out in natural forest environments exposed to elevated ambient sulfur dioxide from surrounding point sources.

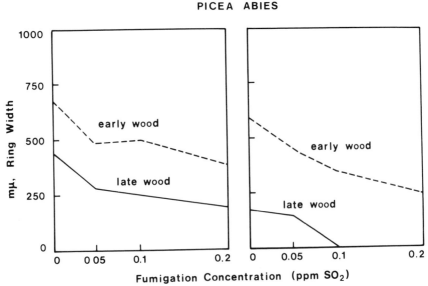

Figure 15-1. The influence of laboratory fumigation with sulfur dioxide on the average radial ring width of early wood and late wood of two spruce clones. (From Keller, 1980.)

Scheffer and Hedgcock (1955) included an investigation of growth impact in their study of smelter influence on adjacent coniferous ecosystems in Washington and Montana. Damage in the Deerlodge National Forest from the Washoe smelter located close to Anaconda, Montana, was investigated in 1910 and 1911. A small number of sapling or pole-size trees were felled within the zone of visible forest damage, in the transition zone and in areas not subject to smelter damage. Radial diameter increments were determined for the sample trees. Species given primary attention included subalpine fir, lodgepole pine, and Douglas fir. Retardation of diameter growth of all three species was indicated for all years from 1892 through 1910 in the zone of visible injury. Growth suppression of transition zone trees was not well defined, but suggestive of some degree of intermediate retardation. The authors concluded, however, that the differences in growth rates of the different zones surrounding the Washoe smelter were indicative, but not proof of, a sulfur dioxide effect. They were unable to separate the influence of the sulfur gas from differences due to natural forces that may have also influenced growth in the various zones. Conifer forests of the upper Columbia River Valley, Washington, subject to sulfur dioxide from a smelter located at Trail, British Columbia, were studied from 1928 through 1936. Height increment and current annual increase in length of the terminal shoot were employed to evaluate sulfur dioxide impact on ponderosa pine growth in forest zones exhibiting varying degrees of foliar symptoms. In zone 1, 60-100% of the foliar tissue exhibited sulfur dioxide injury symptoms, while zones 2 and 3 had 30-60% and 1-30%, respectively. Measured trees included saplings, height range 3-7 m (8-20 feet), and pole size trees, height range 8-17 m (25-50 feet). Height increments were taken during the fall by bending small trees and felling large ones. The results of these measurements are presented in Figure 15-2. The authors concluded that there was no indication of sulfur dioxide influence on height growth in any zone prior to 1926. After 1926, however, height growth in zones 1, 2, and 3 showed generally less response to precipitation differences relative to the response outside the symptomatic zones. It was concluded that there was a definite retardation of height growth in 1928 resulting from 1927 injury (a 40% increase in sulfur dioxide discharge began in 1926) in zones 1 and 2, and approximately a year later in zone 3, and that the retardation was sustained in all three zones through the end of observations in 1936. Annual radial diameter growth was also determined for the ponderosa pines of the various zones (Figure 15-3). Diameter growth revealed initial retardation as early as 1921 in zone 1. Retardation in zones 2 and 3 began in 1924. All zones exhibited sustained suppression of diameter growth through the conclusion of the experiment in 1934. The average percentage reductions in diameter growth were 38, 24, and 17% for zones 1, 2, and 3, respectively. The authors judged that height growth curves and diameter growth curves were generally similar, except that the diameter measurements appeared to reveal the suppression sooner. It is extremely unfortunate that we do not know the ambient concentrations of sulfur dioxide in the atmospheres of the various zones along with the diurnal and seasonal fluctuations of these concentrations.

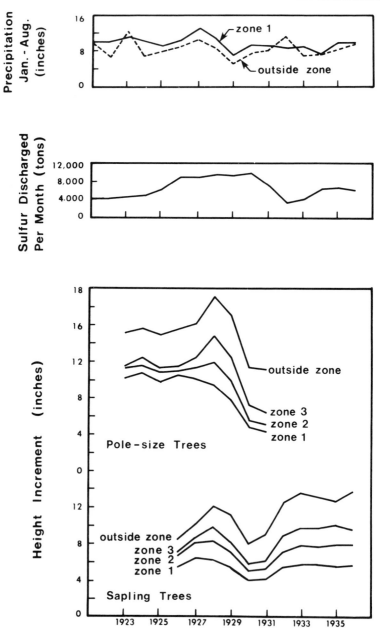

Figure 15-2. Annual height increment of ponderosa pine in the upper Columbia River Valley, Washington, subject to sulfur dioxide stress from a smelter in Trail, British Columbia. (From Scheffer and Hedgcock, 1955.)

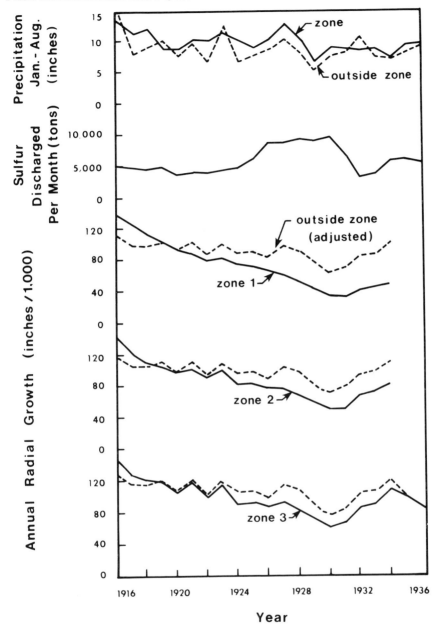

Figure 15-3. Annual radial increment of ponderosa pine in the upper Columbia River Valley, Washington, subject to sulfur dioxide stress from a smelter in Trail, British Columbia. Actual measurements (solid line) are compared with theoretical growth (dotted line) that could have occurred in the absence of sulfur dioxide injury. (From Scheffer and Hedgcock, 1955.)

Samuel N. Linzon, Air Management Branch of the Ontario, Canada Department of Energy and Resources Management, conducted a ten year (1953-1963) assessment of the Sudbury smelter district of Ontario on the growth of surrounding forests (Linzon, 1971). During the study period three large smelters were discharging approximately two million tons of sulfur dioxide to the atmosphere annually. The investigation was concentrated on eastern white pine growth because of its susceptibility to sulfur dioxide injury and because of its commercial importance. Forty-two permanent sample plots containing eastern white pine ranging in age from 65 to 85 years were established. Fume area plots were established at increasingly greater distances northeast of Sudbury in line with the prevailing wind, while control plots were located beyond the influence of smelter effluent. Degree of foliar injury was employed to segregate the fume area into inner, intermediate, and outer zones. Severe tree damage occurred up to 48 km (30 miles) northeast of Sudbury and the inner zone comprised an area of approximately 1865 km^2 (720 miles^2). The intermediate and outer zones contained 4144 km^2 (1600 miles^2) and 7770 km^2 (3000 miles^2), respectively. Radial increment cores were taken from 20 dominant living white pine trees on each sample plot. Annual radial growth for the 1940-1960 period was measured on each core. The results indicated a gradual decline in the growth of white pine in areas adjacent to the smelters, while a constant growth pattern was indicated in other areas. Height and diameter measurements of the pines were employed to construct a local volume table for each plot. Throughout the inner fume zone there was a net average annual loss in total volume of 0.03 m^3 (0.10 feet^3) per white pine tree in the 18-30 cm (7-12 inch) diameter class due to the combination of mortality and reduced growth. In the control plots each tree added 0.09 m^3 (0.30 feet^3) in total volume per year. The total reduction of volume per tree per year in the inner zone, therefore, was 0.12 m^3 (0.40 feet^3). Unfortunately the suppression of volume growth of stressed, but not killed, trees was not clear from this study. The relative significance of weevil infestation, *Cronartium ribicola* infection, sulfur dioxide influence, and other stress factors in causing mortality was not clear. It is obvious, however, that the growth of eastern white pine is dramatically influenced over a very large area by the Sudbury smelter complex.

Stemple and Tryon (1973) conducted a unique investigation to assess the influence of sulfur dioxide and fly ash from coal burning railroad locomotives on surrounding forest vegetation. A deciduous forest ecosystem, dominated by white oak, and situated on a 267 m (800 foot) hill with a railroad at its base was examined in northern West Virginia. Sample plots were established in the "damage area" delineated by a layer of fly ash on the soil and a control area not influenced by the railroad. Measurements were made of site index (Figure 15-4), tree height, and annual radial increment. The site quality of the land adjacent to the railroad was drastically reduced. The authors judged that the coal tonnage hauled past the damaged area was directly correlated with the sulfur dioxide and other effluents impacting the area. Annual radial increment of the white oaks varied inversely with the annual coal tonnage moved through the study forest.

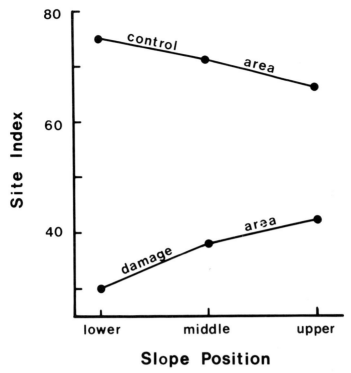

Figure 15-4. Site index for white oaks in Preston County, West Virginia, subject to effluent from coal-burning locomotives compared with an unimpacted control site. (From Stemple and Tryon, 1973.)

John M. Skelly and colleagues of the Virginia Polytechnic Institute and State University, Blacksburg, Virginia, have examined the growth of a variety of forest trees subject to effluent from the U.S. Army Radford Ammunition Plant in southwest Virginia. The Radford Arsenal initiated operations in 1939 and has had three significant production peaks that have corresponded with the major United States defense efforts in World War II, the Korean Conflict, and the Vietnam Conflict. The fact that the ammunition plant was located in an isolated location coupled with the realization that major effluent release was periodic and associated with massive national defense efforts presented the investigators with an opportunity to correlate surrounding forest growth with the quantity of air pollutants (sulfur dioxide and nitrogen oxides) released from the facility. Initial efforts to evaluate the influence of past pollutant levels on the growth of two important commercial tree species involved eastern white pine and yellow poplar (Stone and Skelly, 1974). Sample plots were established in the prevailing downwind direction. Annual ring widths of 43 eastern white pine and 50 yellow poplar, all in dominant or codominant crown classes, were determined to the nearest 0.05 mm with a dendrochronograph. A highly significant inverse relationship was found by linear regression analyses to exist between the fluctuating production levels of the arsenal and the annual ring widths of white pine and yellow poplar (Figure 15-5). Severely symptomatic 13-year-old eastern white pine in the vicinity of the arsenal have been shown to average only 66% of the height of trees lacking symptoms (Skelly et al., 1972). In an effort to better resolve the relationship in intensity of foliar symptoms of white pine to growth suppression, a study was conducted in a stand 1.6 km (1 mile) downwind of the closest arse-

(a)

(b)

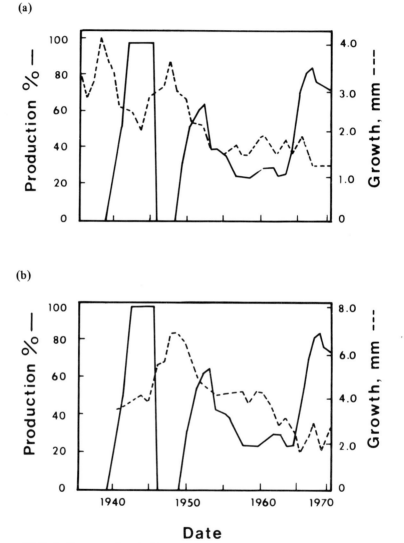

Date

Figure 15-5. Influence of periodic production levels, and associated effluent containing sulfur dioxide and nitrogen oxides, from a military arsenal on the average annual ring width of 43 white pine trees (a) and 50 yellow poplar trees (b). (From Stone and Skelly, 1974.)

nal emission source of nitrogen oxides. The trees of this stand ranged in age from 35 to over 135 years and exhibited foliar symptoms ranging from severe to asymptomatic. Analysis of regression correlations indicated no significant growth rate differences between symptom classes during pollution peaks. Growth of asymptomatic trees was judged to be reduced as much as that of injured trees during peak release episodes (Phillips et al., 1977a). The growth response of young loblolly pine has also been examined in the vicinity of the arsenal (Phillips et al., 1977b). Three stands of plantation loblolly ranging in age from 15 to 18 years and located 1.4-2.7 km (0.9-17 miles) northeast or northwest of the main power facility were examined. A significant inverse relationship was demonstrated between annual radial increment growth and arsenal production levels in two of the loblolly pine stands. In both of these stands additional analyses suggested theoretical reductions in diameter growth without the presence of visible injury.

Legge et al. (1977) have conducted field studies on the growth of lodgepole pine-jack pine hybrids, white spruce, and aspen subjected to sulfur dioxide and hydrogen sulfide from a natural gas processing plant in Whitecourt, Alberta, Canada. Growth of lodgepole pine-jack pine hybrids was quantified by examining basal area increments. Significant decreases in basal area increment for hybrid stands after 1965 were judged to be the result of unfavorable moisture conditions (excess water) rather than to sulfur gas emissions.

It is important to realize that all studies reviewed in this section have considered the amount of rainfall occurring over their study periods. Failure to do this would, of course, meaningfully reduce the confidence of experimental results as precipitation exerts such a profound influence on tree growth. Perhaps the most disappointing aspect of the sulfur dioxide studies is infrequent presentation of dose information. Ambient concentrations and exposure durations were rarely monitored or reported. For historical studies it is clearly frequently impossible to provide this information. For contemporary investigations, however, inclusion of this information would greatly increase the utility of the data. Horntvedt (1970) has reported the results of a field study of the response of radial increment growth of spruce subjected to ambient sulfur dioxide. Reductions of 45, 27, and 18% in diameter growth over a 9-year period corresponded directly to average annual ambient sulfur dioxide levels of 543, 143, and 114 μg m^{-3}, respectively.

2. Acid Precipitation

Initial delineation of the acid precipitation problem (Chapter 9) and early efforts to describe attendant environmental ramifications have come from Scandinavia (Bolin, 1971). It is not surprising, therefore, to realize that a considerable effort has been made in Sweden and Norway to evaluate the influence of acid rain on forest growth (Dahl and Skre, 1971; Sundberg, 1971).

Jonsson and Sundberg (1972) and Jonsson (1976) developed a model to examine the statistical correlation of Scotch pine and Norway spruce growth, as measured by annual ring widths for the period 1910-1965, with the intensity of increasing acid precipitation initiated in 1950 in southern Sweden (below 61°N latitude). Fifty percent of the forest growth in Sweden occurs in the latter area. The authors statified their analysis of over 4000 trees by site classes and by region of differing susceptibility to acid rain stress. The latter classification was based on regional intensity of acid rain, chemistry of regional lakes and rivers, and distribution of soil types. The latter criterion was deemed especially important and regions with basepoor tills or sand deposits and with aquatic resources of low pH and cation concentration were judged to be particularly liable to stress from acid precipitation. The authors made a sincere effort to consider nonpollution factors that may have influenced forest growth. Special attention was given to differences in climatic factors and silvicultural practices in the study regions. The model predicted reduced forest growth in more susceptible regions of approximately 0.3-0.6% for the period 1951-1965 relative to less susceptible areas. The analysis did not enable the authors to conclude that acid rain was the cause of this reduced growth. Likewise acid precipitation was not eliminated as a cause and the analysis did not support any alternative explanations for the poorer growth observed. Jonsson and Sundberg concluded their study by presenting predictions of future Swedish forest growth given various scenarios of sulfur deposition.

Abrahamsen et al. (1976a,b) have examined trends in Norwegian forest growth and acid rain in a manner very similar to the Jonsson and Sundberg Swedish study. Comparisons of tree growth, as indicated by ring widths, were examined between regions presumed or known to have different inputs of acid precipitation and between sites of differing susceptibilities to acid stress due to soil characteristics. No consistent regional growth differences were observed. In fact "somewhat better" development of pine was observed in one region (Sørlandet area) following 1950. Less productive sites, poor vegetation types, and shallow soils did not appear to be more sensitive to soil acidification. The authors concluded that no clear effects of acid precipitation on diameter growth of spruce and pine were detected by the regional tree-ring analyses.

Artificial applications of acid rain to Norway spruce and lodgepole pine seedlings have been conducted at the Sønsterud Forest Nursery in Norway (Tveite and Teigen, 1976). After 3 years of field applications no negative impacts of acid application were detected. Lodgepole pine exhibited a 20% stimulation of height growth after 3 years following application of 50 mm water per month at pH 4 and 3. A stimulation of height growth of approximately 15% was recorded for Norway spruce given a similar treatment.

In the United States the forest regions subject to the most acidic precipitation are located in the Northeast. In a comprehensive study of production and biomass of the northern hardwood forest conducted at the Hubbard Brook Experimental Forest in New Hampshire, Whittaker et al. (1974) observed a significant decline in growth from 1956-1960 to 1961-1965. Wood volume growth declined 17% between these two periods. Net ecosystem production was estimated as 350 g m^{-2} year^{-1} aboveground and 85 belowground for 1956-1960 and 238 and 52 g m^{-2} year^{-1} for 1961-1965. A widespread drought occurred throughout the Northeast during 1961-1965, but examination of wood volume growth patterns, which for some trees could be traced for two centuries at Hubbard Brook, revealed no previous decrease similar to the precipitous decrease recorded in 1961-1965. The authors noted that the period of growth decrease was coincident with a period of increasing acidity in precipitation and inferred that this may have been responsible for the decrease in productivity.

Cogbill (1976, 1977) selected two United States mountainous, remote forest sites to assess the effect of acid precipitation on tree growth by tree-ring analysis. A northern hardwood forest site in the White Mountain National Forest, New Hampshire, and a red spruce site in the Great Smoky Mountains National Park, Tennessee, were studied. Average acidity in the precipitation in New Hampshire was pH 4.1 and in Tennessee was pH 4.4 at the time of the study. The initiation date of increased acidity of precipitation was unkonwn for both sites but was presumed to be prior to 1955 for New Hampshire and approximately 1955 for Tennessee. The tree-ring chronologies for the three northern hardwood species and the spruce are presented in Figure 15-6. No clear indication of regional decrease in tree growth was found. The variation of relative tree growth responses, after subtraction of variation due to climate, showed no recent trend. Cogbill concluded that no correlation of forest growth and acid rain could be established for eastern North America.

Laboratory applications of simulated acid precipitation have revealed no effect or a slightly stimulatory influence on the growth of seedlings of northeastern United States forest species (Wood and Bormann, 1974, 1975).

3. Ozone

The ability of ozone to reduce the growth of agricultural plants has been appreciated for approximately 25 years (Todd and Garber, 1958). The influence of ambient oxidants on citrus yield was dramatically demonstrated by comparing trees grown outside with those growin in environments with carbon-filtered air (Thompson, 1968). Heagle et al. (1972) exposed two varieties of sweet corn to 0, 5, or 10 pphm (98,196 μg m^{-3}) ozone for 6 hr day^{-1} from emergence to harvest in field exposure chambers. Ear weight, kernel number, and kernel dry weight of plants receiving 10 pphm were significantly reduced in the Golden Midget variety. In 1979, Heagle et al. reported the results of chronic dose exposure of a commercial field corn hybrid to ozone. Plants were exposed in open-top field chambers to four doses of ozone 25 days after planting to maturity. Threshold concentrations of ozone causing foliar injury, 0.02-0.07 ppm (39-137 μg m^{-3}), were lower than concentrations required to decrease kernel yield, 0.11-0.15 ppm (216-294 μg m^{-3}). Despite the fact that ozone is the most widespread gaseous air contaminant influencing United States forests today, there is relatively little information concerning the potential influence ambient concentrations of this gas may have on forest growth. Growth of plane trees in ambient greenhouse air in Washington, D.C., was observed by Santamour (1969) to be only 75% of the height growth in filtered air. Jensen (1973) observed that the growth of 1-year-old sycamore seedlings was reduced by ozone doses of 30 pphm (588 μg m^{-3}) for 8 hr day^{-1} for 5 days week^{-1} for 5 months.

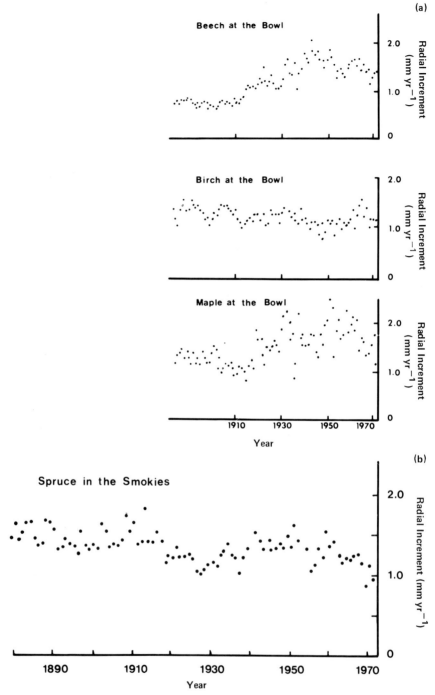

Figure 15-6. Tree ring chronologies for American beech, yellow birch, and sugar maple of the Bowl Natural Area, White Mountain National Forest, New Hampshire (a), and red spruce of the Great Smoky Mountains National Park, Tennessee (b). (From Cogbill, 1977.)

The only documentation of correlation of growth parameters of large trees growing under field conditions with ambient ozone levels has been provided by the comprehensive oxidant study conducted in the San Bernardino National Forest in California (Miller, 1977). Radial growth of ponderosa pine was compared for periods of low pollution (1910-1940) and high pollution (1941-1971) (Table 15-2). The average annual rainfall between these periods was 111 cm and 117 cm year^{-1}, respectively. The 0.20 mm difference in average annual growth between the two periods was attributed to air pollution. Average 30-year-old trees grown in the two periods were estimated to have diameters of 30.5 cm (1910-1940) and 19.0 cm (1941-1971). An average 30-year-old tree grown in contemporary air was estimated to reach 7.0 m height, 19 cm diameter, and be capable of producing one log 1.8 m long with a volume of 0.047 m^3. An average 30-year-old tree grown in the absence of oxidants was estimated to be 9.1 m height, 30.5 cm in diameter, and produce one log 4.9 m long with a volume of 0.286 m^3 (Figure 15-7).

Table 15-2. Average Annual Radial Growth of 19 Ponderosa Pine Trees in Two Levels of Oxidant Air Pollutants in the San Bernardino National Forest, California

High pollution		Low pollution	
Age[a] (years)	Average radial growth (cm) 1941-1971	Age[a] (years)	Average annual radial growth (cm) 1910-1940
20	0.20	60	0.52
21	0.33	55	0.49
29	0.22	55	0.61
22	0.33	57	0.34
25	0.30	64	0.40
35	0.23	63	0.55
27	0.29	60	0.44
28	0.31	65	0.46
35	0.26	60	0.75
22	0.43	71	0.67
39	0.21	63	0.71
35	0.34	71	0.65
29	0.37	66	0.78
33	0.37	63	0.53
35	0.34	60	0.33
35	0.37	70	0.38
36	0.35	61	0.32
36	0.33	62	0.37
34	0.36	59	0.37

Source: Miller (1977).
[a] Age at 1.4 m above ground in 1971.

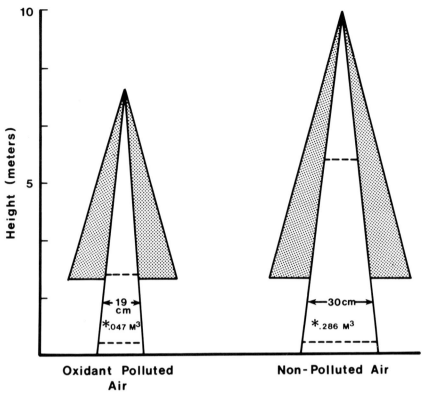

Figure 15-7. Calculated average growth of 30-year-old 15-cm San Bernardino National Forest, California, ponderosa pines in polluted and nonpolluted air based on radial growth samples from 1941-1971 and 1910-1940. The asterisk indicates wood volume in log with 15-cm top (min. merchantable diameter). (From Miller, 1977.)

4. Fluorides

It is apparent from research conducted on western United States conifers that elevated atmospheric fluoride can reduce the growth of trees contaminated from industrial sources. Treshow et al. (1967) recorded a reduction of up to 50% in the annual radial growth of Douglas fir subject to ambient fluoride. This suppression was noted irrespective of foliar symptoms. Extreme reductions in annual diameter growth of symptomatic ponderosa pine, presumably due to fluoride accumulation, have been recorded by Shaw et al. (1951) and Lynch (1951).

In his comprehensive study of lodgepole pine subjected to elevated fluorides from the Anaconda Aluminum facility in Columbia Falls, Montana (Chapter 12), Clinton F. Carlson included tree growth assessment in his observations. A 1973 investigation revealed that statistically significant growth losses attributed primarily to fluorides (secondarily to insects) were observed in 14 of 17 unmanaged stands for the period 1968-1973 in the vicinity of the aluminum plant (Carlson and Hammer, 1974a,b). Radial growth reductions, estimated from a prediction model, ranged from 0.005 mm (0.002 inch) to 1.8 mm (0.071 inch). This radial growth loss was equivalent to an average volume loss of 3.7 m^3 ha^{-1} $year^{-1}$ (57 board feet $acre^{-1}$ $year^{-1}$) for the measured stands. Extrapolation to the entire lodgepole pine ecosystem impacted by the industry suggested a total growth loss estimate of approximately 38,505 m^3 (8.5 million board feet) $year^{-1}$ (Carlson and Hammer, 1974a,b). Radial growth measurements were also made on Douglas fir and western white pine in the vicinity of the Anaconda plant. Ten-year periodic radial growth for two periods, 1958-1967 and 1968-1977, was measured on increment cores. Current 10-year periodic radial growth decreased dramatically with increasing foliar fluoride. Impact, defined as radial growth loss plus mortality, was calculated by Carlson (1978) to be 156,000 m^3 (5.5 million cubic feet) of usable wood for the period 1968-1977 on the 5360 ha (13,245 acres) influenced by the industry.

5. Trace Metals

Most of the evidence implicating trace metal pollutants in tree growth suppression comes from studies involving seedling-size plants. Carlson and Bazzaz (1977) grew 2- to 3-year-old seedlings of American sycamore in an agricultural soil treated with various amounts of lead and cadmium chloride. Various growth parameters were found to be synergistically affected by the lead-cadmium treatment (Figure 15-8). Diameter growth, foliage growth, and new stem growth of sycamore appeared to be synergistically affected by the combined lead plus cadmium treatment at lead/cadmium soil concentrations greater than 250/25 μg g^{-1}.

Lamoreaux and Chaney (1977) exposed first year seedlings of silver maple to cadmium concentrations they judged might be common in soils near industrial areas or in soils treated with sewage sludge or effluent. Seedlings were grown in sand amended with 0, 5, 10, or 20 μg g^{-1} cadmium chloride on a weight basis of rooting medium. Leaf, stem, and root dry weight were significantly reduced by all cadmium treatments.

Two- and three-year-old seedlings of red maple, white pine, and Norway spruce were grown in sand and irrigated with nutrient solution supplemented with either 0, 0.5, 1, 2, 4, 8, or 16 μg $liter^{-1}$ cadmium (Mitchell and Fretz, 1977). White pine seedlings were also treated with 32 and 64 μg $liter^{-1}$ cadmium. All three species were exposed to 0, 6.25, 25, 50, 100, 200, and 400 μg $liter^{-1}$ zinc. Exposure of seedlings to these trace metals in an artificial soil was also examined. Highly significant correlations between a root growth index and cadmium and zinc concentrations in the nutrient solution suggested that increasing metal levels resulted in poorer root development. Seedlings in the cadmium- and zinc-amended soils developed symptoms similar to the nutrient solution study except that injury was less severe at a given treatment level.

Kelly et al. (1979) grew first year seedlings of white pine, loblolly pine, yellow poplar, yellow birch, and choke cherry in a greenhouse environment in a natural forest soil collected in northwestern Indiana. Natural concentrations of trace metals in the soil were 0.6, 11.4, 2.0, and 20.6 μg g^{-1} cadmium, lead, copper, and zinc, respectively. The soil was amended with cadmium chloride to produce cadmium levels of 0, 15, and 100 μg g^{-1}. Shoot elongation and root and

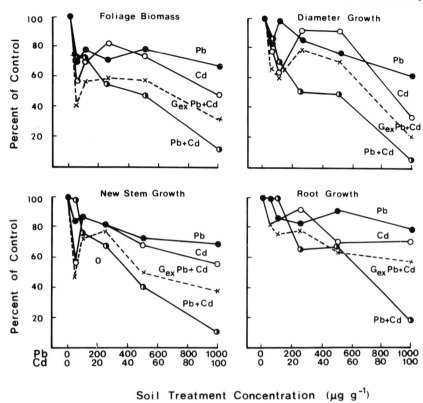

Figure 15-8. Growth parameters of sycamore seedlings grown in an agricultural soil amended with various combinations of lead and cadmium chloride. Treatments included lead alone (●), cadmium alone (○), or lead plus cadmium combined (◑). Values of expected growth for the combined treatment (Gex Pb + Cd) were calculated by multiplication of the reduction in growth due to separate heavy metal treatments and are indicated by dashed lines. (From Carlson and Bazzaz, 1977.)

shoot dry weights were reduced by increasing levels of cadmium. Shoot height for all species was reduced by the 100 $\mu g\,g^{-1}$ cadmium treatment, although yellow poplar and yellow birch height after 17 weeks was not significantly influenced. Heights of white pine, loblolly pine, and choke cherry were not significantly different between the 0 and 15 $\mu g\,g^{-1}$ treatments at the end of 17 weeks (Figure 15-9).

Working with larger trees, Symeonides (1979) has demonstrated that tree growth reduction in the vicinity of a Swedish smelting complex was positively correlated with tree core concentrations of copper and lead.

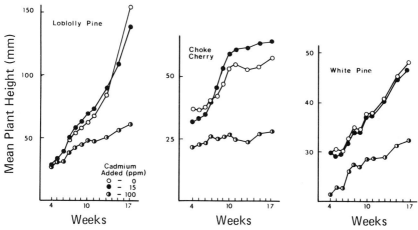

Figure 15-9. Height growth of loblolly pine, choke cherry, and white pine seedlings grown in a natural forest soil amended with various levels of cadmium. (From Kelly et al., 1979.)

6. Dust

Bohne (1963) presented radial increment evidence suggesting that poplar tree growth was reduced by coarse particles from a cement plant 1.7 km (1 mile) distant. Heavy accumulations of cement kiln dust have been observed to reduce early growing season elongation of coniferous twigs and foliage (Darley, 1966). Terminal shoot growth of limestone dust contaminated eastern hemlock was shown to be reduced by Manning (1971).

Brandt and Rhoades (1973) measured annual ring widths to assess the influence of limestone dust on the growth of four forest tree species in southwestern Virginia growing in clean and dusty sites. A reduction in radial growth of at least 18% was exhibited by red maple, chestnut, oak, and red oak. Growth of yellow poplar, on the other hand, was increased by 76%.

In general it can be concluded that coarse dusts may have an adverse effect on the growth of forest vegetation, mainly attributable to crust formation on foliage.

B. Altered Succession and Species Composition

As a result of the considerable varietal and species variation in relative susceptibility to the various Class II interactions, it is reasonable to suppose that differential tolerance to air pollution influence at the species level may be reflected in altered patterns of succession and species composition at the ecosystem level.

Ecologists recognize two major types of processes that influence ecosystem succession. Autogenic processes are those resulting from biological factors within the system. In forest ecosystems autogenic processes would include site alterations caused by the vegetation, influence of one plant species on another (Fisher,

1980) and impact of native insect or disease microorganisms. Allogenic processes, on the other hand, are abiotic factors that influence succession from without the system. Geochemical and climatic forces are especially important examples of allogenic factors that influence forest ecosystems. Idealized ecosystem development characteristically is portrayed as an orderly change of biological progression occurring in a more or less constant environment (Odum, 1969; Woodwell, 1974). It has been generally assumed that autogenic processes dominate allogenic processes in terrestrial ecosystem succession. This generalization, however, is quite inconsistent with data generated by recent imaginative studies with forest ecosystems. The importance of fire (an allogenic force) in influencing presettlement forest ecosystems in the North Central states of the United States has been substantial (Loucks, 1970; Frissell, 1973; Heinselman, 1973). The significance of wind stress (an allogenic force) has been suggested to exert substantial control over successional development of forest ecosystems in New England (Stephens, 1955, 1956; Raup, 1957; Henry and Swan, 1974). Forest management practices imposed by man, for example, clear-cutting, may simulate the influence of natural allogenic forces on forest development and interrupt progress toward a steady state condition (Bormann and Likens, 1979). Conversely, other forest management procedures, for example, fire control, may eliminate a controlling allogenic force and permit succession to proceed toward an unnatural steady state condition. Class II stresses imposed on forest ecosystems by air pollutants may be considered a twentieth century allogenic process of potential importance to forest ecosystem development. As in the case of clear-cutting, this human related force might be expected to alter the attainment of steady-state conditions. Air pollution stress would appear to have certain unique qualities that may make it an allogenic influence of particular importance. Length of exposure to this force precludes evolutionary adjustment and its influence, in certain areas, may be quite continuous rather than cyclic as are windstorms and fires. What is the evidence available to support the importance of air pollution as an allogenic force of significance in forest ecosystem development?

In 1968, prior to sophisticated understanding of most Class II interactions, Treshow provided an excellent review of the impact of air contaminants on plant populations. Treshow's review, along with a variety of additional papers from the late 1960s, for example, Niklfeld (1970), Hajdúk and Ružička (1968), and Trautmann et al. (1970), have indicated alterations in successional pattern or species composition in forest ecosystems subject to air pollution exposure.

The forests of the San Bernardino Mountains in southern California have been subject to oxidant stress from the Los Angeles metropolitan complex for 30 years. Intensive investigations conducted in the San Bernardino National Forest over the years have provided valuable insight and perspective on a variety of forest air pollution relationships. A summary of current research being performed in this area is provided in Section D. In 1970, Cobb and Stark concluded that if air pollution from the Los Angeles basin continued to increase, there will be a conversion from well stocked forests dominated by ponderosa pine to poorly stocked stands of less susceptible tree species in the San Bernardino Mountains.

Miller (1973) has provided a thorough discussion of this oxidant-induced forest-community change. Ponderosa pine is one of five major species of the "mixed conifer type" that covers wide areas of the western Sierra Nevada and the mountain ranges, including the San Bernardino Mountains, in southern California from 1000 to 2000 m (3000-6000 feet) elevation. Other species represented include sugar pine, white fir, incense cedar, and California black oak. The response of these five major tree species to oxidant air contaminants in the San Bernardino National Forest has been variable. Ponderosa pine exhibits the most severe foliar response to elevated ambient ozone. A 1969 aerial survey conducted by the U.S.D.A. Forest Service indicated 1.3 million ponderosa (or Jeffrey) pines on more than 405 km^2 (100,000 acres) were stressed to some degree. Mortality of ponderosa pine has been extensive. Actual death is typically attributed to bark beetle infestation of air pollution stressed trees. White fir has suffered slight damage, but scattered trees have exhibited severe symptoms. Sugar pine, incense cedar, and black oak have exhibited only slight foliar damage from oxidant exposure. A 233 ha (575 acre) study block was delineated in the northwest section of the San Bernardino National Forest in order to conduct an intensive inventory of vegetation present in various size classes and to evaluate the healthfulness of the forest. Ponderosa pines in the 30 cm (12 inch) diameter class or larger were more numerous than any other species of comparable size in the study area. These pines were most abundant on the more exposed ridge crest sites of the sample area. Mortality of ponderosa pine ranged from 8 to 10% during 1968-1972. The loss of a dominant species in a forest ecosystem clearly exerts profound change in that system. Miller concluded from his investigation that the lower two-thirds of the study area will probably shift to a greater proportion of white fir. It was judged that incense cedar will probably remain secondary to white fir. Sugar pine was presumed to be restricted by lesser competitive ability and dwarf mistletoe infection. The rate of composition change was deemed dependent on the rate of ponderosa pine mortality. The upper one-third of the study area, characterized as more environmentally severe due to climatic and edaphic stress, supports less vigorous white fir growth. Following loss of ponderosa pine in this area, sugar pine and incense cedar may assume greater importance. Miller judged, however, that natural regeneration of the latter species may be restricted in the more barren, dry sites characteristic of the upper ridge area. California black oak and shrub species may become more abundant in these disturbed areas. Additional and intensive research on forest composition in the San Bernardino National Forest has been reported (Miller, 1977). Tree population dynamics were examined on 18 permanent plots established in 1972 and 1973 and on 83 temporary plots established in 1974 to investigate forest development as a function of time since the most recent fire. Generally, the data still support the hypothesis that forest succession toward more tolerant species such as white fir and incense cedar occurs in the absence of fire. In the presence of fire, pine may be favored by seedbed preparation and elimination of competing species. These more recent studies suggest a larger number of forest subtypes may exist within the forest ecosystem than initially realized.

The changes in forest composition caused by oxidants in this southern California forest have created a management concern, as well as ecological change, because the forest is intensively used as a recreational resource and the loss of ponderosa pine is judged to reduce aesthetic qualities of the forest.

Other examples, not as dramatic as the San Bernardino example, can be found. Hayes and Skelly (1977) have monitored total oxidants and associated oxidant injury to eastern white pine in three rural Virginia sites between April 1975 and March 1976. Varieties of pine categorized as sensitive and intermediate to oxidant stress were judged to be under stress. The authors speculated that susceptible eastern white pine in the Blue Ridge and Southern Applachian Mountains may be rendered less competitive by air pollution stress. Shifts in species composition away from white pine importance may be occurring in certain eastern regions. Brandt and Rhoades (1973) in their investigation of limestone dust impact in deciduous forests in southwestern Virginia predicted changes in species composition resulting from dust influence. Dusty sites had reduced seedling and sapling density of red maple, chestnut oak, and red oak. This observation, along with documentation of reduced mean basal area and lateral growth of these trees, led the authors to suggest that yellow poplar, more resistant to stress caused by dust accumulation, would increase in importance in these hardwood stands.

Treshow and Stewart (1973) have conducted one the few studies truly concerned with air pollution impact on an entire vegetative community. Portable fumigation chambers were placed over representative plants in intermountain grassland, oak, aspen, and conifer communities. Ozone fumigations were conducted to establish injury thresholds for 70 common plant species indigenous to these communities (Table 15-3). Generally, injury was evident at varying concentrations above 15 pphm (294 μg m^{-3}). Species that were found to be most sensitive to ozone in the grassland and aspen communities investigated included some dominants which were considered key to community integrity. The most dramatic example was aspen itself. Single 2-hr exposure to 15 pphm ozone caused severe symptoms on 30% of the foliage exposed. White fir seedlings require aspen shade for optimal juvenile growth. The authors judged that significant aspen loss might restrict white fir development and alter forest succession. In a companion study, Harward and Treshow (1975) pursued their interest in evaluating ozone impact on aspen communities by evaluating the growth and reproductive response of 14 understudy species to ozone. Plants were fumigated in greenhouse chambers throughout their growing seasons. It was concluded from these fumigations that plant sensitivities varied sufficiently to make probable major shifts in composition in aspen communities following only a year or two of exposure to ozone above concentrations of 7-15 pphm (137-294 μg m^{-3}). The authors observed that comparable doses are widespread in the vicinity of urban areas and that widespread impacts on plant community stabilities may be common in nature.

The efforts of Michael Treshow and colleagues highlight the importance of examining shrub and herb strata when assessing air pollution impact on forest ecosystems. Nyborg (1978) has made an interesting and related observation.

Table 15-3. Injury Thresholds for 2-hr Field Fumigations with Ozone in Grassland, Oak, Aspen, and Conifer Intermountain Plant Communities

Species	Injury threshold (pphm ozone)	
Grassland-oak community species		
Trees and shrubs		
Acer grandidentatum	over	40
Acer negundo	over	25
Artemesia tridentata		40
Mahonia repens	over	40
Potentilla fruticosa		30
Quercus gambelii		25
Toxicodendron radicans	over	30
Perennial forbs		
Achillea millefolium	over	30
Ambrosia psilostachya	over	40
Calochortus nuttallii	over	40
Cirsium arvense		40
Conium maculatum	over	25
Hedysarum boreale		15
Helianthus anuus	over	30
Medicago sativa		25
Rumex crispus		25
Urtica gracilis		30
Vicia americana	over	40
Grasses		
Bromus brizaeformis		30
Bromus tectorum		15
Poa pratensis		25
Aspen and conifer community species		
Trees and shrubs		
Abies concolor		25
Amelanchier alnifolia		20
Pachystima myrsinites	over	30
Populus tremuloides		15
Ribes hudsonianum		30
Rosa woodsii	over	30
Sambucus melanocarpa	over	25
Symphoricarpos vaccinioides		30
Perennial forbs		
Actaea arguta		25
Agastache urticifolia		20
Allium acuminatum		25
Angelica pinnata	under	25
Aster engelmanni		15
Carex siccata		30
Cichorium intybus		25

Table 15-3. (continued)

Species	Injury threshold (pphm ozone)
Cirsium arvense	under 40
Epilobium angustifolium	30
Epilobium watsoni	30
Eriogonum heraclioides	30
Fragaria ovalis	30
Gentiana amarella	over 15
Geranium fremontii	under 25
Geranium richardsonii	15
Juncus sp.	over 25
Lathyrus lanzwertii	over 25
Lathyrus pauciflorus	25
Mertensia arizonica	30
Mimulus guttatus	over 25
Mimulus moschatus	under 40
Mitella stenopetala	over 30
Osmorhiza occidentalis	25
Phacelia heterophylla	under 25
Polemonium foliosissimum	30
Rudbeckia occidentalis	30
Saxifraga arguta	under 30
Senecio serra	15
Taraxacum officinale	over 25
Thalictrum fendleri	over 25
Veronica anagallis-aquatica	25
Vicia americana	over 25
Viola adunca	over 30
Annual forbs	
Chenopodium fremontii	under 25
Collomia linearis	under 25
Descurainia californica	25
Galium bifolium	over 30
Gayophytum racemosum	30
Polygonum douglasii	over 25
Grasses	
Agropyron caninum	over 25
Bromus carinatus	under 25

Source: Treshow and Stewart (1973).

While most commercially important forest trees develop well in soil with a pH as low as 3.5, a variety of forest shrubs exhibits a gradient in tolerance of low soil pH. In Alberta, Canada, Nyborg suggested that measurements on forest soils acidified by windblown elemental sulfur showed that when soil pH was 5 to 4 the number of species in the understory was reduced, when pH was 4 to 3 only a few species grew, and when the pH was less than 3 there was no undergrowth!

McClenahen (1978) has provided a most interesting study with quantitative data on the impact of polluted air on the various strata of a forest ecosystem. Forest vegetation was measured in seven stands on similar sites in a 50-km area of the upper Ohio River Valley. The stands were situated along a gradient of polluted air containing elevated concentrations of chloride, sulfur dioxide, fluoride, and perhaps other contaminants. Species richness (number of different species), evenness (dominance index—low values indicate dominance by one or a few species), and Shannon diversity index were typically reduced within the overstory, subcanopy, and herb strata near industrial sources of air contaminants. Increasing air pollutant exposure reduced canopy stem density, but abundance of vegetation in other strata tended to increase along the same gradient. The relative importance of sugar maple was greatly reduced in all strata with increasing pollutant dose, while yellow buckeye appeared tolerant of poor air quality. In the shrub layer the importance of spicebush increased with increasing pollutant exposure.

In southern California the predominant native shrubland vegetation consists of chaparral and coastal sage scrub. The former occupies upper elevations of the coastal mountains, extending into the North Coast ranges, east to central Arizona, and south to Baja California; while the latter occupies lower elevations on the coastal and interior sides of the coast ranges from San Francisco to Baja California. Westman (1979) applied standard plant ordination techniques to these shrub communities to examine the influence of air pollution. The reduced cover of native species of coastal sage scrub documented on some sites was statistically indicated to be caused by elevated atmospheric oxidants. Sites of high ambient oxidants were also characterized by declining species richness.

Influence of air pollution stress on succession and ecosystem species composition probably varies with the age and successional status of the forest. Harkov and Brennan (1979) have observed that most woody plants susceptible to ozone injury are generally early successional plant species. Most trees intermediate or tolerant of ozone stress are typically mid- or late-successional types. It is not unreasonable to propose, as Harkov and Brennan did, that late successional forest communities may be the most resistant to compositional change as a result of chronic air pollution exposure. Mature ecosystems are also typified by other characteristics that may increase their resistance to air pollution stress (Table 15-4). Low net production may reduce potential importance of restrictions imposed by air contaminants on photosynthesis. Closed and slow nutrient cycling may make nutrient capital less liable to loss by air pollutant influence.

C. Altered Forest Pest Influence

Native phytophagous forest insects and microorganisms that function as tree pathogens represent critically important autogenic influences on forest ecosystem structure and function. Interaction of these stresses with forest trees throughout their life cycles exerts powerful control on reproduction, stand density, plant distribution, and tree competition.

Table 15-4. Characteristics of Ecosystem Development

Ecosystem attributes	Developmental stages	Mature stages
Community energetics		
1. Gross production/community respiration (P/R ratio)	Greater or less than 1	Approaches 1
2. Gross production/standing crop biomass (P/B ratio)	High	Low
3. Biomass supported/unit energy flow (B/E ratio)	Low	High
4. Net community production (yield)	High	Low
5. Food chains	Linear, predominantly grazing	Weblike, predominantly detritus
Community structure		
6. Total organic matter	Small	Large
7. Inorganic nutrients	Extrabiotic	Intrabiotic
8. Species diversity— variety component	Low	High
9. Species diversity— equitability component	Low	High
10. Biochemical diversity	Low	High
11. Stratification and spatial heterogeneity (pattern diversity	Poorly organized	Well organized
Life history		
12. Niche specialization	Broad	Narrow
13. Size of organism	Small	Large
14. Life cycles	Short, simple	Long, complex
Nutrient cycling		
15. Mineral cycling	Open	Closed
16. Nutrient exchange rate, between organism and environment	Rapid	Slow
17. Role of detritus in nutrient regeneration	Unimportant	Important
Selection pressure		
18. Growth form	For rapid growth ("r-selection")	For feedback control ("K-selection")
19. Production	Quantity	Quality
Overall homeostasis		
20. Internal symbiosis	Undeveloped	Developed
21. Nutrient conservation	Poor	Good
22. Stability (resistance to external perturbations)	Poor	Good
23. Entropy	High	Low
24. Information	Low	High

Source: Odum (1969).

We have reviewed the considerable evidence concerning the influence of air pollutants on insects and disease microbes and the abnormal physiology they induce in individual tree hosts (Chapters 12 and 13). It is extraordinarily unfortunate, but we do not have substantial information on the interaction of air pollutants with insect pests and microbial pathogens at the population level! It is further true that our understanding of native insect and microbial influences on forest ecosystems over extended time intervals, in the absence of air pollutants, is not as sophisticated as it should be.

It is tempting to make some generalizations from the information that is available. It is undoubtedly true that the activities of some insects and some pathogens are increased while others are decreased in forest ecosystems subject to air contaminant stress. In the instance of oxidant induced predisposition of western conifers to bark beetle infestation, enhancement of this important insect in ponderosa pine populations does not appear to occur as predisposed and infested trees do not appear to function as efficient brood trees for insect population build-up. In the case of infection by the ubiquitous *Armillariella mellea* fungus, it is judged that chronic stress imposed by air pollutants may enhance the significance of this fungus in forest tree populations.

D. Case Study: Response of a Forest Ecosystem to Air Pollution

Our appreciation of the influence of Class II relationships resulting from air pollution influence on forest ecosystems will not significantly advance unless and until comprehensive investigations of natural ecosystems under ambient air contaminant influence are planned and conducted. The most comprehensive North American effort to study forest ecosystem response to ambient atmospheric contaminants has been performed in southern California under the leadership of Paul R. Miller, Research Plant Pathologist, U.S.D.A. Forest Service, Pacific Southwest Forest and Range Experiment Station, Riverside, California, and sponsored by U.S. Environmental Protection Agency and U.S.D.A. Forest Service. The southern California forest study and the ecosystem model developed to systematically investigate Class II air pollution impacts is thoroughly described in Taylor (1974), National Academy of Sciences (1977), and Miller (1977). In the author's judgment it is essential to establish similar studies in eastern as well as other western forests.

Concentration of human activities, meteorology, and topography interact to produce elevated atmospheric concentrations of ozone, nitrogen oxides, and peroxyacetylnitrate in southwestern California. Adverse impact from these oxidants has been especially high in the coniferous forests of southern California. In the San Bernardino National Forest (Figure 15-10), mortality of ponderosa and Jeffrey pines has been considerable over the past decade. Since 1972, twelve investigators representing various research disciplines developed an integrated study to evaluate Class II interactions in the forests of the San Bernardino Mountains.

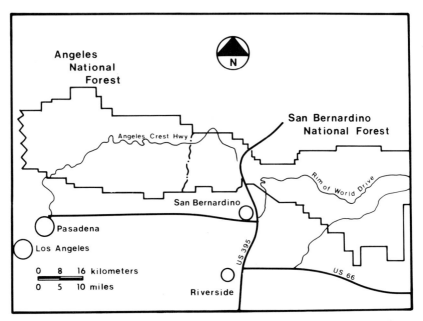

Figure 15-10. Location of the Los Angeles urban complex, an efficient producer of atmospheric oxidants, in relation to the San Bernardino and Angeles National National Forests. (From Wert et al., 1970.)

Figure 15-11. Community-level interactions in a mixed conifer ecosystem. 1, Competition between woodpeckers and small mammals; 2, Climate control of oxidant concentration in different forest communities; 3, Effect of precipitation and temperature on soil moisture and soil temperature in different forest communities; 4, Predation of bark beetles by woodpeckers in different forest communities; 5, Effect of cone crop abundance on cone insect populations in different forest communities; 6, Effect of cone crop abundance on small-mammal populations in different forest communities; 7, Fruiting bodies of nonpathogenic fungi as food for small mammals in different forest communities; 8, Smog-caused mortality and morbidity in different forest communities; 9, Fruiting bodies of pathogens as food for small mammals in different forest communities; 10, Effect of temperature and evaporative stress on species composition in different forest communities; 11, Relationship between soil characteristics and population density of burrowing small mammals in different forest communities; 12, Relationship between soil microarthropods and plants; 13, Relationship between soil characteristics and microarthropod population; 14, Bark beetle mortality caused by natural enemies in different forest communities; 15, Effect of bark beetles on tree mortality and vigor in different forest communities; 16, Relationship between soil characteristics and forest community composition and growth; 17, Relationship between soil characteristics and species distribution and behavior of nonpathogenic fungi; 18, Relationship between soil characteristics and species distribution and behavior of pathogens; 19, Influence of forest community type on populations of natural enemies of bark beetles; 20, Woodpecker distribution and density in different forest communities; 21, Effect of pathogens on tree vigor and mortality in different forest communities; 22, Relationship between nonpathogenic fungi and forest community composition and growth. (From Taylor, 1974.)

In 1975 a systems simulation modeling process was initiated. Initially an inventory of ecosystem components and processes was performed. Organism, tree, community, and stand level interactions were summarized and have been presented graphically in Taylor (1974). Figure 15-11 presents the community level interactions identified. Only interactions judged to have especially important roles in regulating forest structure and function were selected for immediate study. The basic unit for modeling purposes was defined as the forest stand. Stands could be comprised of 10-200 trees on land areas from 100-25,000 m^2. Subsystems treated at the stand level included tree population dynamics, oxidant flux, canopy response, stand tree growth, stand moisture dynamics, tree seedling establishment, cone and seed production, litter production, litter decomposition, and small mammal population dynamics. Time resolution varied with subsystem and ranged from hours to several years.

Eighteen major study plots were selected along an east-to-west gradient of air pollutants (Figure 15-12). Monitoring stations were established along the pollution concentration gradients to provide accurate dose information for the major study plots. Operation of this project for several years has produced an enormous amount of data. Data digestion and model refinement will ultimately permit assessment of oxidant impact on southern California forests. Model pre-

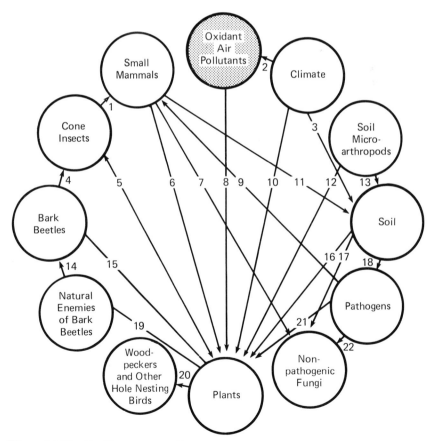

Figure 15-11. Caption on preceding page.

Figure 15-12. Air mass contaminated with photochemical oxidants moving into the forests of the San Bernardino Mountains, California. (Photograph courtesy Prof. Joe R. McBride, University of California, Berkeley.)

dictions may be able to be extended beyond the study ecosystem. Even in the absence of this achievement, however, the project was a success for the organization and perspective it has given to air pollution-forest ecosystem studies.

E. Summary

Perhaps the single most important characteristic of an ecosystem is its productivity. Compared to a variety of other terrestrial ecosystems, the productivity of forest ecosystems is relatively high. Productive forests are critically important, not only for the obvious relationship between wood volume and commercial products in managed forests, but also for the regulation and maintenance of quality for associated ecosystems, amenity functions, and general climatic and terrestrial stability. It is disconcerting to realize, therefore, that there is substantial and impressive evidence to indicate that two widespread air contaminants, sulfur dioxide and ozone, are capable of reducing forest growth. The more localized release of fluoride can also reduce the amount of forest biomass.

Evidence from a variety of studies examining forest growth in the vicinity of large point sources of sulfur dioxide has indicated significantly reduced growth. Generally, the correlation of growth impact with degree of foliar injury caused by sulfur dioxide is not high. Growth retardation occurs in the absence of any visible indication of stress. Most sulfur dioxide studies have accounted for precipitation influence on forest growth over the study periods. Evidence for ozone suppression of forest growth was provided by the comprehensive oxidant impact study on ponderosa pine forest ecosystems in California. Localized reduction of forest growth may also occur in environments subject to elevated levels of fluoride and potentially to a variety of heavy metals.

Acid precipitation studies have not demonstrated a significant influence on forest growth. Investigations conducted in the most seriously impacted temperate forest regions in Scandinavia and the United States have not provided convincing evidence that acid rain either reduces or increases forest growth.

There are two serious deficiencies of forest growth: air pollution stress research. The first relates to the paucity of ambient air quality determinations in growth studies. This makes establishment of dose thresholds or correlations of dose with growth influence nearly impossible. The second serious limitation relates to the inability to partition reduced growth to the various Class II interactions that may actually be responsible for it. For example, what percentage of reduced growth may be due to reduced nutrition, reduced photosynthesis, increased insects or disease, or increased foliar damage?

Future investigations of forest growth, as impacted by air quality, must also include better accounts of growth influencing factors other than precipitation and air pollutants. Better awareness of additional climatic factors, impacts of insect and disease influence, and management strategies must be indicated.

There is increasing appreciation of the importance of allogenic forces on forest ecosystem succession. The significance of fire and wind stress on forest development is substantial in certain environments. It is concluded that air pollutant impact also exerts critically important control over forest succession and species composition. Long-term, continual stress tends to decrease the total foliar cover of vegetation, decrease the species richness, and to increase the concentration of dominance by favoring a few, tolerant species. The importance of air contaminants is probably most significant during early and midsuccessional stages of forest development.

Very unfortunately we are unable, with data available at this time, to evaluate the importance of insect and disease alterations resulting from air pollution interactions at the population level. This topic must be considered of highest priority for future research.

References

Abrahamsen, G., R. Horntvedt, and B. Tveite. 1976. Impacts of acid precipitation on coniferous forest ecosystems. *In*: L. S. Dochinger and T. A. Seliga (Eds.), 1st International Symp. Acid Precipitation and the Forest Ecosystem. U.S.D.A. Forest Service, Upper Darby, Pennsylvania, pp. 991-1009.

Abrahamsen, G., K. Bjor, R. Horntvedt, and B. Tveite. 1976. Effects of acid precipitation on coniferous forest. *In*: F. H. Braekke (Ed.), Impact of Acid Precipitation on Forest and Freshwater Ecosystems in Norway. SNSF Project Research Resport No. 6, Oslo, Norway, pp. 37-63.

Bohne, H. 1963. Schädlihkeit von Staub aus Zementwerken für Waldbestände. Allg. Forstz. 18:107-111.

Bolin, B., Ed. 1971. Air Pollution Across National Boundaries: The Impact on the Environment of Sulfur in Air and Precipitation. Report of the Swedish Preparatory Committee for the U.N. Conference on Human Environment. Stockholm, Sweden, 96 pp.

Bormann, F. H., and G. E. Likens. 1979. Catastrophic disturbance and the steady state in northern hardwood forests. Am. Sci. 67:660-669.

Brandt, C. J., and R. W. Rhoades. 1973. Effects of limestone dust accumulation on lateral growth of forest trees. Environ. Pollut. 4:207-213.

Carlson, C. E. 1978. Fluoride induced impact in a coniferous forest near the Anaconda aluminum plant in northwestern Montana. Unpublished Ph.D. Thesis. Univ. Montana, Missoula, Montana, 165 pp.

Carlson, R. W., and F. A. Bazzaz. 1977. Growth reduction in American sycamore (*Plantanus occidentalis* L.) caused by Pb-Cd interaction. Environ. Pollut. 12: 243-253.

Carlson, C. E., and W. P. Hammer. 1974a. Impact of fluorides and insects on radial growth of lodgepole pine near an aluminum smelter in northwestern Montana. U.S.D.A. Forest Service, Northern Region Rept. No. 74-25, 14 pp.

Carlson, C. E., and W. P. Hammer. 1974b. Impact of fluorides and insects on radial growth of lodgepole pine near an aluminum smelter in northwestern Montana. U.S.D.A. Forest Service, Northern Region Rept. No. 74-26, 14 pp.

Cobb, F. W., and R. W. Stark. 1970. Decline and mortality of smog-injured ponderosa pine. J. For. 68:147-149.

Cogbill, C. V. 1976. The effect of acid precipitation on tree growth in eastern North America. *In*: L. S. Dochinger and T. A. Seliga (Eds.), 1st Internat. Symp. Acid Precipitation and the Forest Ecosystem. U.S.D.A. Forest Service, Upper Darby, Pennsylvania, pp. 1027-1032.

Cogbill, C. V. 1977. The effect of acid precipitation on tree growth in eastern North America. Water, Air, Soil Pollut. 8:89-93.

Dahl, E., and O. Skre. 1971. An investigation of the effect of acid precipitation on land productivity. Konferens om avsvavling, Stockholm, Nov. 11, 1969. Nordforsk, Miljövardssedretariatet, Pub. 1971, 1:27-39.

Darley, E. F. 1966. Studies on the effect of cement-kiln dust on vegetation. J. Air. Pollut. Control Assoc. 16:145-150.

Davis, D. R. 1972. Sulfur dioxide fumigation of soybeans: Effect on yield. J. Air Pollut. Control Assoc. 22:12.

Fisher, R. F. 1980. Allelopathy: A potential cause of regeneration failure. J. For. 78:346-350.

Frissell, S. S., Jr. 1973. The importance of fire as a natural ecological factor in Itasca State Park, Minnesota. Quat. Res. 3:397-407.

Fritts, H. C. 1976. Tree Rings and Climate. Academic Press, New York, 567 pp.

Hajdúk, J., and M. Ružička. 1968. Das Studium der Schäden an Wildpflanzen und Pflanzengesellschaften verursacht durch Luftverunreinigung. In: Air Pollution, Wageningen, Pudoc, pp. 183-192.

Harkov, R., and E. Brennan. 1979. An ecophysiological analysis of the response of trees to oxidant pollution. J. Air Pollut. Control Assoc. 29:157-161.

Harward, M., and M. Treshow. 1975. Impact of ozone on the growth and reproduction of understory plants in the aspen zone of western U.S.A. Environ. Conserva. 2:17-23.

Hayes, E. M., and J. M. Skelly. 1977. Transport of ozone from the Northeast U.S. into Virginia and its effect on eastern white pines. Plant Dis. Reptr. 61: 778-782.

Heagle, A. S., D. E. Body, and E. K. Pounds. 1972. Effect of ozone on yield of sweet corn. Phytopathology 62:683-687.

Heagle, A. S., R. B. Philbeck, and W. M. Knott. 1979. Thresholds for injury, growth, and yield loss caused by ozone on field corn hybrids. Phytopathology 69:21-26.

Heck, W. W., A. S. Heagle, and E. B. Cowling. 1977. Air pollution: Impact on plants. In: New Directions in Century Three: Strategies for Land and Water Use. Proc. 32nd Annual Meeting, Soil Conservation Soc. Amer., Aug. 7-10, 1977, Richmond, Virginia, pp. 193-202.

Heinselman, M. L. 1973. Fire in the virgin forests of the Boundary Waters Canoe Area, Minnesota. Quat. Res. 3:329-383.

Henry, J. D., and J. M. A. Swan. 1974. Reconstructing forest history from live and dead plant material: An approach to the study of forest succession in southwest New Hampshire. Ecology 55:772-783.

Horntvedt, R. 1970. SO_2 injury to forests. J. For. Utiliz. 78:237-286.

Jensen, K. F. 1973. Response of nine forest tree species to chronic ozone fumigation. Plant Dis. Reptr. 57:914-917.

Jensen, K. F., L. S. Dochinger, B. R. Roberts, and A. M. Townsend. 1976. Pollution responses. In: J. P. Miksche (Ed.), Modern Methods in Forest Genetics, Springer-Verlag, New York, pp. 189-215.

Jones, H. C., N. L. Lacasse, W. S. Liggett, and F. Weatherford. 1977. Experimental Air Exclusion System for Field Studies of SO_2 Effects on Crop Productivity. U.S. Environmental Protection Agency, Publica. No. EPA-600/7-77-122, Washington, D.C., 67 pp.

Jonsson, B. 1976. Soil acidification by atmospheric pollution and forest growth. In: L. S. Dochinger and T. A. Seliga (Eds.), Proc. 1st Internat. Symp. Acid

Precipitation and the Forest Ecosystem. U.S.D.A. Forest Service, Upper Darby, Pennsylvania, pp. 837-842.

Jonsson, B., and R. Sundberg. 1972. Has the acidification by atmospheric pollution caused a growth reduction in Swedish forests? Royal College of Forestry, Res. Note No. 20, Stockholm, Sweden, 48 pp.

Keller, T. 1978. Wintertime atmospheric pollutants—Do they affect the performance of deciduous trees in the ensuing growing season? Environ. Pollut. 16: 243-247.

Keller, T. 1980. The effect of a continuous springtime fumigation with SO_2 and CO_2 uptake and structure of the annual ring in spruce. Can. J. For. Res. (in press).

Kelly, J. M., G. R. Parker, and W. W. McFee. 1979. Heavy metal accumulation and growth of seedlings of five forest species as influenced by soil cadmium levels. J. Environ. Qual. 8:361-364.

Kramer, P. J., and T. T. Kozlowski. 1979. Physiology of Woody Plants. Academic Press, New York, 811 pp.

Kress, L. W., and J. M. Skelly. 1980. The interaction of O_3, SO_2, and NO_2 and its effect on the growth of two forest tree species. (In press).

Lamoreaux, R. J., and W. R. Chaney. 1977. Growth and water movement in silver maple seedlings affected by cadmium. J. Environ. Qual. 6:201-205.

Legge, A. H., D. R. Jaques, R. G. Amundson, and R. B. Walker. 1977. Field studies of pine, spruce and aspen periodically subjected to sulfur gas emissions. Water, Air, Soil Pollut. 8:105-129.

Linzon, S. N. 1971. Economic effects of sulfur dioxide on forest growth. J. Air Pollut. Control Assoc. 21:81-86.

Loucks, O. L. 1970. Evolution of diversity, efficiency and community stability. Am. Zool. 10:17-25.

Lynch, D. F. 1951. Diameter growth of ponderosa pine in relation to the Spokane pine-blight problem. Northwest Sci. 25:157-163.

Manning, W. J. 1971. Effects of limestone dust on leaf condition, foliar disease incidence and leaf surface microflora of native plants. Environ. Pollut. 2:69-76.

McClenahen, J. R. 1978. Community changes in a deciduous forest exposed to air pollution. Can. J. For. Res. 8:432-438.

Miller, P. R. 1973. Oxidant-induced community change in a mixed conifer forest. In: J. Naegle (Ed.), Air Pollution Damage to Vegetation, Adv. Chem. Series No. 122, Am. Chem. Soc., Washington, D.D., pp. 101-117.

Miller, P. R., Ed. 1977. Photochemical Oxidant Air Pollutant Effects on a Mixed Conifer Forest Ecosystem. A progress report. U.S. Environmental Protection Agency, Publica. No. EPA-600/3-77-104, Corvallis, Oregon, 338 pp.

Mitchell, C. D., and T. A. Fretz. 1977. Cadmium and zinc toxicity in white pine, red maple and Norway spruce. J. Am. Soc. Hort. Sci. 102:81-84.

Nash, T. H., H. C. Fritts, and M. A. Stokes. 1975. A technique for examining non-climatic variation in widths of annual tree rings with special reference to air pollution. Tree-ring Bull. 35:15-24.

National Academy of Sciences. 1977. Ozone and Other Photochemical Oxidants. Chapter 12. Ecosystems. NAS, Washington, D.C., pp. 586-642.

Niklfeld, H. 1970. Pflanzensoziologische Beobachtungen in Rauchschadengebiet eines Aluminiumwerkes. Zentbl. Ges. Forstw. 84:318-329.

Nyborg, M. 1978. Sulfur pollution in soils. *In*: J. O. Nriagu (Ed.), Sulfur in the Environment. Part II. Ecological Impacts. Wiley, New York, pp. 359-390.

Odum, E. D. 1969. The strategy of ecosystem development. Science 164:262-270.

Phillips, S. O., J. M. Skelly, and H. E. Burkhart. 1977a. Eastern white pine exhibits growth retardation by fluctuating air pollutant levels: Interaction of rainfall, age, and symptom expression. Phytopathology 67:721-725.

Phillips, S. O., J. M. Skelly, and H. E. Burkhart. 1977b. Growth fluctuation of loblolly pine due to periodic air pollution levels: Interaction of rainfall and age. Phytopathology 67:716-728.

Raup, H. M. 1957. Vegetational adjustment to the instability of the site. *In*: Proc. 6th Technical Meeting of the Internat. Union for the Protection of Nature, June 1956, Edinburgh, pp. 36-48.

Santamour, F. S., Jr. 1969. Air Pollution Studies on *Plantanus* and American elm seedlings. Plant Dis. Reptr. 53:482-485.

Scheffer, T. C., and G. C. Hedgcock. 1955. Injury to Northwestern Forest Trees by Sulfur Dioxide from Smelters. U.S.D.A. Forest Service Tech. Bull. No. 1117, Washington, D.C., 49 pp.

Shaw, C. G., G. W. Fischer, D. F. Adams, M. F. Adams, and D. W. Lynch. 1951. Fluorine injury to ponderosa pine: A summary. Northwest Sci. 15:156.

Skelly, J. M., L. D. Moore, and L. L. Stone. 1972. Symptom expression of eastern white pine located near a source of oxides of nitrogen and sulfur dioxide. Plant Dis. Reptr. 56:3-6.

Sprugel, D. G., J. E. Miller, R. N. Muller, H. J. Smith, and P. B. Xerikos. 1979. Effect of SO_2 fumigation on yield and seed quality in field-grown soybeans. Argonne National Laboratory Publica. No. ANL-ERC-72-22, Argonne, Illinois, 27 pp.

Stemple, R. B., and E. H. Tryon. 1973. Effect of coal smoke and resulting fly ash on site quality and radial increment of white oak. Castanea 38:396-406.

Stephens, E. P. 1955. Research in the biological aspects of forest production. J. For. 53:183-186.

Stephens, E. P. 1956. The uprooting of trees: A forest process. Soil Sci. Soc. Am. Proc. 20:113-116.

Stone, L. L., and J. M. Skelly. 1974. The growth of two forest tree species adjacent to a periodic source of air pollution. Phytopathology 64:773-778.

Sundberg, R. 1971. On the estimation of pollution-caused growth reduction in forest trees. The IASPS Symposium on Statistical Aspects of Pollution Problems. Boston, Massachusetts.

Symeonides, C. 1979. Tree-ring analysis for tracing the history of pollution. Application to a study in northern Sweden. J. Environ. Qual. 8:482-486.

Taylor, O. C. 1974. Oxidant Air Pollutant Effects on a Western Coniferous Forest Ecosystem. Annual Progress Report 1973-1974. Statewide Air Pollution Research Center, Univ. California, Riverside, 111 pp.

Thompson, C. R. 1968. Effects of air pollutants on lemons and navel oranges. Calif. Agr. 22:2-3.

Todd, G. W., and M. J. Garber. 1958. Some effects of air pollutants on the growth and productivity of plants. Bot. Gaz. 120:75-80.

Trautmann, W., A. Krause, and R. Wolff-Straub. 1970. Veränderungen der Bodenvegetation in Kiefernforsten als Folge industrieller Luftverunreinigungen im Raum Mannheim-Ludwigshafen. Schraftinr. Reihe Vegetationsk. 5:193-207.

Treshow, M. 1968. The impact of air pollutants on plant populations. Phytopathology 58:1108-1113.

Treshow, M., and D. Stewart. 1973. Ozone sensitivity of plants in natural communities. Biol. Conserva. 5:209-214.

Treshow, M., F. K. Anderson, and F. Harner. 1967. Responses of Douglas-fir to elevated atmospheric fluorides. For. Sci. 13:114-120.

Tveite, B., and O. Teigen. 1976. Acidification experiments in conifer forest. 3. Tree growth Studies. SNSE Project, Research Report No. 7, Oslo, Norway.

Weinstein, L. H., and D. C. McCune. 1979. Air pollution stress. *In*: H. Mussell and R. Staples (Eds.), Stress Physiology and Crop Plants. Wiley, New York, pp. 328-341.

Wert, S. L., P. R. Miller, and R. N. Larsh. 1970. Color photos detect smog injury to forest trees. J. For. 68:536-539.

Westman, W. E. 1979. Oxidant effects on Californian coastal sage scrub. Science 205:1001-1003.

Whittaker, R. H. 1975. Communities and Ecosystems. Macmillan, New York, 385 pp.

Whittaker, R. H., R. H. Bormann, G. E. Likens, and T. G. Siccama. 1974. The Hubbard Brook Ecosystem Study: Forest biomass and production. Ecol. Mono. 44:233-252.

Wood, T., and F. H. Bormann. 1974. The effects of an artificial acid mist upon the growth of *Betula alleghaniensis* Britt. Environ. Pollut. 7:259-268.

Wood, T., and F. H. Bormann. 1975. Short-term effects of an artificial acid rain upon the growth and nutrient relations of *Pinus strobus* L. *In*: L. S. Dochinger and T. A. Seliga (Eds.), Proc. 1st Internat. Symp. Acid Precipitation and the Forest Ecosystem. U.S.D.A. Forest Service, Upper Darby, Pennsylvania, pp. 815-825.

Woodwell, G. M. 1974. Success, succession and Adam Smith. BioScience 24:81-87.

SECTION III

FOREST ECOSYSTEMS ARE INFLUENCED BY AIR CONTAMINANTS IN A DRAMATIC MANNER—CLASS III INTERACTIONS

16

Forest Ecosystem Destruction: A Localized Response to Excessive Air Pollution

Under conditions of excessive dose, that is, atypically high atmospheric concentrations of one or more contaminants for extended (or continuous) time periods, the impact on forest ecosystems may be very severe and dramatic. This response is designated a Class III interaction. The reaction of vegetation in this case is characterized by severe morbidity and mortality caused directly by air pollutants.

Atmospheric burdens of contaminants are generally of sufficient magnitude to cause Class III interactions *only* in those portions of forest ecosystems in the vicinity of major point sources of atmospheric contaminants. Stationary sources of primary importance include energy production facilities, for example, electric generating plants, gas purification plants; metal related industries, for example, copper, nickel, lead, zinc, or iron smelters and aluminum production plants; and a variety of other industries, for example, cement plants, chemical plants, and pulp mills. The forest area impacted by these facilities is typically confined to a zone of a few kilometers immediately surrounding the plant and for a distance of several kilometers in the downwind direction. The extent of the latter influence is quite variable and primarily controlled by source strength of the effluent, local meteorology, regional topography, and susceptibility of the area vegetation.

Terrestrial ecosystems respond to gradients of natural environmental stress by predictable changes in structure. As environmental conditions become more damaging and restrictive, the size of dominant plants becomes smaller and a progression from systems dominated by trees to shrubs to grasses is typical. This pattern of structural change can readily be observed along temperature and wind gradients in mountainous regions, along atmospheric and soil salt gradients in maritime regions, and along moisture gradients in zones of differing annual precipitation input. It has been proposed that environmental stresses imposed on terrestrial ecosystems by anthropogenic activities may induce similar alterations in ecosystem structure (Curtis, 1956). Woodwell (1970) has developed this hypothesis and has indicated that evidence from studies dealing with the ecosystem effects of ionizing radiation, persistent pesticides, and of eutrophication is especially supportive. Woodwell generalized that a common reaction of forest ecosystems to environmental stresses is a "systematic dissection of strata layer by layer." Moving along a gradient of increasing stress: trees are eliminated first, then taller shrubs, then lower shrubs, then herbs, and finally bryophytes. Change in forest structure as described would be the most obvious but not the sole alteration resulting from stress. Associated with gross simplification would also be altered ecosystem functions including reduced rate of energy fixation, reduced biomass, increased nutrient loss, and reduced animal populations. As a result of the considerable influence intact forest ecosystems exert on surrounding geology and climate, gross simplification would also cause appreciable change in local erosion and sedimentation, hydrology, and meteorology.

Woodwell (1970) and Whittaker and Woodwell (1978) have indicated that air pollution stress results in forest ecosystem response comparable to other severe anthropogenic stresses. As indicated, Class III interactions are generally associated with industrial point sources of atmospheric contaminants. The two most important and pervasive of the latter are sulfur dioxide and fluoride. What is the evidence that these pollutants destroy and simplify forest ecosystems? A comprehensive review of this question has been provided by Miller and McBride (1975).

A. Sulfur Dioxide

Smelting of metal bearing ores and combustion of high sulfur fossil fuels for energy generation results in excessive production of sulfur dioxide. Since many of these facilities are located in forested regions, examples of Class III interactions are considerable.

1. Copper Basin, Tennessee

This area centered in Ducktown, Tennessee, consists of approximately 243 km^2 (60,000 acres) and was originally covered with southern deciduous forest. Mining operations were initiated in the basin in 1850 and smelting operations were most active between 1890 and 1895. By 1910 gross forest simplification resulting from excessive sulfur dioxide had created three new vegetative zones surrounding Ducktown (Haywood, 1905; Hedgcock, 1914; Hursh, 1948). In a 27 km^2 (10.5 mile^2) area closest to the source vegetation was devastated and largely eliminated. All trees and shrubs were destroyed and only a few, isolated islands of sedge grass occurred in the outer portions of this zone. A belt of grassland ecosystem, 68 km^2 (17,000 acres) in size, surrounded the barren zone. The principal grassland species was broomsedge. A transition zone of somewhat indefinite boundary and consisting of approximately 120 km^2 (30,000 acres) was located beyond the grassland. Few trees were located along the inner edge of the transition zone. Sassafras, red maple, sourwood, and post oak were common in the middle of the transition forest. The uninfluenced forest beyond the impact of the smelter consisted principally of mixed oaks, hickory, dogwood, sourwood, black tupelo, and some eastern white pine. The distance of vegetative impact extended 19-24 km (12-15 miles) to the north and approximately 16 km (10 miles) to the west of the smelter. Eastern white pine damage was recorded 32 km (20 miles) from the industry (Hursh, 1948).

Sheet and gully soil erosion has been excessive in the acutely damaged inner zone. Micrometeorological changes in the inner zone relative to the surrounding forest have been substantial: summer air temperature averages 1-2°C higher, while the winter air temperature averages 0.3-1°C lower, the soil temperature is 11°C higher in the summer; the wind velocities are five to 15 times higher and rainfall is consistently lower (Hepting, 1971).

2. Sudbury, Ontario

In the Sudbury area, three large nickel and copper smelters have discharged several thousand tons of sulfur dioxide daily into the surrounding atmosphere for roughly 8 decades. Sulfur dioxide emissions from this area approximate 10% of the North American sulfur dioxide total and 25% of the smelter total. Extensive simplification of the mixed Boreal forest ecosystem surrounding this region has occurred primarily via the mortality of eastern white pine throughout a 1865 km^2 (720 square mile) area to the northeast of the Falconbridge, Copper Cliff, and Coniston smelters. This region has been extensively studied and reviewed by Gorham (1970) and Linzon (1978). Class III relationships exist with black spruce and balsam fir in addition to white pine. Acute impact on the latter species has been recorded in excess of 40 km (25 miles) from the source of sulfur dioxide. Red oak, red maple, and red-berried elder are more tolerant and may exist in disturbed forests as close as 1.6 km (1 mile) of the smelters. Morbidity and mortality of forest trees in the Sudbury region continued to spread at least until the construction of the world's tallest smokestack of 403 m (1250 feet) at Copper Cliff in 1972 and the closing of the Coniston smelter also in 1972.

Soil erosion has followed the destruction of surrounding forest ecosystems. Rainfall has been made highly acidic, commonly less than pH 3.0 in 1971. Elevated nickel and copper concentrations in soils have been recorded to distances of 50 km (31 miles). Eroded sediment has contaminated area lakes and water courses with resulting long distance transport of trace metal loads (Hutchinson and Whitby, 1976).

3. Wawa, Ontario

Forest destruction in the vicinity of an iron smelter in Wawa, northern Ontario, exhibits a more discrete pattern, one smelter relative to three, and a shorter history of impact resulting in less equilibration of vegetative response than in Sudbury. The Wawa smelter initiated operations in 1939 (significantly expanded in

1949) and has released as much as 100,000 tons of sulfur dioxide annually to the surrounding atmosphere. Vegetative impact is primarily confined to a strip northeast from the plant in the direction of the prevailing wind. Symptoms of sulfur dioxide damage may be observed for at least 32 km (20 miles) to the northeast. The mixed Boreal forest in the Wawa area consists mainly of white spruce, black spruce, balsam fir, jack pine, white cedar, larch, and white pine in the dominant layer. Mountain maple and *Pyrus decora* are common in the understory.

Gordon and Gorham (1963) have systematically studied the response of the forest to sulfur dioxide from Wawa and have presented evidence consistent with Woodwell's pattern of forest ecosystem response to stress. These investigators established a series of vegetative sampling plots along a transect running from the smelter to 58 km (36 miles) northeast of Wawa. The variety of the flora declined acutely from approximately 20-40 species per 40 m^2 quadrat beyond 16 km (10 miles) of the facility to 0-1 species within 3 km (2 miles). The authors recognized four zones of destroyed or simplified forest ecosystem along the transect. In the zone of "very severe" impact, within 8 km (5 miles) of the source, both tree and continuous shrub layers were nonexistent. Some elder remained alive, but was symptomatic. In the zone of severe damage, extending from 16-19 km (10-12 miles), the forest canopy was still lacking. Some symptomatic mountain maple occurred along with plentiful but slightly injured elder. In the zone of considerable damage, 19-27 km (12-17 miles) from the source a discontinuous tree canopy was present. Tree mortality in this region was high. Only a few white birch and white spruce persisted. Tall shrubs were vigorous and *Pyrus* and mountain maple were common. Elder was present in large numbers in the understory. In the zone of moderate damage, from 27-37 km (17-23 miles), the tree canopy was continuous but symptoms of stress were obvious. Table 16-1 presents a summary of ground flora species recorded at increasing distances from the smelter.

Gordon and Gorham concluded that the forest ecosystem was "peeled off in layers" as the smelter source was approached from the northeast. The tree stratum was intact at 37 km (23 miles) distance. It was discontinuous within 37 km and absent within 27 km (17 miles). The shrub stratum dominated from 27 km to 19 km (12 miles). Herbs were dominant from 19 km to 8 km (5 miles) and there was no continuous plant cover within 8 km. Erosion at Wawa was extensive in the devegetated area (Figure 16-1).

A large number of additional examples involving severe morbidity or mortality of forest species in the immediate vicinity of industrial point sources of sulfur containing contaminants may be cited (Table 16-2). Unfortunately, most reports do not include useful and accurate suggestions of the size of forest areas impacted. If these were available and could be totaled the sum would be impressive.

Table 16-1. Number of Species Recorded, Tree Symptoms, and Soluble Soil Sulfate at Increasing Distances from an Iron Smelter along a Transect in the Direction of the Prevailing Wind

Distance NE of smelter (miles)	Number ground flora species		Tree damage (aerial estimate)	Soluble sulfate in soil (meq 100 g^{-1} ignition loss)
	Quadrat 20 m × 2 m	Accessory		
0.56	0	2	Very severe	31
1.0	1	3	Very severe	24
1.2	1	2	Very severe	18
1.6	1	2	Very severe	16
2.1	2	1	Very severe	15.5
2.5	1	1	Very severe	14.2
3.2	3	2	Very severe	12.2
4.7	2	8	Very severe	8.2
5.8	5	10	Severe	6.5
6.4	9	6	Severe	5.0
6.6	15	5	Severe	6.9
7.3	8	14	Severe	3.9
9.5	15	23	Severe	4.1
10.8	26	14	Severe	4.8
11.7	20	16	Considerable	7.6
14.0	39	16	Considerable	6.1
15.7	27	15	Considerable	5.7
17.3	32	17	Moderate	4.2
19.2	27	10	Moderate	9.2
21.9	20	12	Moderate	9.0
25.5	36	16	No damage	9.4
27.9	27	19	No damage	9.1
30.1	27	17	No damage	8.0
36.1	27	10	No damage	4.8

Source: Gordon and Gorham (1963).

B. Fluorides

The release of particulate sodium and aluminum fluoride and gaseous carbon tetrafluoride and hydrogen fluoride from aluminum ore reduction facilities and phosphate fertilizer plants has resulted in Class III relationships in a variety of forest ecosystems surrounding these industries (Miller and McBride, 1975).

Two significant United States examples of fluoride damage are located in Columbia Falls, Montana, and Franklin Park (Spokane), Washington. In Columbia Falls, the impact of the Anaconda Company aluminum reduction plant on the surrounding forest has been previously discussed (Chapter 12). Severe morbidity and mortality are concentrated in ponderosa pine and lodgepole pine on 8 km^2 (2000 acres) surrounding the plant. In Franklin Park, the Kaiser aluminum ore reduction operation has caused similar severe stress on ponderosa pine forests

Figure 16-1. Forest destruction in the vicinity of a smelter operation in Wawa, Ontario, Canada. Years of effluent discharge (sulfur dioxide) has caused mortality in various forest strata in the downwind plume direction. At 30 km (18 miles) northeast of the smelter the white spruce, balsam fir, and white birch forest exhibits normal structure (a). As the smelter is approached in the plume zone, the various strata are sequentially destroyed. At 16 km (10 miles) from the smelter the tree stratum is almost completely killed with only scattered white spruce and white birch remaining alive; the shrub layer is vigorous and consists of

(c)

(d)

mountain maple, showy mountain ash and invading red-berried elder (b). Beginning at approximately 13 km (8 miles) from the smelter the tree stratum is completely destroyed and mortality and morbidity in the shrub layer is extensive; climbing buckwheat is vigorous in the herb stratum (c). Within 6 km (4 miles) of the smelter almost all vegetation is destroyed and only scattered; symptomatic climbing buckwheat remains (d). (Photographs courtesy of Prof. Eville Gorham, University of Minnesota, Minneapolis.)

Table 16-2. Additional Examples of Boreal and Temperate Forest Ecosystems That Have Exhibited Class III Relationships with Sulfur Contaminants from Industrial Point or Area Sources

Location	Source	Severely injured or killed species	Reference
Redding, California, USA	Smelter	Pine, oak	Haywood (1905, 1910)
Anaconda, Montana, USA	Smelter	Douglas fir, lodgepole pine	Scheffer and Hedgcock (1955)
Missoula, Montana, USA	Pulp mill	Ponderosa pine, Douglas fir	Carlson (1974)
Superior, Arizona, USA	Smelter	Paloverde	Wood and Nash (1976)
Jackson, Mississippi USA	Nuclear electric generating facility (cooling tower SO_4^{2-})	Red oak, white pine, sassafras, white ash	Rochow (1978)
Colstrip, Montana, USA	Electric generating complex	Ponderosa pine	U.S.D.A. Forest Service (1978)
Trail, B.C., Canada	Smelter	Ponderosa pine, Douglas fir, western larch, lodgepole pine	Scheffer and Hedgcock (1955)
Anyox, B.C., Canada	Smelter	Western red cedar, western hemlock, Pacific silver fir, Sitka spruce	Errington and Thirgood (1971)
Yellowknife, NW Territories, Canada	Smelter	Black spruce, white spruce, paper birch, poplar, willow	Hocking et al. (1978)
Industrial Pennines, Great Britain	Mixed industrial	Scotch pine	Farrar et al. (1977)
Ruhr Valley, West Germany	Mixed industrial	Scotch pine	Knabe (1970)
Rouen, France	Mixed industrial	Scotch pine	Décourt (1977)

over a 130 km^2 (50 mile^2) area surrounding the plant. Because foliar fluoride analyses have been employed to document damage in both of these examples, confidence in assigning the destruction to atmospheric fluorides released by the plants is considerable (Carlson, 1972; Adams et al., 1952). Mortality of 81 ha (200 acres) of Douglas-fir has been documented adjacent to a phosphate reduction facility in Georgetown Canyon, Idaho (Treshow et al., 1967). In their review, Miller and McBride (1975) indicated that additional examples of localized forest destruction caused by fluorides have occurred in the states of Washington, Oregon, and Montana.

Examples of Class III forest relationships with fluoride are numerous in Europe (Scurfield, 1960). Robak (1969) observed that none of the larger aluminum plants in Norway has been able to avoid some damage to the more sensitive conifers existing within 2 km of their smoke outlets. In some cases Scotch pine forests have been destroyed up to a distance of 10-13 km (6-8 miles) from the fluoride source. Gilbert (1975) has indicated that forest destruction by fluorides in Norway is consistent with the Woodwell hypothesis concerning relative susceptibility of the various forest strata. In Hungary, an oak-pine forest has been destroyed to a distance of 800 m (875 yards) from an aluminum smelter (Keller, 1973). In Raushofen, Austria, approximately 800 ha (2000 acres) of mixed forest in the vicinity of an aluminum plant has been destroyed (Jung, 1968).

Hällgren and Nyman (1977) have investigated the influence of gaseous pollutants from an iron-sintering plant in Vitafors, Sweden, on surrounding Scotch pine forests. The Vitafors industry releases significant amounts of both sulfur dioxide and hydrogen fluoride. Foliar symptoms were not useful in distinguishing damage caused by sulfur dioxide and fluoride. Scotch pine foliage was examined for elevated sulfur and fluoride content. Foliar levels representative of emission free areas (baseline) were judged to be 0.07-0.12% sulfur and 10 μg g^{-1} fluoride. Elevated fluoride was detected to a distance of approximately 5 km (3 miles) from the industry. Foliar sulfur concentrations did not appear to be correlated with distance from the source. This study emphasizes the difficulty of making specific phytotoxicity judgments concerning industrial effluents as these frequently contain more than one pollutant.

C. Other Pollutants

A variety of additional examples of Class III interactions could be cited involving industrial release of numerous other pollutants. Release of coarse dust, chlorine, ammonia, and other materials has resulted in local forest destruction. The significance of these incidents is not great, however, as they are usually infrequent in occurrence and commonly highly localized.

The most important deficiency in our understanding of Class III relationships concerns oxidant influence on forest ecosystems. Oxides of nitrogen, ozone, and peroxyacetylnitrate are generated and released from area sources, frequently large urban complexes or regions of concentrated industrial-commercial-transportation activity, and as such are transported into surrounding forests in association with large air mass movements rather than in small, discreet plumes characteristic of point source industrial release of sulfur dioxide and fluoride. The oxidants are also rapidly metabolized by the biota and soils and their detection in association with the vegetation is not possible, as they are neither persistent nor accumulative. As a result of these limitations we have less than acceptable appreciation of the ability of ambient oxidants to cause acute morbidity and mortality in natural forests. The most convincing evidence for Class III oxidant involvement concerns the severe impact of ozone on ponderosa pine of the Western Montane forest along the western side of the Sierra Mountains in California (Chapter 15) (Miller and McBride, 1975). An approximate dose of more than 0.08 ppm (157 μg m^{-3}) for 12 hr has been judged to cause moderate to severe damage to ponderosa pine.

In the eastern United States considerable interest has been focused on eastern white pine as varieties of this species readily exhibit foliar symptoms in response to elevated oxidant concentrations. This species is a major component of four forest types and an associate in 14 other types with a range extending over 28,350 km^2 (7 million acres) from the Lake States to the Appalachian Mountains (U.S.D.A. Forest Service, 1973). Documentation of field injury is available (Linzon, 1965; Berry and Ripperton, 1963; Berry, 1961; Skelly and Jonston, 1978). Fumigation experiments have suggested that other widespread eastern pine species, for example, Virginia pine and jack pine, may be even more susceptible to oxidant injury than white pine (Chapter 14). It has been suggested that if oxidant concentrations in eastern forests reach daily peaks in the range of 0.30-0.60 ppm (588-1176 μg m^{-3}), widespread impact on forest ecosystems with susceptible pine species may occur (Miller and McBridge, 1975). Experiments designed to explore this possibility remain a research priority.

D. Summary

Under conditions of high air pollution dose, forest trees may be severely injured or killed. This may result in severe perturbations to ecosystem structure and function (Table 16-3). Responses of forest trees and ecosystems to excessively damaging doses of air contaminants are designated Class III relationships. A very large number of examples of Class III relationships have been documented in North America and Europe. Typically these situations involve the impact of sulfur dioxide or fluoride released from industrial sources on surrounding forests. Generally an elliptical pattern of forest stress occurring in the prevailing downwind direction is recorded. The extent of the influence downwind and the area of forest impact is very variable and controlled by emission strength, local meteorology, and topography and tree species susceptibility. The fact that many industries are located in valleys appears to exacerbate the damage by restricting plume expansion and reducing atmospheric mixing. In the western United States oxidants are imposing a Class III stress on portions of the Western Montane forests of southern California. A similar relationship for several eastern United

Table 16-3. Interaction of Air Pollution and Temperate Forest Ecosystems under Conditions of High Air Contaminant Load, Designated Class III Interaction

Forest soil and vegetation: Activity and response	Ecosystem consequence and impact
1. Severe morbility, excessive foliar damage	1. Dramatic change in species composition, reduced biomass, increased erodebility, nutrient attrition, altered microclimate and hydrology
2. Mortality	2. Forest simplification or destruction

States pine species and oxidants has been suggested but not clearly documented in the field.

In extreme Class III situations, irrespective of the specific pollutant, forest communities react, first, by losing sensitive species, second, by losing the tree stratum, and third, by maintaining cover in resistant shrubs and herbs widely recognized as seral or successional species. In less extreme Class III situations, the loss of sensitive species may be followed by maintenance of a tree stratum by resistant species. In this case, the forest consists of a fewer number of dominant plants and is characterized as being "simplified." The report of McClenahen (1978) provides a very nice example of air pollution induced forest simplification in the upper Ohio River Valley.

A fundamental question that has intrigued ecologists for a long time concerns the relationship between ecosystem simplification and stability (Holling, 1973). Are simplified ecosystems less stable? Ecosystems with complex structure can be maintained with relatively less energy. Ecosystem maturity is generally associated with greater complexity (Margalef, 1963). May (1973) has reviewed a variety of theoretical models dealing with population stability in diverse biological communities. He concluded that there is no indication that increasing diversity and complexity occasion enhanced community stability. Further, as Longford and Buell (1969) have indicated, temperate forest ecosystems have traditionally been viewed as systems that naturally combine characteristics of low diversity with high stability. The loss of individual tree species from forest ecosystems by air pollution stress cannot, therefore, be judged to make the residual systems less stable, only more simple. The impact of species loss to a given ecosystem will be largely dependent on the overall significance of that species to the functions of the system. Whittaker (1965) has indicated that natural communities are mixtures of unequally important species. The loss of an important species from a forest ecosystem will have greater impact on system function than a less important species. Productivity appeared, to Whittaker, to be the best predictor of relative species importance. When evaluating responses of forest ecosystems to Class III interactions we should evaluate the consequences of the loss of particular species accordingly. By this criterion, the loss of ponderosa pine in some mixed conifer stands in California and the loss of white pine in eastern forests would be significant in certain forest types and insignificant in others.

In addition to the importance of tree mortality for ecosystem considerations, species loss must also be evaluated in terms of management objectives imposed by people on the stressed ecosystem. If a severely impacted or destroyed species has commercial or aesthetic importance, its loss will be significant. In the case of gross forest destruction, regional impacts on nutrient cycling, soil stabilization, sedimentation and eutrophication of nearby aquatic systems, and climatic and hydrologic influences are important. Biomass reduction results in a corresponding reduction in the total inventory of nutrient elements held within a system and loss of the dominant vegetation destroys cycling pathways and mechanisms of nutrient conservation (Chapter 8). Research on the Northern Hardwood forest has clearly established that retention of nutrients within a forest ecosystem is dependent on constant and efficient cycling between the various components of the intrasystem cycle and that deforestation impairs this retention (Likens et al., 1977). Extensive nutrient loss can pollute downstream aquatic resources and can result in depauperization of a site and have long-term consequences with regard to future plant growth potential. Sundman et al. (1978) have reported significant failures in reforestation after clearcutting of coniferous forests in northern Finland during recent decades. Increases in soil instability and erosion follow extensive mortality of dominant vegetation, particularly in regions with steep or unstable slopes. Increased erodibility was found to follow deforestation at the Hubbard Brook Experimental Forest in New Hampshire (Bormann et al., 1969). This mature forest ecosystem, when undisturbed, was little affected by erosion, with an average annual particulate matter export of only 2.5 tons km^{-2} $year^{-2}$. Deforestation and repression of growth for 3 years increased export to a maximum of 38 tons km^{-2} $year^{-1}$ (Bormann et al., 1974). Soil erosion has been extensive at both Copper Hill, Tennessee, and Wawa, Ontario. In areas subject to fire stress, for example, southern California, an increase in the abundance of dead trees and increased shrub cover would act to increase the fire hazard. Fire incidence has become more frequent in a coastal western hemlock forest severely stressed by sulfur dioxide from a copper smelter in Anyox, British Columbia (Errington and Thirgood, 1971).

Class III interactions between forest ecosystems and air pollutants are the most dramatic of all responses. The forest areas involved in Class III interactions, however, are much less than in the Class I and II interactions.

References

Adams, D. F., D. J. Mayhew, R. M. Gnagy, E. P. Rickey, R. K. Koppe, and I. W. Allan. 1952. Atmospheric pollution in the ponderosa pine blight area. Spokane County, Washington. Ind. Eng. Chem. 44:1356-1365.

Berry, C. R. 1961. White pine emergence tipburn, a physiogenic disturbance. U.S.D.A. Forest Service, Southeast For. Exp. Sta. Paper No. 130, 8 pp.

Berry, C. R., and L. A. Ripperton. 1963. Ozone, a possible cause of white pine emergence tipburn. Phytopathology 53:552-557.

Bormann, F. H., G. E. Likens, and J. S. Eaton. 1969. Biotic regulation of particulate and solution losses from a forest ecosystem. BioScience 19:600-610.

Bormann, F. H., G. E. Likens, T. G. Siccama, R. S. Pierce, and J. S. Eaton. 1974. The effect of deforestation on ecosystem export and the steady-state condition at Hubbard Brook. Ecol. Monogr. 44:255-277.

Carlson, C. E. 1972. Monitoring fluoride pollution in Flathead National Forest and Glacier National Park. U.S.D.A. Forest Service, Div. State and Private Forestry, Missoula, Montana, 25 pp.

Carlson, C. E. 1974. Sulfur damage to Douglas-fir near a pulp and paper mill in western Montana. U.S.D.A. Forest Service, Div. State and Private Forestry Publica. No. 74-13, 41 pp.

Curtis, J. T. 1956. The modification of mid-latitude grasslands and forests by man. In: W. L. Thomas (Ed.), Man's Role in Changing the Face of the Earth. Univ. Chicago Press, Chicago, Illinois, pp. 721-736.

Décourt, N. 1977. Premier inventaire des effets de la pollution atmosphérique sur le massif forestier de Roumare. Biologie et Forêt, pp. 435-447.

Errington, J. C., and J. V. Thirgood. 1971. Search through old papers helps reconstruct recovery at Anyox from fume damage and forest fires. Northern Miner Annu. Rev., pp. 72-75.

Farrar, J. F., J. Relton, and A. J. Rutter. 1977. Sulfur dioxide and the scarcity of Pinus sylvestris in the industrial Pennines. Environ. Pollut. 14:63-68.

Gilbert, O. L. 1975. Effects of air pollution on landscape and land-use around Norwegian aluminum smelters. Environ. Pollut. 8:113-121.

Gordon, A. G., and E. Gorham. 1963. Ecological aspects of air pollution from an iron-sintering plant at Wawa, Ontario. Can. J. Bot. 41:1063-1078.

Gorham, E. 1970. Air pollution from metal smelters. Naturalist 21:12-15, 20-25.

Hällgren, J. E., and B. Nyman. 1977. Observations on trees of Scots pine (Pinus silvestris L.) and lichens around a HF and SO_2 emission source. Stud. For. Suec. No. 137, 40 pp.

Haywood, J. K. 1905. Injury to vegetation by smelter fumes. U.S.D.A., Bur. Chem. Bull. No. 89, 23 pp.

Haywood, J. K. 1910. Injury to vegetation and animal life by smelter wastes. U.S.D.A., Bur. Chem. Bull. No. 113, 63 pp.

Hedgcock, G. G. 1914. Injuries by smelter smoke in southeastern Tennessee. J. Wash. Acad. Sci. 4:70-71.

Hepting, G. H. 1971. Air pollution and trees. In: W. H. Matthews, F. E. Smith, and E. D. Goldberg (Eds.), Man's Impact on Terrestrial and Oceanic Ecosystems. MIT Press, Cambridge, Massachusetts, pp. 116-129.

Hocking, D., P. Kuchar, J. A. Plambeck, and R. A. Smith. 1978. The impact of gold smelter emissions on vegetation and soils of a sub-arctic forest-tundra transition ecosystem. J. Air Pollut. Control Assoc. 28:133-137.

Holling, C. S. 1973. Resilience and stability of ecological systems. Annu. Rev. Ecol. System. 4:1-23.

Hursh, C. R. 1948. Local climate in the copper basin of Tennessee as modified by removal of vegetation. U.S.D.A. Circular No. 774, 38 pp.

Hutchinson, T. C., and L. M. Whitby. 1976. The effects of acid rainfall and heavy metal particulates on a boreal forest ecosystem near the Sudbury smelting region of Canada. In: L. S. Dochinger and T. A. Seliga (Eds.), Proc. 1st Internat. Symp. Acid Precipitation and the Forest Ecosystem, U.S.D.A. Forest

Service, Genl. Tech. Rept. No. NE-23, Upper Darby, Pennsylvania, pp. 745-765.

Jung, E. 1968. Bestandesumwandlungen im Rauchschadensgebiete von Ranshofen. Miedzynarodowej Konf. Wplyw. Zanieczyszczen Powietrza Na Lasy, 6th, Katowice, 1968, pp. 407-413.

Keller, T. 1973. Report on the IUFRO meeting "Air pollution effects on forests." Sopron, Hungary, Oct. 9-14, 1972. Eur. J. For. Pathol. 3:56-60.

Knabe, W. 1970. Distribution of Scots pine forest and sulfur dioxide emissions in the Ruhr area. Staub Reinhalt. Luft. 30:43-47.

Langford, A. N., and M. F. Buell. 1969. Integration, identity and stability in the plant association. Adv. Ecol. Res. 6:83-135.

Likens, G. E., F. H. Bormann, R. S. Pierce, J. S. Eaton, and N. M. Johnson. 1977. Biogeochemistry of a Forested Ecosystem. Springer-Verlag, New York, 146 pp.

Linzon, S. N. 1965. Semimature-tissue needle blight of eastern white pine and local weather. Ont. Dept. Forestry, Res. Lab. Inform. Rept. No. O-X-1.

Linzon, S. N. 1978. Effects of airborne sulfur pollutants on plants. In: J. O. Nriagu (Ed.), Sulfur in the Environment, Part II, Ecological Impacts. Wiley, New York, pp. 109-162.

Margalef, R. 1963. On certain unifying principles in ecology. Am. Natur. 97: 357-374.

May, R. M. 1973. Stability and Complexity in Model Ecosystems. Princeton Univ. Press, Princeton, New Jersey, 235 pp.

McClenahen, J. R. 1978. Community changes in a deciduous forest exposed to air pollution. Can. J. For. Res. 8:432-438.

Miller, P. R., and J. R. McBride. 1975. Effects of air pollutants on forests. In: J. B. Mudd and T. T. Kozlowski (Eds.), Responses of Plants to Air Pollution. Academic Press, New York, pp. 195-235.

Robak, H. 1969. Aluminum plants and conifers in Norway. In: Air Pollution Proc. 1st European Congr. on the Influence of Air Pollution on Plants and Animals. Centre for Agric. Publish. and Documentation, Wageningen, The Netherlands, pp. 27-31.

Rochow, J. J. 1978. Measurements and vegetational impact of chemical drift from mechanical draft cooling towers. Environ. Sci. Technol. 12:1379-1383.

Scheffer, T. C., and G. G. Hedgcock. 1955. Injury to northwestern forest trees by sulfur dioxide from smelters. U.S.D.A. Forest Service, Tech. Bull. No. 1117, 49 pp.

Scurfield, G. 1960. Air pollution and tree growth. For. Abstr. 21:339-347, 517-528.

Skelly, J. M., and J. W. Jonston. 1978. A status report of the deterioration of eastern white pine due to oxidant air pollution in the Blue Ridge Mountains of Virginia. Proc. Amer. Phytopathol. Soc. 5:398.

Sundman, V., V. Huhta, and S. Niemela. 1978. Biological changes in northern spruce forest soil after clear-cutting. Soil Biol. Biochem. 10:393-397.

Treshow, M., F. K. Anderson, and F. Harner. 1967. Responses of Dougas fir to elevated atmospheric fluorides. For. Sci. 13:114-120.

U.S.D.A. Forest Service. 1973. Silvicultural Systems for the Major Forest Types of the United States. Agr. Handbk. No. 445. U.S.D.A. Forest Service, Washington, D.C., 114 pp.

U.S.D.A. Forest Service. 1978. Forest Insect and Disease Conditions in the United States–1976. U.S.D.A. Forest Service, Washington, D.C., 40 pp.

Whittaker, R. H. 1965. Dominance and diversity in land plant communities. Science 147:250-260.

Whittaker, R. H., and G. M. Woodwell. 1978. Retrogression and coenocline distance. *In*: R. H. Whittaker (Ed.), Ordination of Plant Communities. Dr. W. Junk, The Hague, The Netherlands, pp. 51-70.

Wood, C. W., and T. N. Nash. 1976. Copper smelter effluent effects on Sonoran desert vegetation. Ecology 57:1311-1316.

Woodwell, G. M. 1970. Effects of pollution on the structure and physiology of ecosystems. Science 168:429-433.

17
Synopsis and Prognosis

Large areas of the temperate forest ecosystem are currently experiencing major perturbation from air pollution. The influence of a variety of air contaminants on biogeochemical cycling, patterns of succession and competition, and individual tree health, designated Class II interactions in this book, are causing significant forest change in the temperate zone. At the ecosystem level the major perturbations include decreased productivity, biomass and diversity; at the community level, reduced growth; and at the population level, altered species composition. Early and midsuccessional forests are concluded to be at particular risk. Temperate forests have historically been subjected to major change resulting from the activities of human beings. For centuries the major influence was gross destruction for agricultural, fuel, or other wood-product purposes. In the present century reduced need for agricultural land and increased forest management have reduced the adverse impact on forests in temperate latitudes. Human activities of primary contemporary importance to forest structure and function have included the introduction of exotic arthropod and microbial tree pests into forest systems lacking evolutionary exposure to these destructive agents, enhancement of native and natural stresses by cultural practices, and the creation of artificial forests of one or a few commercially important species. In the past several decades, however, we have accumulated sufficient evidence to indicate that an additional major anthropogenic modifier of temperate forest ecosystem development is air pollution.

The interactions between air pollution and forest systems have been described in this book as falling into one of three Classes. Due to the enormous complexities of atmospheric chemistry and transport and forest ecosystem structure and function these Classes of interaction may not occur as discreet entities in time or space. It is perhaps more accurate to view the interactions of forests and air pollutants as a continuum of responses that vary greatly depending on the specific forest and the specific air contaminant of interest. The value of the three Class perspective, however, is that it facilitates the appreciation of a comprehensive relationship and permits generalizations concerning ecosystem response as well as individual species reaction.

Class II interactions are judged to be the most significant as they are capable of inducing major alterations in the patterns and processes of forest ecosystems, frequently subtle and insidious in character, and are widespread in occurrence. The ability of forest ecosystems to act as sources of air contaminants and as sinks for air pollutants, Class I interactions, follows Class II relationships in importance. Forests do have important roles in global major element cycles. Gross forest destruction is currently concentrated in the tropical zones and it is concluded that destruction of tropical forests will greatly exceed the importance of changes in temperate forests relative to global nutrient cycling over the next decade. Temperate forests are clearly efficient producers of a variety of air contaminants, most notably pollen, particulate and volatile hydrocarbons, and combustion products associated with forest fires. These can be of extreme local and regional importance. Forests have generated these contaminants for a long period, however, and it is presumed that evolutionary adjustments in the biota have been made. The capability of forests to serve as sinks or repositories for air pollutants is very significant. The improvement of local air quality by large forests in urban or industrial environments is important. The regional and continental importance of forest soils as sinks for carbon monoxide and trace metals is even more important. The Class I concept infers no adverse impact on forest ecosystems, however, and it is judged that the threshold of transition between Class I and Class II interactions has been crossed for several pollutants in numerous areas throughout the temperate zone. Class III interactions involving severe impact of air pollutants on forests are dramatic, but judged relatively unimportant because of their extremely localized nature.

A. Relative Importance of Specific Air Pollutants to Forest Ecosystems

The following air pollutants are judged to be most importantly involved in widespread interactions with temperate forest ecosystems: carbon oxides, sulfur oxides and secondary products, nitrogen oxides and secondary products, hydrocarbons, photochemical oxidants, and particulates.

1. Carbon Oxides

An increase in global temperature resulting from increased atmospheric carbon dioxide would have profound influence on the development of temperate forest ecosystems. Primary influence on forest development would be manifest in altered growth, plant competition, nutrient cycling, and arthropod and microbial pest relationships. If forest destruction is ultimately shown to be significantly involved in the increasing atmospheric burden of carbon dioxide, the primary contribution will be from the loss of tropical and not temperate forest systems.

The soil compartment of forest ecosystems is judged to function in a key manner in holding atmospheric carbon monoxide levels at relatively constant concentration. This removal may be the single most important contribution that forest ecosystems make to ambient air quality.

The next several years will see a continuation of increasing input of carbon dioxide into the atmosphere from anthropogenic sources primarily from fossil fuel combustion processes. Carbon monoxide, on the other hand, will continue to decrease at least in the United States. The reduction in carbon monoxide release has been, and will continue to be, largely the result of increased installation of catalytic converters on motor vehicle engines and reduced burning of solid wastes. The reduced incidence of carbon monoxide generated by forest wildfires has also contributed to regional reductions in ambient carbon monoxide. Increased use of prescribed burning as a management tool could, on the other hand, increase local burdens.

2. Sulfur Oxides and Secondary Products

The interactions between sulfur dioxide, atmospheric sulfates, and acid precipitation and temperate forest ecosystems are varied and profound. At low dose these atmospheric pollutants may stimulate forest growth. The evidence to support sulfur dioxide and acid precipitation involvement in numerous Class II interactions, however, is much more substantial. The potential for widespread influence, on tree reproduction and metabolism, nutrient cycling, and insect and microbial pest relationships is very high. The demonstrated involvement of sulfur dioxide in regional forest destruction is very impressive.

In the United States sulfur oxide emissions were slightly reduced during the period 1970 to 1977. This was largely due to increased recovery of sulfur at primary nonferrous smelters and to combustion of low sulfur fossil fuels. Over the next decade, however, the emission of sulfur dioxide and the associated formation of sulfates and acid precipitation is expected to increase as utility and industrial sectors make a major shift in fuel source from oil to coal. Large forest areas downwind of power generating facilities will be subject to sulfur pollutants as sulfur dioxide in plumes is converted to sulfates over long distances, commonly in the range of 100 to 500 km. If the heavy haze, regularly observed over the Arctic in the springtime, is proved to contain particles derived from pollution sources in the middle latitudes, these transport distances may have to be revised upward to in excess of 10,000 km.

3. Nitrogen Oxides and Secondary Products

Nitric oxide and nitrogen dioxide have the potential to supply forest ecosystems with nitrogen and in this context could be involved in a fertilizing Class I relationship. Limited evidence also suggests some potential for the involvement of these oxides in Class II interactions. The typically low ambient concentrations of nitrogen oxides, however, is judged to result in extremely modest direct involvement of these gases with forest ecosystems. It is surely true that the influence of secondary products, including nitrates, acid precipitation, and photochemical oxidants, are more importantly impacting forests than the oxides themselves. Substantial evidence has been presented to indicate that the former may be involved in widespread influence on tree reproduction and metabolism, nutrient cycling, and pest relationships.

Nitrogen oxides emissions have generally increased during the last decade in the United States. Approximately equal amounts are generated by transportation and stationary boiler sources. The general economic slowdown and motor vehicle control devices have caused a leveling off of these emissions over the last few years. As in the case of sulfur dioxide, however, major increases in nitrogen oxides emission are anticipated over the next decade primarily due to the utility sector shift to coal. Coal-fired boilers release three to six times as much nitrogen oxides as do oil and gas fueled boilers.

4. Hydrocarbons

Aside from pollen, hydrocarbons are the most important air pollutants generated by forest ecosystems. Increased appreciation of the role of these hydrocarbons in photochemical oxidant formation is needed. Current evidence indicates a role of relative unimportance for forest hydrocarbons in atmospheric oxidant chemistry and in associated adverse impact on human and forest vegetation health.

Anthropogenic hydrocarbons, on the other hand, clearly function as primary precursors in the generation of oxidant pollutants in major urban areas with roles of primary significance to surrounding forest ecosystems. As indicated in the nitrogen oxide discussion, these oxidants have been implicated in a variety of Class II and III relationships.

In the United States hydrocarbon emissions have been decreased by the use of oxidation catalysts that burn hydrocarbons emitted from motor vehicle engines. Despite continued increases in automotive use, there has been a net reduction in hydrocarbon release. Decreased burning of solid waste has also contributed to reduced hydrocarbon emission. This downward trend is expected to continue and may even accelerate if higher gasoline costs reduce automotive travel. Hydrocarbon emissions from industrial sources may increase if facilities are expanded or if leaks and spills occur that result in major evaporative losses.

5. Photochemical Oxidants

The diverse group of photochemical oxidants including various aldehydes, peroxyacetylnitrates, and most importantly ozone has been shown to be related to forest ecosystems in a large number of critically important ways. The hydrocarbons released from forest foliage may be important in the formation of ozone in rural atmospheres. The soil and vegetation of forest ecosystems may also remove substantial quantities of ozone from polluted environments. These Class I interactions, however, are judged to be secondary in importance to the extremely large number of Class II interactions that have been demonstrated for ozone. For selected tree species it is possible that ambient ozone concentrations downwind of major urban centers exert significant depressive effects on tree reproduction and photosynthesis and thereby adversely impact competition and productivity. While photochemical oxidants have not, and will not, cause dramatic localized forest destruction as in the case of sulfur dioxide and fluorides, the large forest areas subject to Class II interactions with this gas make this pollutant of enormous importance to those concerned with forest ecosystem health. There is increasing evidence that photochemical oxidants, or their precursors, can be transported over considerable distances from their areas of origin.

In the United States, the trend of ozone pollution has been relatively constant or slightly increased over the past decade in most areas of the country. In view of the increasing trend predicted for nitrogen oxides along with the decreasing trend for hydrocarbons over the next several years, the patterns of ambient ozone may be expected to continue relatively unchanged.

6. Particulates

The extremely diverse group of solid and liquid particles that contaminate the troposphere are intricately involved with forest ecosystems. The pollen and fungal spores produced in forested areas have enormous medical and economic significance to a very large segment of the population of the temperate zone. The large surface-to-volume ratio that forest trees present to the atmosphere makes this vegetative form a relatively efficient particle interceptor. Increased appreciation of the capability that forested regions have for air filtration in urban and industrial areas is needed.

With respect to Class II interactions, sulfate and trace metal particles are determined to be of primary importance. The ability of heavy metal particles to interfere with nutrient cycling and tree metabolism, with associated potential for reductions in forest productivity, are viewed as critically serious problems of potentially wide area significance.

Over the past decade in the United States, particulate emissions have generally been down. This was largely achieved by the installation of control equipment on industrial sources and large stationary combustion sources fueled by coal. Also small industrial and power generating facilities consumed lower quantities of coal during this period. Reduced burning of solid waste, agricultural burning, and fewer forest wildfires also contributed to the declining trend. Over the next decade, however, significant increases in particulate emissions may occur. The most important source of new contamination will be increased coal combustion. If efficient stack emission control devices are employed, and if chemical and physical coal cleaning procedures are developed and implemented, the increases associated with coal utilization can be moderated. It is important to realize, however, that volatile trace metals and those associated with particles with diameters less than 1 μm are not effectively removed by electrostatic precipitators. As previously indicated, increased coal combustion will release increased quantities of sulfur dioxide and nitrogen oxides. The secondary formation of sulfate and nitrate aerosols in industrial and power generating facility plumes will further contribute to the atmospheric particulate load. There are additional sources for increased particulate loads in the near future. Diesel-powered cars and light trucks are rapidly increasing in the United States. Relative to motor vehicles with standard gasoline engines equipped with catalysts, light-duty diesel engines emit greater amounts of solid material consisting of unburned hydrocarbons and other partial combustion products associated with micro size carbon particles.

B. Relative Importance of Air Pollutants for Specific Forest Regions: United States Case Summary

Regional differences in air quality are very great due to differences in source strength, climate, and topography. These differences combined with diverse forest ecosystem types result in differential significance of specific pollutants and differential importance of Class I, II, and III interactions in various areas of the temperate zone. A summary of the air pollution impact on the forests of the coterminous United States is presented by examining the influence of major contaminants in five major eastern and western forest ecosystems.

1. Eastern United States

The major forest types of the eastern United States are dominated by deciduous species and are conveniently divided into the northern, central, and southern hardwood regions (Figure 17-1). Conifers have considerable significance in portions of these regions with spruce and balsam fir especially important in northern New England; white, red, and jack pine especially important in the Lake States; and longleaf, loblolly, and slash pine especially important in the Southeast.

a. Boreal Forest

The true Boreal forest is not significant in the coterminous United States. The northern portions of the Northern Hardwood forest, for example, in portions of Maine, have an aspect of the Boreal forest due to the large amount of white spruce, balsam fir, and coniferous swamps and bogs. This forest does, of course, occupy a position of prominence in Canada and a similar Boreal forest covers large areas of Eurasia. The climate of this forest is extreme and the growing season short. The Boreal forest has the lowest species diversity and lowest net primary productivity of any major North American forest type.

In eastern North America, primary air pollution concerns with the Boreal forest relate to the impact of acid precipitation. As detailed in Chapter 9, major sulfate deposition and acid precipitation is concentrated in the northeastern United States and in eastern Canada. Heavy metal input to the eastern Boreal forest is also a principal concern because of the atmospheric input of trace metals from smelters to the west and from industrial and urban sources in the United States to the south.

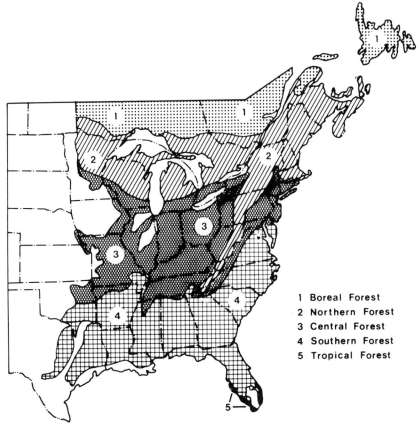

Figure 17-1. Major forest ecosystems of the eastern United States. (From Society of American Foresters, 1954.)

b. Northern Forest

The Northern Hardwood forest is dominated by yellow birch, American beech, and sugar maple. Coniferous species of great importance in selected areas include eastern hemlock and white, red, and jack pines.

As in the case of the Boreal forest, the contaminants of primary concern in this area are secondary products of sulfur pollution and heavy metal particles. Midwestern sources of sulfur dioxide contribute sulfates and acid precipitation to extensive areas of the Northern Hardwood forest. The Ohio River basin alone, with scores of large coal-burning power plants and numerous industrial boilers, generates polluted air masses that are regularly transported, depending on the wind, northeastward to the Northern Hardwood forests of Pennsylvania, New York, and New England and northwestward to the Northern Hardwood forests of Wisconsin and Minnesota or due north into Ontario (Galvin et al., 1978). Certain southern and western portions of this forest are also subjected to elevated concentrations of photochemical oxidants for occasional periods during the growing season. This may have particular significance because of the especially high susceptibility of coniferous components of the Northern forest to ozone. Heavy metal contaminants are widely input to the Northern forest from numerous urban and industrial sources. Class I and II interactions with acid precipitation, heavy metals, and ozone may be particularly widespread in the Northern forest.

c. Central Forest

This forest is dominated by various species of oak and hickory with yellow poplar an important associate in various portions of the distribution. The forest is characterized by a favorable climate, extended growing season, and relatively high species diversity and productivity.

Large portions of the Central forest are characterized by relatively high atmospheric sulfate concentrations and acid precipitation is judged to be of importance in the region. The portion of the Central forest east of the Appalachian Mountains and north of Virginia is subject to ambient photochemical oxidant doses second only to those of southern California. The area is also characterized by excessive trace metal and a variety of other particulate inputs from the intensive industrial-urban zone between Washington, D.C., and Boston, Massachusetts.

As in the case of the Northern forest, it is judged that Class I and II interactions, especially with secondary sulfur products, oxidants, and heavy metals, are extensive in the Central forest. Major conversion to coal as an industrial fuel and source of power generation may cause substantial increases of sulfur contaminants to be introduced into numerous regions throughout the Central forest.

d. Southern Forest

This extensive area is conveniently divided into two regions: an oak-pine area in the northern portion and a longleaf-loblolly-slash pine region in the southern (coastal) portion. The northern region is dominated by a variety of oaks and hickories and conifers are restricted to the poorer soils and drier sites. While longleaf pine forests dominate the landscape of the southern portion, the region is a mosaic of vegetative types with variable mixtures of pine and deciduous species.

Of all eastern forest areas, the Southern forest is judged to be the least impacted by atmospheric contaminants. Examples of Class II interactions can be found in the vicinity of major industrial and power generating facilities. Regional air masses, however, are generally characterized by appreciably less sulfur, oxidant, and particulate contaminant loads than those that traverse the Northern and Central forest regions. The importance of Class I and II interactions, therefore, are judged to be less significant in this area relative to other Eastern forests. This is fortunate as the prevalance of susceptible species and successional nature of much of this forest make it especially vulnerable. Increased coal combustion in this region may seriously degrade air quality.

e. Tropical Forest

This forest is restricted to the margins of the southern Florida peninsula and to the keys of south Florida. Certain Florida counties on the eastern side of the peninsula are characterized by high levels of oxidant and particulate pollutants.

The significance of these contaminants to the mahogany, fig, and mangrove species of the tropical forest is not clear.

2. Western United States

The major forest types of the western United States are dominated by coniferous species and are conveniently divided into five major forest ecosystems (Figure 17-2).

a. Western Montane Forest

The extensive Western Montane forest is largely characterized by ponderosa, lodgepole, and western white pines along with western larch.

Numerous Class III relationships exist in this forest in the immediate vicinity of various metal smelters and power generating facilities. Sulfur dioxide and fluoride emissions are primarily involved in these situations. Elevated sulfate inputs do occur in the forests in the eastern portion of the distribution. In southern California this forest type is subject to excessive Class II interactions with oxidants (Chapter 15). Additional areas of Class I and II interactions are presumed to be associated with other major industrial and urban areas scattered throughout the range of this forest.

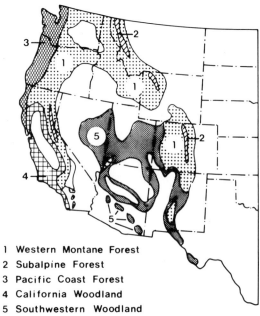

1 Western Montane Forest
2 Subalpine Forest
3 Pacific Coast Forest
4 California Woodland
5 Southwestern Woodland

Figure 17-2. Major forest ecosystems of the western United States. (From Miller and McBride, 1975.)

b. Subalpine Forest

The Subalpine forest consists largely of spruce-fir communities. Engelmann spruce and subalpine fir are the dominants throughout most of the range of this limited system. The research dealing with air pollution impact on this forest is limited and the potential for Class I and II relationships is not clear. The presumption is one of minimal impact.

c. Pacific Coast Forest

The most common species in this forest is Douglas fir which is seral and forms pure, even-aged stands following fire disturbance. Dominant climax species include western hemlock, Sitka spruce, and western red cedar.

This forest is subject to major air mass migrations from the Pacific Ocean and thereby generally has relatively clean air. Recent evidence has indicated elevated atmospheric sulfates in the southern portion of the range. Generally, however, Class I and II interactions are judged to be minimal. Some may occur downwind of principal industrial-urban sites. Class III relationships do occur in the immediate locality of major industrial facilities.

d. California Woodland

Dominant coniferous species of the California woodland include sugar and ponderosa pines. This forest is subject to extensive urban and suburban development and certain areas are characterized by high doses of oxidant and particulate pollution. Oxidant levels are greatest in the central and southern portions of the Sierra Nevada, near cities such as Los Angeles, Bakersfield, and Fresno. Class I and II interactions are important in portions of this forest.

e. Southwestern Woodland

This relatively large forest ecosystem is generally characterized by relatively low density pinon pine and juniper species. Isolated Class III interactions occur in association with industrial sites. Research on potential Class I and II interactions has been minimal but is justified particularly in regard to the elevated sulfur dioxide and particulate loads that have characterized southern Arizona atmospheres. Increased coal combustion in the Southwestern woodland would increase the need for additional research.

C. Research Needs

Air pollution over the next decade has the potential to cause significant reductions in forest productivity and shifts in species composition in numerous areas throughout the temperate zone. This potential, resulting from a variety of Class II interactions, must be given consideration in deliberations concerning clean air laws and regulations, alternative energy strategies, industrial and transportation location, and forest research funding. The ability of forests to influence the quantities of atmospheric pollutants, Class I interaction, is recognized as a second priority topic. The gross destruction of local forest ecosystems by large industrial and power generating facilities, Class III interaction, is recognized as the least important situation.

Scientists can rarely resist the temptation to call for additional research (Kozlowski, 1980) and the author finds he is no exception. During the last decade forest researchers have outlined numerous Class II interactions by largely utilizing relatively young forest plants grown in controlled environment facilities. During the next decade we must make an effort to perform experiments in natural forest ecosystems to confirm our hypothesis that ambient air pollution is reducing forest productivity and altering species composition.

1. Establishment of Comprehensive Air Pollution—
Forest Ecosystem Studies

The very highest research priority is reserved for the establishment of comprehensive investigations to systematically examine Class II interactions in forest ecosystems located in those portions of the temperate zone particularly subject to air pollution stress. These investigations should include analysis of air contaminant influence on soil metabolism and structure, nutrient cycling, tree reproduction, photosynthesis and respiration, important arthropod species and microbial pathogens, foliar symptoms of important vegetation in all forest strata, and a careful examination of forest productivity and alterations in successional trends and species dominance. These studies will be of extended term. They will require the participation of numerous scientific disciplines, minimally including pathology, entomology, meteorology, soil science, soil microbiology, ecology, and systems analysis. Continuous meteorological and air quality monitoring will be required. Air pollutants measured should include sulfur dioxide, nitrogen oxides, hydrocarbons, ozone, and particulates, the latter to include determination of sulfates, nitrates, and trace metals. Precipitation acidity will be routinely determined. The objective of these comprehensive studies will be to clarify and quantify various Class I and II interactions. The ecosystems will be evaluated for their ability to resist (inertia) and respond (resilience) to disturbance from air pollution stress. Model development for the various interactions will hopefully allow future projections given various air quality scenarios and allow extrapolation of findings to other ecosystems.

In the United States the only research program presently addressing this need is the oxidant study in progress on the San Bernardino National Forest in California (Chapter 15). It is imperative that additional investigations be initiated as soon as possible. The studies should be established in those areas judged to be under the greatest stress and they should be initiated in association with integrated and comprehensive forest ecosystem studies currently in progress. Priority forest ecosystems in the United States include the (1) Northern Hardwood forest, (2) Central Hardwood forest, and (3) Western Montane forest (San Bernardino project in progress). Appropriate locations, in terms of existing research facilities or abundant ancillary information, for the Northern forest are the Hubbard

Brook Experimental Forest in New Hampshire; the Isle Royale National Park, Michigan, and the Itasca Forest, Minnesota. In the Central forest the Camp Branch Forest watershed in east-central Tennessee and the Coweeta Hydrologic Laboratory in western North Carolina would be appropriate. With regard to location, the Wayne National Forest in Ohio would appear to represent an interesting research opportunity. In addition to the San Bernardino Forest study, the Andrews Experimental Forest, Oregon, and the Bitterroot National Forest, Idaho, would be other strategically located sites for the Western Montane forest. The Hubbard Brook, Coweeta and H. J. Andrews Experimental Forests have the advantage of being established Biosphere Reserves (U.S.D.A. Forest Service, 1979).

2. Specific Research Priorities

Several topics have especially high research priority for both the comprehensive ecosystem studies and investigations conducted in the field, greenhouse, or laboratory. It is important to realize, as in the case of epidemiological investigations concerning air pollution impact on human health, field experimentation must be supported by a strong laboratory (controlled environment) research program. Particularly important research areas include the following:

1. Dose-response information on visible (symptomatic) response with experiments appropriately designed to accommodate and consider the influence of genetic factors, environmental factors, and interaction of air contaminants.
2. Development of accurate, relatively simple, and reproducible methodologies to identify and inventory visible (symptomatic) injury in the field.
3. Analysis of the ability of air pollution stress to predispose, aggravate, or reduce stresses caused by insect, microbial, edaphic, and climatic agents or human management strategies.
4. Dose-response information on invisible (asymptomatic) response including an evaluation of the ability of air pollution exposure to influence tree metabolism, reproduction, competition, and growth along with the associated ecosystem parameters of productivity, succession, and species composition.
5. Determine the physiological and biochemical bases of air pollution stress on forest vegetation.
6. Determine the ability of forest vegetation and forest soils to act as a sink and source for atmospheric contaminants.
7. Develop suitable models of air pollution interactions so that future trends may be predicted and projected and information extrapolated from one ecosystem to another.
8. Develop reliable and economically sound cultural procedures for protecting valuable trees and determine the usefulness of the use of resistant varieties to reduce air pollution significance.

D. Conclusion

The integrity and healthy function of forest ecosystems are intimately associated with the health and welfare of human beings. During the twentieth century we have made enormous advancements in our understanding of forests and their values to us. We have learned to appreciate, manage, and manipulate forests. We have created new forests and have damaged others. We have recognized many of the strengths and weaknesses of our production, stewardship, and regulatory policies.

The recognition that we are capable of stressing forests in so many ways has sobered our sense of accomplishment derived from the sophisticated appreciation of forest form and function, the ability to increase forest productivity, and the strategies for preserving forests in perpetuity that we have developed.

Research indicates that air pollution is one of the most significant contemporary anthropogenic stresses imposed on temperate forest ecosystems. Gradual and subtle change in forest metabolism and composition over wide areas of the temperate zone over extended time, rather than dramatic destruction of forests in the immediate vicinity of point sources over short periods, must be recognized as the primary consequence of air pollution stress. This realization will permit rational evaluation of the total costs of various technologies. Failure to perform this evaluation is unthinkable.

References

Galvin, P. J., P. J. Samsun, P. E. Coffey, and D. Romano. 1978. Transport of sulfate to New York State. Environ. Sci. Technol. 12:580-584.

Kozlowski, T. T. 1980. Impacts of air pollution on forest ecosystems. BioScience 30:88-93.

Miller, P. R., and J. R. McBride. 1975. Effects of air pollutants on forests. In: J. B. Mudd and T. T. Kozlowski (Eds.), Responses of Plants to Air Pollution. Academic Press, New York, pp. 195-235.

Society of American Foresters. 1954. Forest Cover Types of North America. Soc. Am. For., Washington, D.C., 67 pp.

U.S.D.A. Forest Service. 1979. Selection, Management, and Utilization of Biosphere Reserves. Genl. Tech. Rept. No. PNW-82, Pacific Northwest Forest and Range Experiment Station, Portland, Oregon, 308 pp.

Appendix

Common and scientific names of woody plants cited in this book

A

Ailanthus, (tree-of-heaven)	(*Ailanthis altissima*)
Alder, European	(*Alnus glutinosa*)
Alder, thinleaf (mountain)	(*Alnus tenuifolia*)
Apple	(*Malus* species)
Apricot	(*Prunus armeniaca*)
Apricot, Chinese	(*Prunus armeniaca* var. Chinese)
Arborvitae, (white cedar)	(*Thuja occidentalis*)
Arborvitae, oriental	(*Thuja orientalis*)
Ash, European	(*Fraxinus excelsior*)
Ash, European mountain	(*Sorbus aucuparia*)
Ash, green (red)	(*Fraxinus pennsylvanica*)
Ash, white	(*Fraxinus americana*)
Aspen, large-tooth (big)	(*Populus grandidentata*)
Aspen, quaking (trembling)	(*Populus tremuloides*)
Avocado	(*Persea americana*)
Azalea	(*Rhododendron canescens*)
Azalea, campfire	(*Rhododendron kaempheri*)
Azalea, Chinese	(*Rhododendron mollis*)
Azalea, Hinodegiri	(*Rhododendron hinodegiri*)
Azalea, Korean	(*Rhododendron poukhanensis*)
Azalea, snow	(*Rhododendron kurume*)

B

Basswood	(*Tilia americana*)
Barberry	(*Berberis* species)
Barberry, Japanese	(*Berberis thunbergii*)
Beech, American	(*Fagus grandifolia*)
Beech, European	(*Fagus sylvatica*)
Beech, purple-leaved	(*Fagus sylvatica atropurpurea*)
Birch, European (weeping)	(*Betula pendula*)
Birch, European white	(*Betula alba*)
Birch, gray	(*Betula populifolia*)
Birch, sweet (black)	(*Betula lenta*)
Birch, water	(*Betula occidentalis fontinalis*)
Birch, western paper	(*Betula papyrifera commutata*)
Birch, white (paper)	(*Betula papyrifera*)
Birch, yellow	(*Betula alleghaniensis*)
Blueberry	(*Vaccinium* species)
Blueberry, lowbush	(*Vaccinium angustifolium*)
Box, Japanese (boxwood)	(*Buxus sempervirens*)
Boxelder	(*Acer negundo*)
Bridal wreath	(*Spiraea vanhoutii*)
Buckeye, yellow	(*Aesculus octandra*)
Buck-brush	(*Ceanothus velutinus*)
Buckwheat, climbing	(*Polygonum cilinode*)
Buffalo-berry	(*Lepargyraea canadensis*)

C

Ceanothus, redstem	(*Ceanothus sanquineus*)
Cedar, eastern red	(*Juniperus virginiana*)
Cedar incense	(*Libocedrus decurrens*)
Cedar, northern white	(*Thuja occidentalis*)
Cedar, western red	(*Thuja plicata*)
Cherry, black	(*Prunus serotina*)
Cherry, Bing	(*Prunus avium* var. Bing)
Cherry, bitter	(*Prunus emarginata*)
Cherry, choke	(*Prunus demissa*)
Cherry, Lambert	(*Prunus avium* var. Lambert)
Cotoneaster, rock	(*Cotoneastern horizontalis*)
Cotoneaster, spreading	(*Cotoneaster divaricata*)
Cottonwood, black	(*Populus trichocarpa*)
Cottonwood, eastern	(*Populus deltoides*)
Cottonwood, narrowleaf	(*Populus angustifolia*)
Currant, northern black	(*Ribes hudsonianum*)
Currant, sticky	(*Ribes viscosissimum*)
Cypress	(*Cupressus* species)
Cypress, Lawson's	(*Chamaecyparis lawsoniana*)

D

Dogwood, gray	(*Cornus racemosa*)
Dogwood, red osier	(*Cornus stolonifera*)
Dogwood, white (flowering)	(*Cornus florida*)
Douglas-fir, coast	(*Pseudotsuga menziesii* var. *menziesii*)
Douglas-fir, Rocky Mountain	(*Pseudotsuga menziesii* var. *glauca*)

E

Elder	(*Sambucus nigra*)
Elder, black bead	(*Sambucus melanocarpa*)
Elder, blueberry	(*Sambucus cerulea*)
Elder, red-berried	(*Sambucus pubens*)
Elderberry, (Pacific red elder)	(*Sambucus callicarpa*)
Elm, American (white)	(*Ulmus americana*)
Elm, Chinese	(*Ulmus parvifolia*)
Elm, mountain	(*Ulmus montana*)
Elm, rock	(*Ulmus Thomasi*)
Elm, Siberian	(*Ulmus pumila*)
Euonymus, winged	(*Euonymus alatus*)
Euonymus, dwarf winged	(*Euonymus alatus compactus*)

F

Fig	(*Ficus* species)
Filbert, European	(*Corylus avellana*)
Fir, balsam	(*Abies balsamea*)
Fir, cephalonica	(*Abies cephalonica*)
Fir, common silver	(*Abies pectinate*)
Fir, Douglas	(*Pseudotsuga menziesii*)
Fir, grand	(*Abies grandis*)
Fir, Japanese	(*Abies homolepis*)
Fir, Pacific silver	(*Abies amabilis*)
Fir, silver	(*Abies alba*)
Fir, subalpine	(*Abies lasiocarpa*)
Fir, white	(*Abies concolor*)
Firethorn	(*Pyracantha coccinea*)
Firethorn, Laland's	(*Pyracantha coccinea laland*)
Forsythia	(*Forsythia viridissima*)
Forsythia, Lynwood gold	(*Forsythia intermedia spectabilis*)

G

Ginkgo	(*Ginkgo biloba*)
Grape, concord (European)	(*Vitis vinifera*)
Grape, wild	(*Vitis riparia*)
Gum, black	(*Nyssa sylvatica*)
Gum, sweet	(*Liquidambar styraciflua*)

H

Hawthorn	(*Crataegus* species)
Hawthorn, black	(*Crataegus douglasii*)
Hawthorn, red	(*Crataegus columbiana*)
Hazel, beaked	(*Corylus cornuta rostrata*)
Hazel, California	(*Corylus cornuta californica*)
Hazel, witch	(*Hammamelis virginiana*)
Hemlock, eastern	(*Tsuga canadensis*)
Hemlock, western	(*Tsuga heterophylla*)
Hickory	(*Carya* species)
Holly, American (female)	(*Ilex opaca femina*)
Holly, American (male)	(*Ilex opaca mascula*)
Holly, English	(*Ilex aquifolium*)
Holly, Hetz Japanese	(*Ilex crenata Hetzi*)
Honeylocust	(*Gleditsia triacanthos*)
Honeysuckle, blue-leaf	(*Lonicera korolkowi*)
Honeysuckle, tatarian	(*Lonicera tatarica*)
Hornbean, European	(*Carpinus betulus*)
Horsechestnut	(*Aesculus hippocastanum*)
Hydrangea	(*Hydrangea paniculata*)

J

Juniper, common	(*Juniperus communis*)
Juniper, Rocky Mountain	(*Juniperus scopulorum*)
Juniper, shore	(*Juniperus conferta*)
Juniper, Utah	(*Juniperus osteosperma*)
Juniper, Western	(*Juniperus occidentalis*)

K

Kinnikinnick	(*Arctostaphylos uva-ursi*)

L

Larch, European	(*Larix decidua*)
Larch, Japanese	(*Larix leptolepis*)
Larch, western	(*Larix occidentalis*)
Laurel, mountain	(*Kalmia latifolia*)
Laurel, sheep	(*Kalmia angustifolia*)
Ligustrum	(*Ligustrum lucidum*)
Lilac, Chinese	(*Syringa chinensis*)
Lilac, common	(*Syringa vulgaris*)
Lime, summer	(*Tilia grandifolia*)
Lime, winter	(*Tilia parvifolia*)
Linden, American	(*Tilia americana*)
Linden, European cut-leaf	(*Tilia platyphyllos lacinata*)
Linden, littleleaf	(*Tilia cordata*)

Locust, black   *(Robinia pseudoacacia)*

M

Mahogany   *(Swietenia* species)
Mahogany, curl-leaf   *(Cercocarpus ledifolius)*
Mahogany, mountain   *(Cerocarpus montanus)*
Mahonia, trailing   *(Mahonia repens)*
Mangrove   *(Rhizophora* species)
Maple, bigtooth   *(Acer grandidentatum)*
Maple, coliseum   *(Acer cappadocicum)*
Maple, Douglas   *(Acer glabrum douglassi)*
Maple, fan   *(Acer palmatum)*
Maple, hedge   *(Acer campestre)*
Maple, Manitoba   *(Acer negundo interius)*
Maple, mountain   *(Acer spicatum)*
Maple, Norway   *(Acer platanoides)*
Maple, red   *(Acer rubrum)*
Maple, redvein   *(Acer argutum)*
Maple, Rocky Mountain   *(Acer glabrum)*
Maple, silver   *(Acer saccharinum)*
Maple, sugar   *(Acer saccharum)*
Mock-orange, Lewis   *(Philadelphus lewisi)*
Mock-orange, sweet   *(Philadelphus coronarius)*
Mock-orange, virginalis   *(Philadelphus virginalis)*
Mountain-ash, American   *(Sorbus americana)*
Mountain-ash, European   *(Sorbus aucuparia)*
Mountain-ash, showy   *(Pyrus decora)*
Mountain-ash, Sitka   *(Sorbus sitchensis)*
Mountain-ash, western   *(Sorbus scopulina)*
Mountain-laurel   *(Ceanothus sanguineus)*
Mulberry   *(Morus* species)
Mulberry, Texas   *(Morus microphylla)*

N

Ninebark, Pacifica   *(Physcocarpus capitatus)*

O

Oak, black   *(Quercus velutina)*
Oak, bur   *(Quercus macrocarpa)*
Oak, California black   *(Quercus kelloggii)*
Oak, chestnut   *(Quercus prinus)*
Oak, chinkapin   *(Quercus muehlenbergii)*
Oak, common   *(Quercus pendunculata)*
Oak, English   *(Quercus robur)*
Oak, Gambel   *(Quercus gambelii)*

Oak, northern red	(*Quercus rubra*)
Oak, pin	(*Quercus palustris*)
Oak, post	(*Quercus stellata*)
Oak, scarlet	(*Quercus coccinea*)
Oak, sessile	(*Quercus petraea*)
Oak, shingle	(*Quercus imbericaria*)
Oak, white	(*Quercus alba*)
Oak, willow	(*Quercus phellos*)
Ocean-spray	(*Holodiscus arieafolius*)
Olive	(*Olea* species)
Oregon grape	(*Odostemon aquifolium*)

P

Pachistima	(*Pachistima myrsinites*)
Pagoda, Japanese	(*Sophor japonica*)
Palm, coconut	(*Cocos nucifera*)
Peach	(*Prunus* species)
Pear, Barlett	(*Pyrus communis* var. Bartlett)
Pear, wild	(*Pyrus* species)
Pieris, Japanese	(*Pieris japonica*)
Pine, Austrian	(*Pinus nigra*)
Pine, black	(*Pinus austriaca*)
Pine, Coulter	(*Pinus coulteri*)
Pine, digger	(*Pinus sabiniana*)
Pine, eastern white	(*Pinus strobus*)
Pine, hoop	(*Araucaria cunninghamii*)
Pine, jack	(*Pinus banksiana*)
Pine, Japanese red	(*Pinus densiflora*)
Pine, Jeffrey	(*Pinus jeffreyi*)
Pine, knobcone	(*Pinus attenuata*)
Pine, limber	(*Pinus flexilis*)
Pine, loblolly	(*Pinus taeda*)
Pine, lodgepole	(*Pinus contorta*)
Pine, longleaf	(*Pinus palustris*)
Pine, Monterey	(*Pinus radiata*)
Pine, Mugho	(*Pinus mugo*)
Pine, pitch	(*Pinus rigida*)
Pine, ponderosa	(*Pinus ponderosa*)
Pine, red	(*Pinus resinosa*)
Pine, sand	(*Pinus clausa*)
Pine, Scotch	(*Pinus sylvestris*)
Pine, shortleaf	(*Pinus echinata*)
Pine, slash	(*Pinus elliottii*)
Pine, sugar	(*Pinus lambertiana*)
Pine, Swiss stone	(*Pinus cembra*)

Pine, pinyon	(*Pinus edulis*)
Pine, pond	(*Pinus serotina*)
Pine, Torrey	(*Pinus torreyana*)
Pine, Virginia	(*Pinus virginiana*)
Pine, western white	(*Pinus monticola*)
Plane, London	(*Platanus acerifolia*)
Plum	(*Prunus* species)
Poison-ivy	(*Toxicodendron radicans*)
Poison-ivy, western	(*Toxicodendron radicans rydbergii*)
Poplar, balsam	(*Populus balsamifera*)
Poplar, Carolina	(*Populus canadensis*)
Poplar, European white	(*Populus alba*)
Poplar, hybrid	(*Populus maximowiezii* x *trichocarpa*)
Poplar, Lombardy	(*Populus nigra* var. *italica*)
Poplar, tulip (yellow)	(*Liriodendron tulipifera*)
Privet, amur north	(*Ligustrum amurense*)
Privet, common	(*Ligustrum vulgare*)
Privet, lodense	(*Ligustrum vulgare* var. *pyramidale*)
Prune	(*Prunus* species)
Pyracantha	(*Pyracantha coccinea*)

R

Redbud, eastern	(*Cercis canadensis*)
Redwood	(*Sequoia sempervirens*)
Rhododendron	(*Rhododendron nova zembla*)
Rhododendron, Carolina	(*Rhododendron carolinianum*)
Rhododendron, catawba	(*Rhododendron catawbiense*)
Rhododendron, red	(*Rhododendron roseum elegans*)
Rhododendron, white	(*Rhododendreon catawbiense album*)
Rockspirea, creambush	(*Holodiscus discolor*)
Rose	(*Rosa* spp.)
Rose, woods	(*Rosa woodsii*)

S

Sagebrush, big (common)	(*Artemisia tridentata*)
Sassafras	(*Sassafras albidum*)
Sequoia, giant	(*Sequoia gigantea*)
Serviceberry	(*Amelanchier* species)
Serviceberry, low	(*Amelanchier spicata stolonifera*)
Serviceberry, Saskatoon	(*Amelanchier alnifolia*)
Serviceberry, Utah	(*Amelanchier utahensis*)
Snowberry, alba	(*Symphoricarpos alba*)
Snowberry, Columbia	(*Symphoricarpos rivularis*)
Snowberry, mountain	(*Symphoricarpos oreophilus*)
Snowberry, vaccinoides	(*Symphoricarpos vaccinioides*)

Sourwood	(*Oxydendrum arboreum*)
Squawbush	(*Rhus trilobata*)
Spicebush	(*Lindera benzoin*)
Spirea, shineyleaf	(*Spirea lucida*)
Spirea, Van Houts	(*Spirea vanhouttei*)
Spruce, black	(*Picea mariana*)
Spruce, Black Hills	(*Picea glauca densata*)
Spruce, blue	(*Picea pungens*)
Spruce, Englemann	(*Picea engelmannii*)
Spruce, Norway	(*Picea abies*)
Spruce, Sitka	(*Picea sitchensis*)
Spruce, white	(*Picea glauca*)
Sumac, fragrant	(*Rhus aromatica*)
Sumac, smooth	(*Rhus glabra*)
Sumac, staghorn	(*Rhus typhina*)
Sweetgum	(*Liquidambar styraciflua*)
Sycamore, American	(*Platanus occidentalis*)

T

Tree-of-heaven, (ailanthus)	(*Ailanthus altissima*)
Tupelo, black	(*Nyssa sylvatica*)

V

Virburnum, arrowwood	(*Virburnum acerifolium*)
Virburnum, burkwoodii	(*Virburnum burkwoodii*)
Viburnum, Korean spice	(*Viburnum carelsi*)
Viburnum, linden	(*Viburnum dilatatum*)
Viburnum, tea	(*Viburnum setigerum*)

W

Walnut, black	(*Juglans nigra*)
Walnut, English	(*Juglans regia*)
Willow, black	(*Salix nigra*)
Willow, weeping	(*Salix babylonica*)

Y

Yew, common tree	(*Taxus baccata*)
Yew, dense	(*Taxus densiformis*)
Yew, Hatfield's pyramidal	(*Taxus media hatifieldi*)
Yew, pacifica	(*Taxus brevifolia*)

Index

Springer Series on Environmental Management
Robert S. DeSanto, Series Editor

Gradient Modeling
Resource and Fire Management
Stephen R. Kessell
Gradient modeling is a promising new approach to resource management simulation and land management planning—a method that applies the tools of remote sensing, computer technology, and gradient analysis to resource management problems.

"The author has done much research in ecosystem modeling for resource management. Most of the text is devoted to Kessell's study in Glacier National Park, and covers field methods and modeling methods in that study. The author also includes methods for model implementation and applications of gradient modeling systems. A lengthy bibliography and index provide an excellent reference resource. Two appendixes give species lists for the studied area, and gradient model nomograms for Glacier Park. Advanced students and professionals in resource management should welcome this work." —*Choice*

". . . deserves to be read and understood by all those concerned with the wise management of land and resources." —*BioScience*

1979/432 pp./175 illus./27 tables/cloth
ISBN 0-387-**90379**-8

Disaster Planning
The Preservation of Life and Property
Harold D. Foster
Almost daily, stories of tragic occurrences—disasters—appear in the news. However, many communities still ignore the possible risks to life and property imposed by an ever-increasing range of hazards, both natural and manmade. **Disaster Planning** offers local administrators, institutions, and organizations a blueprint for increasing community safety through a comprehensive risk management program. Dr. Foster presents a step-by-step approach to delineating safety goals in which urban and regional planners are introduced to numerous innovative techniques and concepts applicable to existing staff and equipment, thereby minimizing expense while maximizing safety. Also, a new safety index based on stress factors and a methodology for deciding the necessity of evacuation are introduced.

1980/275 pp./48 illus./cloth
ISBN 0-387-**90498**-0